Mangrove Tiger

An Ethological Study

By

Jayanta Kumar Mallick

Mangrove Tiger: An Ethological Study

By Jayanta Kumar Mallick

This book first published 2023

Ethics International Press Ltd, UK

British Library Cataloguing in Publication Data

A catalogue record for this book is available from the British Library

Print Book ISBN: 978-1-80441-270-1

eBook ISBN: 978-1-80441-271-8

This book is dedicated to:

My late parents Shaktipada and Anima
My late wife Krishna
My daughter Runa
My son Sumanta
My son-in-law Saptarshi
and
My daughter-in-law Sagarika

Table of contents

Preface vii

Acknowledgements viii

Chapter 1: Introduction 1

Chapter 2: Materials and Methods 24

Chapter 3: Mangroves of the world and Tiger Colonisation 27

Chapter 4: Migrations from Mainland to Islands: The Last Resort .. 52

Chapter 5: Impact of Biotic and Abiotic Environment 77

Chapter 6: Habitats: Plants-Tiger Interactions 94

Chapter 7: Enumeration Dilemmas: State-of-the-Art 118

Chapter 8: Morphological Adaptations 204

Chapter 9: Behavioural Adaptations 254

Chapter 10: Intraspecific Interactions and Relationships 311

Chapter 11: Interspecific Interactions and Relationships 348

Chapter 12: Conservation Status: Global, National and
Local Levels 398

Chapter 13: Critical Conservation Issues 431

Chapter 14: Prospects 437

Selected Bibliography 440

Preface

The tiger makes no secret of its danger, prowling around in a yellow and black stripy catsuit, a trend it shares with other perilous beasts. After all, most people never get to see them in the wild anyway, and for those who want to, we have ample footage and photographs of them in their natural habitat. In the new "human age", we are going to have to prioritise conservation efforts towards the species we do want to save. Habitat conservation is a key to preventing the animals' extinction in the wild. Perhaps the biggest problem with conserving the tigers' habitats is that these cats typically inhabit the crowded regions of the world. Ultimately, the reason for conserving tigers may be less to do with their ecosystem or tourist economy benefits, but simply because they are magnificent creatures. We have a moral and ethical imperative to save them in the wild.

I remember the story of Jim Corbett as a boy creeping up to a plum bush (*Zizyphus mauritiana*) with a tiger on the other side looking at him with an expression which said as clearly as words, "Hello, kid, what the hell are you doing here?", and, receiving no answer, turning around and walking away very slowly without once looking back. Such instances have earned the tiger the sobriquet "large-hearted gentleman".

In the wild its motions remain still, then its tiptoes forward. Its smell - an incredibly strong musky odour, mixed with the scent of meat - spreads with the slightest movement of the air to attack the prey. Afterwards, the unmistakable sound of bones of the prey being crunched by a hefty jaw is often heard. Hours pass away. With a massive bound, the enormous flame-coloured beast leaps away, disappearing deep inside the forest. It is still warm where he had been lying, with fresh blood and chewed up deer bones. It takes a while for my heart to return to its usual pace, but it takes longer for the grin to fade. That heart-thudding glimpse of copper fur "burning bright" will stain my memory forever; it will be truly devastating if mine is the last generation to have this experience.

Being the largest living cat species in the world, the tiger (*Panthera tigris* Linnaeus, 1758), essentially a crepuscular and nocturnal animal, is a highly adaptable carnivorous species, exhibiting tolerance to a wide range of forest types, climatic regimes, altered landscapes and prey bases, particularly in the one and only mangrove habitat, inasmuch as there is no other tigerland like the Sundarbans left on earth. Over the last three decades, expansion of core area (by 28%) in the Indian Sundarbans and notification of a part of the buffer area as an additional wildlife sanctuary (WLS) i.e. West Sundarban in 2013, a vast tract of rich mangrove forests harbouring tigers, shrinking of the area available for fishing from 892.38 km² to c.523 km² are the positive initiatives by the forest department (FD) for ensuring a viable tiger population being dispersed in all the suitable habitats in the core, buffer or transboundary areas.

Contrary to what the name 'Sundarban' literally implies, i.e. 'beautiful forest', as the novelist Amitav Ghosh writes in his book 'The Hungry Tide', "There is no prettiness here to invite the stranger in." This archipelago is neither just a beautiful forest nor is it just home to the Sundari (*Heritiera fomes*) trees. It comprises an ecosystem that forms the frontal Gangetic belt. Over and above the utilitarian (narrow, broad and ethical) aspects of unique mangrove forests and rich biodiversity, particularly the most powerful bulwark against natural catastrophes, Sundarbans is the world's best attraction as home of the (Royal) Bengal tiger.

The Sundarbans is "a single ecosystem divided between the two countries" (Government of India and Government of Bangladesh) and "the Sundarbans ecosystem is greatly influenced by human use." We have to look after this iconic ecosystem if we want amazing animals like the Bengal tiger to have a chance of survival. The more of the Sundarbans that are conserved- via new PAs and reducing illegal poaching- the more resilient it will be to future climatic extremes and rising sea levels.

It is well-established now that tigers evolved in China, and two-million-year-old fossil remains were found in Central Asia, eastern and northern China, Siberia, Japan, Sumatra and Java—not very different from the modern range of the tiger. The continuous unbroken tiger habitat then with low human population and an abundance of prey enabled the tiger to occupy a wide range of landscapes in the areas around the Caspian Sea (Caspian tiger, *Panthera tigris virgata*), Siberia and northern China (Siberian tiger, *P.t. altaica*), central and southern China (South China tiger, *P.t. amoyensis*), the Indian subcontinent (Indian tiger, *P.t. tigris*), mainland

Southeast Asia (Indo-Chinese tiger, *P.t. corbetti*), the island of Bali (Bali tiger, *P.t. balica*), the island of Java (Javan tiger, *P.t. sondaica*), and the island of Sumatra (Sumatran tiger, *P.t. sumatrae*).

All the subspecies are distinguishable on the basis of the differences in the colour and striping of their coats and the skulls. The three island subspecies were once connected with the mainland species when the climatic conditions were much colder and there were ice bridges.

Taxonomically, the Sundarbans mangrove tiger is the "nominotypical subspecies" (*tigris*) known as the Bengal tigers since Bengal (without specific location) is traditionally fixed as the typical locality for the binomial *Panthera tigris*, commonly known as the continental tiger. The British taxonomist Pocock subordinated the Bengal tiger under the trinomial nomenclature *Panthera tigris tigris* in 1929, a native to the Indian subcontinent (Kitchener *et al.*, 2017) and most numerous of all subspecies.

At the end of the last glaciation (c. 115,000 – c. 11,700 years ago), the climate warmed, the sea level rose and the islands became isolated. This isolation led to the evolution of small-sized tigers—island tigers that are half the size of mainland tigers. Today, this phenomenon is also seen in the Sundarbans tigers, which has remained isolated from the mainland tiger habitat for about 500 to 1,000 years ago. Perhaps the small size of Sundarbans tigers is an adaptation in the marshy mangrove habitat for such longer period.

"The presence of a tiger", writes Romila Thapar 'In Times Past', "represented an affiliation with the most powerful and positive force for good in the natural world. Local people believed the tiger was the intermediary between heaven and earth. Tigers have played integral roles in ancient and modern cultures and folklore, being used to represent various characteristics and symbolisms throughout the centuries."

The most ancient records of tigers are found in the cave paintings of central India dated between 100,000 and 30,000 BP (Badam and Sathe, 1991). There are references to the occult powers of tigers in the Atharva Veda, Ramayana and Mahabharata (circa 400 BCE to CE 400). We find the ecological role of tigers in Nature in the Mahabharata.

"If there is no forest, then the tiger gets killed; if there is no tiger, then the forest gets destroyed. Hence, the tiger protects the forest and the forest guards the tiger!"

- Mahābhārata (Kumbhaghonam Edition) - Udyoga Parva: 5.29.57

The Bengal tiger, a 'living heritage', is often familiar as the 'Royal' beast in all its natural glory- an important symbol in the construction of British imperial and masculine identities during the 19th century (Sramek, 2006). E.P. Gee thinks the Royal may have originated from the fact that a tiger was shot by the Duke of Windsor when he was the Prince of Wales. Hence, by virtue of being recognised as 'king of the jungle' with a gruesome reputation, the tiger was sculpted on seals of Indus Valley civilisation of Harappa and Mohenjo-daro (2,500 to 1,700 BC) and later became the symbol of the Chola Empire (300-1279 CE).

Designated as the national symbol (India and Bangladesh) in early 1970s and subsequently as a World Heritage Site, Sundarbans has a dignity to be preserved. Arin Ghosh (1999), the then Chief Wildlife Warden, West Bengal, writes:

> The 'Royal Bengal Tiger', as it is emphatically called in this part of the country, is the pride of Bengal. It has been intricately enmeshed in the psyche and profile in anything Bengali; be it fiction or folklore, its religious beliefs or its serious literary pursuits.

The popular literature of the Sundarban delta presents an intellectual history of religious and cultural encounters in the region, which transformed humans' relationship with nature by forging the ethics and morality of forest resource use between humans and non-humans.

The mangrove forest is a challenge even for the Bengal tiger- with floods, marshy banks, shifting tides, and pneumatophores of mangrove roots that stick out of the ground like spikes (Mallick, 2021). Most importantly, the 'Mangrove tiger', often represented as ruler of the mangrove forests, exhibits certain distinctive morphological adaptations that make them particularly suited to the mangrove habitat. Their striped coats help them blend in well with the sunlight filtering through the treetops to the forest floor. The big cat's seamless camouflage to their surroundings is enhanced because the striping also helps break up their body-shape, making them difficult to be detected by the unsuspecting prey.

Life in the lush mangrove forests can be both beautiful and savage; a paradise on the one hand, when viewed by us from a safe distance, providing camouflage to a daily hell of trials and survival for those wild animals who live there, the subordinates and preys as well as the top ambush predator. The powerful tiger is by nature born leader, dynamic, very confident, perhaps too confident sometimes, noble and fearless, respected for courage, even by those timber mafias and poachers other than the common people. Tigers are unpredictable, always tense and addicted to excitement. They love adventures when hunting the grazing primary consumers and defending their own territory from other would-be killer machines, like being obeyed and not tolerating others to compete with and challenge them. The tiger usually chooses to operate alone in the natural kingdom, a dictator in peak physical age (optimal condition), who dominates over other associates in intra- and inter-specific interactions. Because of this, they play a crucial role in maintaining a healthy habitat.

I had visited the Sundarbans innumerable times since the mid-1970s observing the mangrove tigerland as well as the northern Bengal mainland intensely. During the last two decades, many of my write ups, highlighting the Bengal tigers in their isolated mainland and island habitats with special emphasis to their behavioural ecology and the most effective ways to protect them, have been published in different journals. Here are some of my published articles and academic books with special emphasis on *P.t. tigris* (Mallick, 2002; 2010a, b; 2011; 2012; 2013; 2015; 2019a; 2023; Bahuguna and Mallick, 2010). With the strong belief that saving tigers offers the best bet for recovering the storehouse of the natural gene pool to address future climatic challenges, I have written the book entitled 'Mangrove Tiger' with compassion, clarity and concern. It is a vivid narration on the biological structures and naturalistic behaviour of the subspecies, its critical role in shaping natural ecosystems, and their striking presence in our collective imagination. In spite of our joy and sensation of being present in the mangrove tigerland, the predicted perils thereof are haunting us like nightmares, but we are 'cautious optimists' professing and working for balancing human emancipation and nature conservation, which is critical for making more room for the mangrove tigers. Conservation is a never-ending war that has to be dealt with absolute passion.

Personally, I have had a special affinity with or passion for the Bengal tiger (*Panthera tigris tigris*) ever since I was a young student studying in the Calcutta university. During the early 1970s, I had my first close contact with a captive young Bengal tiger named 'Shankar', a healthy male performer in the Great Oriental Circus, and got stunning images of him while I was working temporarily as a musician on the stage in that circus and had the opportunity to observe his gait inside the ring and within the campus, and *inter alia* jumping through the rings of fire even though tigers are instinctively afraid of fire (a cruel practice indeed banned in 1998). Although I believe that despite the appearance of pseudo-domestication in some trained tigers, the tigers retain their predatory instincts and neural-visceral reflexes, and they can inflict serious wounds using their teeth or claws suddenly and without forewarning, my beloved Shankar was not of that type, rather calm and quiet, a majestic individual. There were no such untoward incidents. However, I used to note down every minute details about Shankar. He was a good actor too! He was the main attraction of a Bengali comedy film named "Once there was a tiger (*Ek je chhilo bagh*)" released in 1973. The shooting used to start after the night show and I was always present at the shooting spot as an inquisitive spectator in order to be familiar with his nocturnal attitudes and behaviour. At midnight, Shankar entered a hut-like structure by climbing on the thatched roof and dropped on the floor by making a hole. Ultimately, he returned to the wild without injuring anybody. After a few months, I left the circus when it shifted to another state and carried on with my academics.

I always wanted to create a new document about the world-famous king in the wild- the Bengal tiger in the greater Sundarbans mangroves, a mangrove swamps on the tropical coasts of India and Bangladesh, locally known as *bādābān* or *bādā*. Very soon my dream was fulfilled. After completing my post-graduation and passing the competitive examination incredibly, I joined the Wildlife Wing Headquarters in West Bengal, a state in the eastern India bordering Bangladesh and, naturally, got the legal right to visit the tigerland every weekend (Saturday being half-holiday then and Sunday a holiday) to lark around the horizontal and vertical landscape to my heart's content, all the way down to the Bay of Bengal via the Haldibari

Camp under National Park (NP) (West) Range because during late 1970s and the 1980s, Haldibari and Mechua were good sites for tiger sighting, although during the early 21st century I enjoyed the ambience of the Burirdabri camp under Basirhat Range on the international border with the Bangladesh Sundarbans, and got the opportunity of sighting an incident of deer-hunting by a resident tiger at night.

Having seen a tiger in the wild is an incredible privilege, which not many people in the world can boast about. And for many of us, the first sighting of a wild tiger is probably the most remarkable one. But sighting a tiger in the mainland cannot be compared with the charm of locating a half invisible tiger, a semi-aquatic feline, soaked in water and coated in mud, just landed after crossing a river. The diffused light makes the evergreen mangrove swamps mysterious, where the trees are short, have a dense canopy, grow very close to each other and hence visibility is poor compared to the mainland. The best bet to see the tiger in the Sundarbans is to sail through the water courses, expecting the tiger to emerge from the forests on the shore, as I did while approaching the Haldibari jetty in the morning. The chances of sighting a big cat through the pattern of light and shade are usually very remote. I was fortunate to have a glimpse of the healthy, vibrant and glowing orange fur and black stripes, the graceful paws, terrifying eyes but it was not directly looking at me; the wild eye leapt across the bush before bounding away. That's my first encounter with the largest feline in the wild.

I knew Dr. Kalyan Chakraborty, the then Field Director, Sundarban Tiger Reserve (FD, STR), used to spend night after night sitting in the wooden watchtower at Haldibari with the facility of a 360 degree view, overlooking a sweetwater pond with frontal, left and right strip of land bereft of any vegetation for such a precious sighting of the prey and predator from a distance. My first tiger sighting took place onboard while approaching the jetty of Haldibari camp. Since then, I have had so many rewarding opportunities to encounter this majestic big cat in the mangroves. It was so special, most memorable, unpredictable, breathtaking and vibrant to me from an ethological perspective. Their presence can be felt by the prey animals, even when you can't see or smell them; you can rest assured about their presence or footprints in the mud of every nook and cranny of their habitat and they can sense your presence because they are silently monitoring your every movement in the world's most mysterious tigerland.

In spite of the fact that the mangrove forest is a challenging habitat with storms, floods, marshy banks, shifting tides, and pneumatophores that stick out of the muddy ground like spikes at a height of 20-30 cm above soil to make hunting very difficult (20% success-rate), the single-patch mangrove forests is the only area in the world known to have the Bengal tiger, an endangered charismatic megafauna, drawing global public attention and conservation efforts. Whereas for the captive tigers, physical wellbeing is enough for their survival, for a wild Sundarbans tiger physical and psychological performances together can only ensure its survival.

Systematic ethological research on the Bengal tiger in the wild, particularly when the habitat is inaccessible mangroves, is a very challenging task. Such research works were initiated as a sequel to the institutional tiger conservation efforts during the early 1970's. At present a variety of topics such as foraging and feeding behaviour, habitat selection, social and reproductive behaviour, chronobiology, chemical communication signals and neurobehaviour are being investigated by the ethologists. These analyses revealed that there has been a gradual increase in the availability of books and scientific journals in the field of tiger behaviour. The number of workers has increased and the fields of investigation in the area of ethology have diversified. An analysis of the development of tiger ethology in the Indian subcontinent indicates that the progress of this discipline as a major field of research and teaching during the 21st century is satisfactory. Further, more emphasis should be given to field oriented investigations aiming at conservation of the tiger along with coexistence and sustainable development as tiger population is close to saturation. If the present trend is maintained, ethology should develop as a major discipline of tiger research in future. In this perspective, ethological research on endangered Mangrove tigers has a great prospect at local, regional, national and global levels. Tiger conservation can support the realisation of UN Sustainable Development Goals in Asia.

Continuous observations and recordings of individual and social behaviours of big cats in the Sundarbans mangroves, biological rhythms, assessment of their daily activities, life cycle in different seasons, ecological adaptations and development using ethological and physiological parameters, and management of their habitats appears to be particularly complex inasmuch as these animals are indeed prone to develop species-specific stress-related behavioural problems: the females usually show stereotypical behaviour whereas the males do not.

This book contains well-researched (literature review/field surveys) and 48 years' experience-oriented information on the origin and adaptation of Bengal tiger (*Panthera tigris tigris*), the only large carnivore species inhabiting the mangrove forests across the world. A published book highlighting the updated information on the migratory behaviour, plant-tiger interaction, morphological and behavioural adaptations of the Bengal tiger in the mangrove ecosystem is not readily available. In this context, this book is a different one. This monograph is ideal for the researchers, postgraduate and graduate students in zoology, botany, ecology and conservation. This comprehensive treatise will serve the professionals, such as foresters, environmentalists, conservationists, resource managers, planners, government agencies, academic institutions, NGOs and naturalists.

Acknowledgements

I would like to register my acknowledgement of and sincerest appreciation to all the individuals, institutions including the Wildlife Wing of the Forest Directorate, Government of West Bengal, for an intriguing work opportunity for 37 years from 1976 to 2012, quite literally a treasure trove, the giant one, to the best of my knowledge and belief. The Wing was created after promulgation of the Wildlife (Protection) Act, 1972 in 1973. Since then, the Wing has not only contributed to the conservation of tigers in the Sundarbans but also the conservation of the entire ecosystem.

As a part of the wildlife management system, I had learnt the intricacies of tiger conservation since the initial stage, particularly from Late Dr. Ram Krishna Lahiri and Late Amal Bhusan Chaudhuri. From 1999, I was preparing a handbook on the mammals of South Asia, which was published in 2010 (see reference), when I had learnt a lot about the tigers. I am also indebted to Mr. Ganesh Behari Thapliyal, who engaged me in the in-house research projects for biodiversity assessment of the protected areas (PAs) in the state and entrusted me to establish a wildlife data bank, which helped me immensely in writing this book.

After retirement from the Wildlife Wing, I was involved in a Sundarban-based project for five years, which was published in 2020 by the Cambridge University Press, UK, entitled 'Primates in flooded habitats: Ecology and conservation'. During all these years, I have had a thorough experience of working in the Sundarbans mangroves in India and Bangladesh, which has enriched my knowledge about the tigerland and its wild habitants and I could include those data in this book. Special thanks go to three editors-Katarzyna Nowak, University of the Free State, South Africa, Adrian A. Barnett, National Institute for Amazonian Research, Brazil, and Ikki Matsuda, Chubu University Academy of Emerging Sciences, Japan for involving me in this novel project.

I also thank all the frontliners and unsung heroes working day and night to protect the mangrove forests and the wild denizens, who helped me a lot in providing relevant field data.

I always remember the crew of MS Manorama, who miraculously saved my life from drowning in the Bay of Bengal due to a sudden storm surge in the evening (a nor'wester) during the return trip from Haldibari camp.

There are countless official colleagues, friends and associates to whom I would like to convey my gratitude for their help throughout the preparatory phase, but for the sake of brevity here I could not mention their names. I deeply regret this.

Finally, salute to everyone on my publishing team of the Ethics International Press Limited for their unstinting support.

Chapter 1
Introduction

"A tiger does not shout its tigritude, it acts."– Wole Soyinka

Tiger Evolution

Through palaeo-genomic analyses, Hu *et al.* (2022) identified a Pleistocene tiger from northeastern China, dated to beyond the limits of radiocarbon dating (greater than 43,500 years ago). Based on the mitochondrial genome, the divergence time of this ancient lineage was estimated to be approximately 268 ka (95% CI: 187-353 ka), doubling the known age of tigers' maternal ancestor to around 125 ka (95% CI: 88-168 ka). One of the Bengal Tiger's oldest ancestors is the sabre-toothed tiger (*Smilodon populator* Lund, 1842) that lived 35 million years ago in South America and *Proailurus lemanensis* (Filhol, 1879), a smaller prehistoric cat, living 25 million years ago in Europe and Asia. Other old tiger fossils date back to 2.16-2.55 million years in China. Scientists believe that a big change was occurring at that time when the tigers had to migrate very far to survive. Some think that it was a rise in sea level, then South China would have been flooded and that is where the first tigers were.

The Bengal tigers are thought to have come into India around 12,000 years ago because no tiger fossils have been found in the area before that time. The Bengal tiger has eventually adapted to its new habitat conditions. The biogeography of the Bengal tiger is presently confined to India as well as parts of Bangladesh, Nepal, Bhutan, southwestern China, Medog, Tibet autonomous region, and Myanmar.

The tigers are present across 11 Asian nations, despite their recent range collapse, occupying diverse habitats including estuarine mangrove forests (the Sundarbans), dry deciduous forests (parts of India), tropical rainforests (Malay Peninsula), and cold, temperate forests (Russian Far East)(Armstrong *et al.*, 2021). Demographic models suggest recent divergence (within the last 20,000 years) between subspecies and strong population bottlenecks (*ibid*) like natural or man-made disasters, geographic isolation, and natural selection. Amur tiger genomes revealed the strongest signals of selection related to metabolic adaptation to cold, whereas Sumatran tigers show evidence of weak selection for genes involved in body size regulation. However, the specific adaptations of the various populations to their habitats remain largely unknown like the tigers occupying the estuarine mangrove forests (the Sundarbans). However, their sampling of the most populous (Jhala *et al*. 2015) and genetically diverse tiger subspecies- the Bengal tiger- was limited across habitats.

Taxonomy

At least six lines of correlated evidence are required for taxonomic certainty-

1. Morphological- Taxa are diagnosably distinct on the basis of several characters (e.g. skull, pelage) in comparison with all other members of a species or genus (excluding hybrids) from throughout their respective geographical ranges.
2. Genetic- Taxa are genetically distinct based on a variety of genetic information, including mtDNA, Y-chromosome markers, Single Nucleotide Polymorphisms SNPs, etc.
3. Biogeographical- Distinct taxa are more likely to be recognised where there are distinct geographical barriers relevant to the taxon, e.g. rivers, seas, mountains, deserts, or where geological events, such as sea-level changes, or volcanic eruptions are broadly coincident with coalescence times, or where recolonisations following climate change are consistent with former refugia.
4. Behavioural- e.g. predisposition to taming.
5. Ecological- e.g. use of distinct habitats with appropriate adaptations.
6. Reproductive- e.g. seasonality or not of reproductive cycles.

Scientific name

The tigers, having binomial (generic and specific) scientific nomenclature *Panthera tigris* (Linn. 1758), with their distinct yellow to orange-red coat, white areas on the chest, neck, and inside of the legs, individually (not uniform) identified by unique face-markings, broken black-stripes and tail-rings as well as pugmarks, are one of the most captivating carnivorous species on our planet and have evolved to thrive in the diverse climates and ecosystems, which have allowed them to become one of the most successful big cats on the planet. Historically, they inhabited much of Asia, and various tiger subspecies with somewhat different characteristics, naturally migrated, and spread out over time, for which a third (trinomial) name 'tigris' is applied.

The present system of classification of the species is a development from that originally proposed by an eighteenth century biologist Carl Linnaeus. The taxonomic classification of the big cats has fluctuated since new morphological and genetic research has been documented. The Felidae (Fischer von Waldheim, 1817), originated in Central Asia in the Late Miocene, has 18 genera encompassing about 40 species. There are 13 genera within the Felinae subfamily, four genera within the Pantherinae subfamily and one within the Acinonychinae subfamily (Pocock, 1917). Initially, Linneaeus classified it as *Felis tigris* (now Basionym of *Panthera tigris*), but later scientists placed it under the genus *Panthera* because of its distinctive elastic hyoid bone, a U-shaped structure situated at the root of the tongue in the front of the neck and between the lower jaw and the largest cartilage of the larynx or voice box.

Felid Evolution

The modern felids of the 'Panthera' lineage diverged from the ancestral cat species around 10.8 million years (Myr) ago, while five large ('Great Roaring Cats' of genus *Panthera* Oken, 1816) and two medium (Clouded leopards of genus *Neofelis* Gray, 1867) felids diverged between 6 and 6.4 Myr ago (Luo *et al.* 2010) and spread in various environments from high mountains, savannahs to the rainforests of Africa, Asia and Central and South America. The split of the *Panthera* lineage was followed by a rapid series of divergence and migration events starting around 3.7 Myr ago that led to the five extant *Panthera* species [Lion (*Panthera leo* Linnaeus, 1758), Jaguar (*Panthera onca* Linnaeus, 1758), Snow Leopard (*Panthera uncia* Schreber, 1775), Leopard (*Panthera pardus* Linnaeus, 1758) and Tiger (*Panthera tigris* Linnaeus, 1758)]. The Asian-derived *Panthera* species subsequently spread into America (Jaguar), Africa (Lion and Leopard) and others remained in Asia (Tiger, Snow Leopard and Clouded Leopard). Paleo-tigers are ancestral forms of the phyletic line that only much later led to the current tiger. The most recent common ancestor lived about 110,000 years ago (Liu *et al.* 2018). They were compelled to spread southwards in search of suitable habitats as successive Pleistocene glacial and other geological events made Northern Asia inhospitable.

Prior to three extinctions [Balinese tigers (*Panthera tigris balica*) since the 1950s, Caspian tigers (*P.t. virgata*) since the mid-1900s, and Javan tigers (*P.t. sondaica*) since the mid-1970s], there were eight different tiger subspecies. There is, however, some debate today that the five remaining subspecies should, in fact, only be three (based on DNA analysis two merged with *P.t. tigris*)- (a) Type locality Bengal, fixed by Thomas, 1911, *Proc. Zool. Soc. Lond.*, 1911: 135 [now coupled with the South Chinese and Corbetts (with three common names- (i) Corbetts named after Jim Corbett, famed hunter of man-eaters and conservationist, (ii) Indochinese and (iii) Malayan tigers], (b) Amur (erstwhile Siberian) and (c) Sumatran.

The tiger population of the Sundarbans is paraphyletic (i.e. descended from a common evolutionary ancestor but not including all the descendants) within the Bengal tiger (Aziz *et al.*, 2022). The tigers inhabiting the Sundarbans mangrove forests were assumed to be migrants from the mainland especially Central India, via Chhota Nagpur Plateau and have been isolated recently; but phylogenetic ancestry of the present residents has remained poorly understood until recently. When an established tiger habitat reaches its full carrying capacity of breeding population, generally the young tigers are forced to move out in search of new territories risking their lives while passing through hostile terrains. They are known not only to move over long distances between fragmented forests through corridors to establish their own territories but also able to reproduce in new areas, thereby contributing to the genetic diversity of subpopulations. Such dispersal and subsequent reproduction is crucial for the maintenance of long-term genetic health in the fragmented tiger population in the Sundarbans mangroves.

Studies suggest that the Bengal tiger crossed the Eastern Himalayas through the corridors of North-east India, and occupied most of the country since the Late Pleistocene between 12,000 and 16,500 years ago. While tigers populated Central India about 10,000 years ago, their population subdivision began only recently owing to habitat fragmentation, historical events, human pressure and land-use patterns. The regional tigers diverged and isolated from each other around 8,000-9,000 years ago due to intra-specific competition, loss of habitat and corridors.

About 6000 years ago (in the mid-Holocene), the shoreline of the North-eastern Indian Peninsula was situated to the west of the present shoreline which was comparatively closer to the foothills of the Eastern Himalayas. Therefore, the present Sundarbans area did not exist for terrestrial life until the shoreline had moved eastward and the excess population of tigers could arrive from Peninsular India. Hence, the gene pattern of the Sundarbans tiger is identical to the big cat population of the Central Indian landscape including the States like Madhya Pradesh, Chhattisgarh, parts of Andhra Pradesh and Maharashtra. In fact, they migrated to the new coastal mangrove habitat on the east, formed between 2500 and 5000 years ago by the silt deposition of the river Ganges and its tributaries/distributaries. Since the Sundarbans tiger is morphologically distinct in terms of skull and body size the population was evaluated further to determine if it is an evolutionary significant unit (Barlow 2009).

Distribution

Historically, tigers inhabited Mesopotamia and portions of South and East Asia. During the 19h century, they were also found in parts of Turkey, China, Afghanistan, Iran, and Russia. Their eastern distribution ranged from Korea, eastern China, and selected areas within the region of Southeast Asia.

The extant subspecies are native to a wide range of highly threatened habitats throughout Asia, including tropical forests, savannas, grasslands, mountainous regions and even mangroves. They are confined to 11 range countries including India and Bangladesh (study area). These habitats provide them with food, water and cover they need to survive and procreate.

The historical distribution of tigers exemplifies the variety of habitat types to which they have adapted. The tigers are found mainly in the forests of tropical Asia, although they historically occurred more widely in the drier and colder climes with a sufficient prey base, particularly native ungulates-odd-toed (Perissodactyla) and even-toed (Artiodactyla), key to their successful reproduction.

In Turkmenistan, Uzbekistan, and Tajikistan, the tiger was found in the drainage basins of rivers and lakes, where they hunted in the "tugai", which consists of thickets of low-stature trees (turanga, tamarisk), shrubs, and dense reed beds. In Kazakhstan, tigers sometimes ascended into montane forests, attaining heights of 2,500 m, in the summer in pursuit of wild boars. There are also records of tigers in the Himalayas at altitudes of almost 4,000 m, although in most areas these big cats remain well below 2,000 m. In China, they occupied grass thickets, montane sub-tropical evergreen forests, and mixed forests dominated by oak and poplars. Tigers in the Russian Far East live in low mountainous terrain dominated by nut pine, birch, oaks, fir, and spruce. Winter is harsh in this region with deep snow and temperatures dropping to -34°C. In Sumatra and Malaysia, tigers are found in lowland humid tropical rainforest, where precipitation exceeds 2,000 mm annually. In the outwash areas south of the Himalayas, tigers inhabit the "terai", a belt of floodplain habitat dominated by marshes, swamps, oxbow lakes, and tall, dense grasslands intermixed with riverine forest. Sal forest, a climax form of moist deciduous forest, occurs on the slopes of the adjacent hills.

In India, tigers inhabit the tropical, wet evergreen, and semi-evergreen forests of Assam, the mangrove swamps of West Bengal and neighbouring Bangladesh, the vast expanses of dry deciduous forest in the central plateau, the tropical moist and dry deciduous forests of the Western Ghats, and the thorn forests of Rajasthan and Gujarat (now extinct, the last reported tiger in Dangs was shot by a poacher near Waghai in 1983). In Ranthambore NP, Rajasthan, tigers also use the ancient temples and fortresses as places to live during the day. The tiger habitats range from the temperate pine-oak forests of the Russian Far East and the rocky mountain slopes of Manchuria to the tall grasslands of Nepal, up to the mangrove swamps of the Sundarbans and, of course, the rainforests of Malaysia and Indonesia in the Southeast Asia.

The extant tigers are discussed below at subspecific levels.

Mainland Asian populations

(1) The Bengal Tiger (*Panthera tigris tigris* Linnaeus, 1758) has reddish-orange coloured fur with broad black vertical stripes on its head and back, which are also found on the skin, so that if it were to be shaved, its distinctive coat-pattern would still be visible. It is the native of mainland Asia and confined to the Indian Subcontinent. At the beginning of the 20th century, the scientists believe that there were more than 100,000 tigers throughout most of Asia and about 40,000 tigers were estimated to inhabit the Indian subcontinent.

The Malayan tiger (*P.t. jacksoni/malayensis*), native to Peninsular Malaysia, and the northern Indo-Chinese Tiger (*P.t. corbetti*) in Cambodia, China, Laos, Eastern Myanmar, Thailand and Vietnam are now considered *P.t. tigris* subspecies (Kitchener *et al.* 2017); the gene flow between the Sundarbans mangrove islands and the mainland populations is highly restricted.

(2) The Amur (Siberian) Tiger (*P.t. altaica*) is lighter and has fewer stripes than other tiger subspecies. It has long and thick hair to help them stay warm in their cold climate of eastern Siberia, Manchuria to Northeast China, and possibly North Korea; 38-45 individuals remain in the mountain forests along the Sino–Russian border (Xiao *et al.* 2016);

(3) The south China (Amoy) Tiger (*P.t. amoyensis*), native to southern China, listed as Critically Endangered, has not been directly observed in the wild since the 1970s; during field studies (camera traps, GPS technology and extensive sign surveys) in 2001, no evidence of tigers or scats were found and is possibly extinct; and

Sunda island populations

(4) The Sumatran Tiger (*P.t. sumatrae* Pocock 1929), darker in colouration than other tiger species with a deep orange to reddish coat and black stripes, has a population of <400 individuals estimated in 2022; it occurs in tropical broadleaf evergreen forests, freshwater swamp forests and peat swamps.

Extinct tigers

Three subspecies of the tigers are now extinct:

(i) The Javan Tiger (*P.t. sondaica* Temminck, 1844), native to the Sunda islands and inhabited lowland tropical forest until the mid-1970s; last sighted in 1976 and, in 2003, listed as extinct (IUCN Red List) but many are hopeful that the Javan tiger still lives;

(ii) The Bali Tiger (*P.t. balica* Schwarz, 1912), native to the tropical forests of Bali island; extinct since 1950s; and

(iii) The Caspian (Turan/Hyrcanian) Tiger (*P.t. virgata* Illiger, 1815) inhabited Western and Central Asia in dry river valleys of the Taklamakan, western slopes of the Tianshan mountains, Amudarya and Syrdarya river valleys, shores of the Caspian sea, Elburz mountains, eastern Turkey, Tigris and Euphrates river valleys; assessed as extinct around the 1950's.

Population dispersal

The tiger's natal philopatry, which is defined as continued residence on the natal home-range past the age of independence from the parents, across various ecological and life-history regimes depends on a number of variables like adult turnover rates, habitat saturation, spatial patchiness of resources, advantages of familiarity with the natal home range, and reliance on extensive home range "improvements". The most interesting consequence of natal philopatry is the continued spatial association of kin into adulthood.

The recent history of the tigers in the Indian subcontinent is consistent with the lack of tiger fossils from India prior to the late Pleistocene and the absence of tigers from Sri Lanka which was separated from the subcontinent by rising sea levels in the early Holocene. The tigers expanded their range even in the drier regions of North-West India by moving into newly opened ecological niches. While Ranthambore in Rajasthan, cut off from other tiger conservation landscapes, is the western-most extant population of wild tigers, the Sundarbans landscape is the easternmost habitat of the Bengal tiger and completely isolated from its northern and south-western counterparts. Difference between the island tigers and mainland tigers is that mainland tigers with higher body mass and bigger sizes are more equipped with catching larger prey while the island tigers are efficient for preying on the smaller ungulates (Seidensticker and McDougal, 1993).

Studies revealed fascinating levels of tolerance and adaptation in tigers as they inhabit and persist in human-dominated landscapes while dispersing in search of suitable habitat. Studies also showed that tigers can exist without being detected by people and go to extreme extent to avoid conflict in order to achieve safe passage over hundreds of kilometres. A wild tigress fitted with a satellite GPS collar was tracked as she dispersed in the Vidarbha region of the Central Indian Tiger Landscape. The tigress had been rescued by FD staff on 12th October 2011 from a water duct she had fallen into in the Nagpur district of Maharashtra and was released on 27th November 2011 in a forest near her site of capture.

It was observed that she remained in the same forest for almost a month until 25th December and then started moving eastwards into a human-dominated landscape. Human density in the district averages 409 people per km^2 and village density here averages 5.3 villages per km^2. In the immediate area of the tigress' vicinity, human density was still as high as 200 people per km^2. The landscape she now inhabited was a mix of forests, agricultural fields, villages, schools, roads etc. The home range of the tigress in such a challenging landscape encompassed a massive 431 km^2. She avoided human detection by being almost entirely nocturnal and resting in foliage and shade by day. The scientists tracking her found that she often rested very close to places where humans were active by day- sometimes as close as 100 metres- but no untoward incidents ever occurred. She was subsisting largely on a diet of wild pigs, though she also killed cattle and even scavenged.

The scientists lost track of her after her collar ran out of battery and stopped functioning, but she was photographed a year later in April 2013 just 40 km from her site of release, indicating that she was still inhabiting the same region.

The area is in the vicinity of the Tadoba-Andhari, Pench and Navegaon-Nagzira TRs, all of which are source sites from which tigers can potentially disperse into the rest of the landscape.

During the study, the tigress demonstrated the ability of her species to cross human habitation, agricultural fields, roads and railway lines in its search for safe habitat when moving from one PA to another. The case of this dispersing tigress outside PAs implies that conflict is imminent and that activities such as mining, construction of canals and deforestation cause fragmentation of forests and increase human-tiger conflict. Yet in areas where large carnivores are forced to use human-dominated landscapes to disperse, conflict remains minimal particularly if precautions and safeguards are taken by people.

Gour *et al.* (2013) established male-biassed dispersal and female philopatry on the basis of genetic data collected from 22 females and 6 males within the core area of Pench TR, Madhya Pradesh. Females showed positive correlation up to 7 km (which corresponds to an area of approximately 160 km). However this correlation is significantly positive only up to 4 km (corresponding to approximately 50 km). Males do not exhibit any significant correlation in any of the distance classes within the forest (up to 300 km). Evidence of female dispersal up to 26 km in this landscape was also observed.

The tiger's home ranges in the Sundarban mangroves are small in the suitable habitats for the breeding, feeding, resting, or sheltering. Here, the habitats have been changed or fragmented and natural resources are depleted due to overexploitation; for example, where the preferred large ungulate prey is scarce. In order to augment the mangrove habitat potential and facilitate the continued dispersal of tigers within the landscapes, priority ecological corridors have to be restored and connectivity maintained involving the multiple stakeholders.

For survival, the tigers need unrestricted and safe passage in its home range including the natural corridors to breed, hunt and establish their own territories. They live in diverse landscapes from rainforests to grasslands, savannahs, montane forests, boreal forests and mangrove forests. But, when the extensive tiger landscapes are historically cut into more isolated fragments resulting in loss of preys and corridors in between, either broken or lost forever, tigers are increasingly wedged in the smaller habitats that can lead not only to inbreeding, but also intraspecific competition and conflict for the scarce prey animals, breeding partners and other biological needs, which is ultimately associated with inflicted injuries or even death. In addition, natural and anthropogenic threats also cause habitat destruction and reduction in viable breeding populations.

Aziz *et al.* (2018) assessed the genetic status of these populations to reveal the effects of dispersal barriers, "Microsatellite analyses exhibit a signature of fine-scale genetic structure, which might have been the consequence of limited tiger dispersal due to wide rivers across the Sundarbans. Similarly, mitochondrial data show a historic pattern of population isolation that might be due to wider rivers across the entire Sundarbans shared by Bangladesh and India. Given the intrinsic nature of the mangrove habitat embedded with numerous rivers, increased commercial traffic and human activities may further impede tiger dispersal across wide rivers, escalating further genetic isolation of the Sundarbans tigers."

Comparatively, dispersing tigers in the Vidarbha landscape are using a much wider swathe of the landscape outside PAs for movement than earlier known. It extends well beyond forested structural corridors. Tigers in this landscape were pushing the boundaries of humans to accept the risk of exploring a human-dominated landscape. Data has also shown extensive use of agricultural lands for movement. Tigers have used whatever small fragment of forest patch or parcel of cultivated land with standing crops, to seek refuge during daytime. The purview of tiger conservation, which till date was thought to be restricted to lands under the jurisdiction of the forest management, now seems to extend beyond such bound boundaries and into a realm where successful conservation effort should necessarily include many stakeholders. The local people, district administration, local NGOs, and various development agencies should work in tandem with forest management. The report provides clues to managers to target proactive and preemptive management interventions for conflict prevention, mitigation, and connectivity conservation.

Population status

While Bangladesh, Bhutan, Indonesia, Malaysia, Nepal, Russia and Thailand harbour several hundred tigers, only a few are found in Myanmar and China. Tigers are now extinct from Vietnam, Laos and Cambodia. The largest population of global wild tigers (Bengal tiger *Panthera tigris tigris*) is found in the Indian subcontinent and is mostly restricted to the Central Indian landscape, whereas the Sundarbans represent the only population adapted to living in the mangrove forest habitat.

First tiger census in India

The estimate of tigers in the country has always been a wild guess. Gee (Wildlife in India, 1966) estimated the existence of 40,000 tigers at the end of the 19th century and about 4,000 tigers in 1965. According to Corbett (1955) the number of tigers was not >2,000. Sankhala (1970) estimated the number of tigers existing in the country to be about 2,500. However, a countrywide Tiger Census was done in the year 1972 to assess the status of the tiger, after the Survival Service Commission of the I.U.C.N. had drawn pointed attention to the threatened existence of the Royal Bengal tiger (*Panthera tigris tigris*) and the need for its complete protection. The result of the 1972 tiger census is given in Table below.

State	Number of tigers
1. Undivided Madhya Pradesh	457
2. Undivided Uttar Pradesh	262
3. Maharashtra	160
4. Assam	147
5. Odisha	142
6. Karnataka	102
7. Undivided Bihar	85
8. Nagaland	80
9. Rajasthan	74
10. West Bengal	73 (excluding 4/5th of Sundarbans)
11. Arunachal Pradesh	62
12. Kerala	60

State	Number of tigers
13. Undivided Andhra Pradesh	35
14. Tamil Nadu	33
15. Meghalaya	32
16. Gujarat	8
17. Tripura	7
18. Manipur	1
Total	1,820

Census techniques

The methodology adopted for this national census has been called the "Cooperation census". The method consists of the following basic operations:

The States having tiger habitats were grouped into five zones each under a Zonal Coordinator:

(i) Northern Zone: Uttar Pradesh and Bihar

(ii) Eastern Zone: West Bengal, Assam, Arunachal Pradesh, Meghalaya, Mizoram, Manipur, and Tripura.

(iii) Central Zone: Madhya Pradesh and Odisha.

(iv) Western Zone: Maharashtra, Rajasthan, Goa and Gujarat.

(v) Southern Zone: Tamil Nadu, Andhra Pradesh, Kerala and Karnataka.

The census was carried out in two stages:

(i) Eastern region from 22nd to 28th April, 1972; and

(ii) Rest of the country from 15th to 31st May, 1972.

Unfortunately, full details of the census regarding age-sex distribution of the tigers are not available and, as such, detailed analysis was not possible.

The above analysis of the various censuses highlights the following weaknesses in the conduct of various wildlife census and monitoring in India:

(1) Standard techniques for the census/estimation of different wildlife populations have not been evolved. Even in the same area the methodology followed for censusing the same population in two successive censuses has been changing depending upon the persons carrying out the census without examining various alternatives and studying the suitability of a particular type of method. As a consequence, the results, which are available, lack initial comparability and statistical reliability.

(2) Attempts at studying the structural aspects of wildlife populations as a part of census estimation are not properly emphasised. In this connection, studies in ageing and sexing of the population have to be developed further.

(3) Studies on 'natality' and 'mortality' are lacking, especially as a part of census study for purposes of studying population dynamics.

(4) Survival rates and age-specific mortalities for populations under study should form an essential part of population estimation.

(5) It is suggested that studies of population dynamics are immediately taken up for all the populations of animals which are rare or threatened and on the verge of extinction, like the black buck in Point Calimere where the results of the two census surveys reveal an almost complete absence of reproductive activity and consequent degeneration of the population. Similar studies will be needed for rare species like 'Hangul' in Kashmir (Dachigam), Swamp deer, the brow-antlered deer or the Sangai etc.

(6) A study of the age and sex structure of the lion and the tiger is also essential as the different census figures available show inherent contradictions.

(7) Census and monitoring operations as a regular management practice is on a different footing. The type of census estimation which is being made in the Palamau winter track census needs to be statistically refined as suggested in the paper so that results can be analysed on more scientific lines. It is suggested that similar continuous censuses may be made a part

of regular wildlife management practice in all wildlife reserves and sanctuaries in the country.

Conservation efforts in the ecologically sensitive Sundarbans

With a key eye for profits, the colonial rulers realised that large parts of the Sundarbans, which were acquired in 1757, covering parts of three provinces- 24-Parganas, Bakarganj and Jessore districts, were densely forested and far too marshy and saline for cultivation.

"Lying between the Bay of Bengal and the inhabited parts of the Delta, this dense forest with its accumulated and perpetually exhaling malaria, urged by the southwest monsoon spread disease and death over the whole country; the tract swarms with tigers and other wild beasts whose ravages cause wide destruction of life and property; it affords convenient shelter for smugglers and river pirates…..To remove or abate this source of so much material suffering, and to afford employment for hundreds of thousands of cultivators is undoubtedly the first object of Government." - Lord Dalhousie (Calcutta Review; 1858)

This receding of the jungles followed the logic of revenue maximisation and rhetoric of improvement. The rationale behind this was that a land covered with impenetrable forest, a hideous den of beasts and reptiles could only be improved by deforestation.

The idea of forests as useful, and as something that merits conservation did not strike the imperialists in the 18th and the early 19th century when Sundarbans, as Hunter (1875) described the region as a 'drowned island', were 'impenetrable forests' and 'thick brushwood', without any restrictions on subsistence based activities like fishing, honey collection, fuel-wood collection etc. The vast forest clearing operation started in 1781 by bringing migrants from Midnapore (Medinipore) and other parts of central Bengal, primarily tribals.

Tigers were treated to be a major threat for land reclamation, since it was estimated by the second half of the 19th century that tigers killed about 1,600 people every year (Chakrabarty 2010). It was only during the 1870s, that the government, out of a sense of extensive forest depletion, had a realisation that such uncontrolled felling need to be checked. The passing of the Forest Act of 1878 reconfirmed the fact that the forest is a property of the government, but included the category of 'reserved' or 'protected' forests within the forest conservation laws.

Like Dietrich Brandis, the First Inspector General of forests in India, Wilhelm Schlich, the Conservator of Forests, Bengal (1872), was concerned over the disappearance of the Sundarbans forests. Schlich understood the importance of the Sundarbans supply of timber, thatching grasses and fuel wood. In 1875 he along with B.H. Badden Powell inaugurated the journal *Indian Forester*. He wrote an article entitled "Remarks on the Sundarbans" and gives a fair representation of the concerns over conservation which underlay much of the period's scientific forestry. He observed that because of the peculiar ecology of the region, regeneration is uncertain, especially where the forest has been extensively cleared. He further wrote, "It is our duty to see that the supply is not exhausted…, the demand is certain to increase, and we must therefore make sure that the increase also is provided for. It has been said that the supply is inexhaustible, but such is not the case. It appears, on the contrary, that the western part of the Sundarbans, which is nearest Calcutta, is already exhausted to a large extent and that fuel cutters proceed more to the east, year after year."

The strategies drew on four principles- (i) State competence, (ii) State ownership, (iii) Forest economy and (iv) Limited user access.

About 70% of the wasteland of Sundarbans (about 2,000 km²) had been cleared during the period between 1830 and 1873. At that time the Sundarbans were identified as the area bounded on the north by the limits of Permanent Settlement in the 24-Parganas, Jessore (later Khulna) and Bakarganj districts and on the south by the sea face stretching from the Hooghly estuary to the Meghna river-mouth. The total area of the Sundarbans was divided into four categories in the Commissioner's 1873 report detailed below (Table quoted by J. Richards and E. Flint).

- Leased/cultivated- 2,792 km² (14.3%)
- Leased forests- 1,562 km² (8.0%)
- Unleased forests- 10,274 km² (52.7%)
- Water- 4,881 km² (25.0%)
- Total Area- 19,509 km².

A Deputy Conservator of Forest was sent to the Sundarbans in 1873, following which a rudimentary structure of forest administration was set up in the area. Toll stations and offices for issuing licences were established. The tidal forest over 2,292 km², lying within Khulna district, was demarcated in 1875 as the Sundarban FD. The next year Schlich succeeded in adding an additional 1,802 km² to the reserved area under the control of the FD. However, the Conservator was unable to persuade Sir Richard Temple, the Lieutenant Governor of Bengal, to transfer the entire unleased area of the Sundarbans to reserved forest status. This colonial policy eventually created today's Sundarban forest.

Part of the Sundarbans was declared as a Reserve Forest (RF) in 1878. The first formal forest policy was adopted in 1894. 4,095 km² of RF in Khulna and 4,480 km² of Protected Forests (PFs) in 24-Parganas were established during 1890 under the Indian Forest Act of 1878. Khulna also possessed 65 km² of PFs. In the Indian Sundarbans, the undivided 24-Parganas district was first declared as PF following a notification dated 7th December, 1878. The officially recognised area of the Sundarbans was diminished by 2,608 km² (13.3%) in about 30 years, i.e. from 19,510 in 1873 to 16,902 km² in 1904. In the same year, 2,401 km² of reclaimed cultivated land area within the Sundarbans was recorded, which represented new lands leased, cleared and cultivated over the ensuing years covering almost three decades.

By designating the 24-Parganas tidal forests as Protected rather than Reserved, the FD left itself an option. It could either lease these lands for clearing and conversion to rice, or it could transfer them to timber production and management as RFs. The area classified as PFs stayed relatively constant from 1890 through the 1930's between 4,400 and 4,500 km². The boundaries of the remaining PFs in 24-Parganas were fixed by Notification No. 4457-For., dated 9th April, 1926. The PFs in the Basirhat subdivision of the district, presently Basirhat Range in STR, were constituted as RFs as per Government Notification No. 15340-For., dated 9th August, 1928. The PFs of Namkhana Range (except Mahisani and Patibania Islands) were also finally declared as RFs under Notification No. 7737-For, dated 29th May, 1943.

In 1936-1937, five blocks (Dubla, Katkar, Dhondal, Dakshmir Mandarbaria, and Talpatti Maidan) were closed to hunting in order to protect the species in the Bangladesh Sundarbans.

Ecological Landscape

This typical habitat spreads over approximately 10,000 km^2 in both India and Bangladesh on the northern Bay of Bengal and has the highest ecosystem service value to serve as one of Earth's most vital carbon sinks. The Sundarbans NP in India has stores of 60 million tonnes of carbon (Mt C) as against 110 Mt C in the Bangladesh Sundarbans, which varies significantly among different vegetation types, age, salinity zones and vegetation functional attributes. The ecological integrity of the Sundarbans tiger is well maintained by the two countries.

Indian Sundarbans

a) Mangrove forests in the southern part of the Sundarban Biosphere Reserve (SBR) designated as NP or Core Area and Critical Tiger Habitat (CTH);

b) Mangrove forests in Sajnekhali WLS on the north and

c) Mangrove forests in 24-Parganas (South) Forest Division (FD) on the West with three wildlife sanctuaries (WLSs) - West Sundarban, Lothian Island and Haliday Island.

d) Lothian was first notified as a sanctuary over 14.67 miles^2 in 1948, where the major fauna were chital, wild pig and occasionally tiger. Sajnekhali was first notified as a sanctuary over 139.92 miles^2 with a population of tigers (not specified) in 1960. Haliday was first notified as a sanctuary over 2.3 miles^2 in 1960. All the PAs were renotified in 1976 as per provisions of the Wildlife (Protection) Act, 1972.

Bangladesh Sundarbans

Mangrove forests extend over 6,017 km^2, of which the forest land area covers 4,143 km^2 including three designated WLSs declared in 1977- West (715 km^2), South (370 km^2) and East (323 km^2) under the Bangladesh Wildlife (Preservation) (Amendment) Act, 1974, totalling 1,408 km^2, and 1,874 km^2 comprises a maze of numerous estuaries, rivers, channels and creeks. Out of the tiger-occupied forests in Bangladesh Sundarbans, a total of 1,263.7 km^2 was selected for camera trapping with emphasis on the above WLSs. Tigers are known to be present throughout the Bangladesh Sundarbans, with higher concentrations in the south and west compared to the north and east (Barlow *et al.* 2008).

Zonation

Indian Sundarbans

The 15 forest Blocks of the STR have been divided into five Zones, i.e., the Northern (Jhilla, Panchamukhani and Pirkhali), the Western (Chottohardi, Matla and Netidhopani), the Eastern (Arbesi, Harinbhanga and Khatuajhuri) and the Southern (Bagmara, Gona and Mayadwip) and the Central (Chamta, Chandkhali and Gosaba).

Although the Indian Sundarbans region consists of 4,267 km^2 (41.48% in the deltaic Hooghly-Matla estuary) including five PAs and RFs, in 2014 tiger occupancy was estimated to be 1,841 km^2 or about 43.14% (increased to 2,313 km^2 or 54.20% in 2018), whereas the prey occupancy was a bit low, i.e. 1,184 km^2 (Chital) and 1,591 km^2 (wild pig) in 2008.

Bangladesh Sundarbans

Areas of the management zones (in ha) (Halder, 2011)

Management zone	East WLS	South WLS	West WLS
Strict protection zone	38,976	43,772	162,060
Rehabilitation zone	518	744	2,296
Recreational zone	1,055	590	2,449
Special use zone	5	15	24
Total	40,554	45,121	166,829

(i) Strict protection zone: This zone is defined as an area in which rich biodiversity is present and it is closed to human settlement. The strict protection zone allows for scientific studies and research and controlled outdoor recreation activities.

(ii) Rehabilitation zone: This is a degraded area for the restoration of natural habitat associated with biodiversity on a long-term basis and rezonation to a stricter protection level. Here natural regeneration is aided with controlled fire, enrichment plantations with indigenous species and sometimes with exotics for the restoration process.

(iii) Recreational zone: The zone in which high recreational, tourism, educational or environmental awareness values are allowed with priority on sustainable conservation and ecotourism.

(iv) Special use zone: This zone is the area containing existing infrastructures and fenced compounds and installations of natural importance such as Bangladesh Naval Base and Bangladesh Port Authority at Nilkamal inside the south sanctuary.

Buffer zone

Buffer zone is defined as a strip of land/water body outside each sanctuary, but adjoining it. This zone is intended to be managed to provide a social fence and at the same time used for controlled economic activities such as seasonal fishing, gathering of fuelwood, small-sized timber, Nipa Palm, honey, wax and other forest product collection and seasonal hunting of game species. The gathering of forest products and fishing shall be strictly prohibited in the buffer zones. Controlled outdoor recreational activities such as fishing, hunting, boating, etc., shall be allowed in this zone. A strip of about 5-km wide around each sanctuary will serve as a buffer zone.

Buffer zone of sanctuaries

Sanctuary	Area (ha.)
East	27,699
South	23,705
West	15,095

Working plans

India has a long tradition of mangrove forest management. The Sundarbans mangroves, partly in India and partly in Bangladesh were the first mangroves in the world to be put under scientific management as early as 1892 (Heining) and the first working plan came into force in 1893-1894 followed by Lloyd, 1904; Farrington, 1906; Trafford, 1911; Curtis, 1933; Choudhury, 1937; Roy Chaudhary, 1948;

Management Plans of STR

1st (Lahiri, 1973), 2nd, 3rd and 4th (the then FD).

Mangroves vegetation type and dominant species

Mandal *et al.* (2019) pin-pointed the mangroves vegetation type and dominant species in the trans-boundary Sundarbans as follows:

Sl.no	Type of distinct mangrove vegetation	Dominant species	
		Bangladesh	India
1	Tall dense forests	*Heritiera fomes*	*Avicennia officinalis*
2	Low dense forest	*Ceriops* spp	*Ceriops* spp
3	Low dense forest at edge of mangroves	*Nypa fruticans*	-
4	Open mangroves thickets	-	*Phoenix paludosa*
5	Scattered mangroves under shrubs	-	*Suaeda* sp
6	Abundant throughout area coverage	*Excoecaria agallocha*	*E. agallocha*

Sl.no	Type of distinct mangrove vegetation	Dominant species	
		Bangladesh	India
7	Vegetation at newly formed land	*Sonneratia apetala*	*Avicennia alba, A. marina*

Conservation significance

The single-patch mangrove forests, commonly known as the Sundarbans, is the only area in the world known to have the Bengal tiger (*Panthera tigris tigris*). It is considered one of the most complex and most delicate ecosystems in the world. The Bengal tiger requires a steady supply of prey to thrive and large contiguous mangrove forests with fair interspersion of undisturbed breeding areas. The dense mangrove trees provide excellent shade and cover for tigers to evade hunters and stalk their prey, in spite of the fact that the mangrove forest is a challenging habitat with storms, floods, marshy banks, shifting tides, and pneumatophores that stick out of the muddy ground like spikes at a height of 20-30 cm above soil to make hunting very difficult (20% success-rate). Tides often wash away the soil, leaving a skeleton of roots.

The average minimum territory size for a female tiger is around 15-20 km², though some studies have suggested it could be closer to 10 km² in some places, whereas territories of the breeding males may be extended up to a range of 60-100 km², including the home ranges of a few breeding females.

Today, the Sundarbans tiger and its mangrove habitats are considered endangered, which is an indicator of the mangrove ecosystem in crisis. Tiger is, therefore, a conservation-dependent species. Tiger densities are high in those landscapes where the biomass of ungulate prey is extraordinary.

Bengal tigers are defying 12,000 years of evolution by thriving alongside people. Under the right conditions, they are learning to live alongside their dangerous neighbours to save themselves from extinction. These results challenge the narrative within some conservation circles that humans and megafauna are incompatible.

In the mangrove landscape there was a trend during the 19th century towards 'trophic downgrading,' a term referring to the disproportionate loss of the mangrove's largest animals. Trophic downgrading was usually worse in forest edges now converted to completely human-dominated landscape because hunters with arms used to conduct frequent trips for targeting those larger species and return with enviable trophies.

The present scenario is an outcome of tougher anti-poaching efforts in the PAs that are closer to human settlements and are more frequently visited by tourists who shoot the big cats and other associated species with cameras and mobile phones. More tigers are living in substantially less area which has troublesome implications for the flagship species. The increase of tiger populations shows good management of tigerlands, but it has lost ground in connecting corridors. That the number of tigers going up in insufficient areas increases chances of tigers coming in conflict with people. Hence, fragmented landscapes should be consolidated and protecting the existing corridors is the only solution for long term conservation of viable populations of the mangrove tigers.

Tigers are a conservation-reliant species and will likely remain so through the 21st century (Sanderson *et al.* 2019). Indeed, probability of success in reducing the risk of extinction can be improved by empowering conservation managers in three important ways:

(1) By increasing the accuracy of the knowledge of the distribution and abundance of tigers and their prey so that conservation managers can precisely identify and counter threats;

(2) By increasing the understanding of the extent and character of the threats that create extinction risks for tigers so that conservation managers can combat those threats; and

(3) By diffusing this improved knowledge conservation managers and conservationists can educate and engage key stakeholders and more effectively mobilise multi-sector support in efforts to secure a future for wild tigers.

Chapter 2
Materials and Methods

"Where tigers thrive, it is a sign that the ecosystem is healthy."

The present study is an attempt to investigate into the tiger populations inhabiting different habitats of the greater Sundarbans in India and Bangladesh by combining the learnings, knowledge, data and information, scattered in different published literature and also grey literature of the FD with the objectives to identify and bridge the knowledge-gaps through field-based rapid assessment in the potential habitats.

It was not possible to conduct a forest beat-wise survey in the Sundarban Landscape. Since Sundarbans is a unique and hostile tiger habitat, a separate protocol was evolved for evaluating tiger, prey, and habitat status for the Sundarban landscape. Transect surveys were not possible due to hostile habitat, lack of logistic infrastructure, stilt root/pneumatophores and muddy forest floor rhythmically flooded by two high tides every day. The surveys were mainly carried out at high tide, but some important areas had to be surveyed at low tide. Several areas were also surveyed two or three times and at different stages of the tidal cycle to assess tidal differences in direct sighting, indirect evidence and distribution.

After conducting a structured literature review, a database was prepared for facilitating field verification during the study period (April, 2019-March, 2023), including three seasons- (i) dry (October to March), (ii) pre-monsoon (April-May) and (iii) monsoon (June to September).

A sample questionnaire survey on a set *pro forma* was also conducted among the fringe villagers, who depend on the forest resources for their livelihood, the frontline staff, tourist guides, representatives of local Eco-development/Forest Protection Committees and non-government organisations.

Reconnaissance survey was carried out after selecting suitable sites based on the following criteria:

i.   Tiger pugmarks; and

ii. Comparatively high elevation areas unlikely to get submerged even during high tides.

Dense vegetation which made access difficult as well as those with excessive human disturbances was excluded. Males were differentiated from females based on the presence of testicles.

Tigers in the mangrove Sundarbans are often extremely difficult to track or enumerate due to their elusive nature. Before camera trapping or radio-collaring or ear-tagging, the non-invasive "sign surveys" relying on indirect evidences such as prey kills, pugmarks, scats [categorised into (a) Old: dry with hair and bones visible; (b) Fresh: dry but intact with shiny surface; and (c) Very Fresh: soft, moist, and smelly], scent marks (spray, rolling), scrape marks (on trunks) and roaring were conducted to determine the distribution patterns of tigers across space, and even reveal changes in their distribution in the mangrove landscape over time, but not used to estimate numbers or understand movement.

The field survey was conducted with the help of the forest check posts, protection camps and mobile squads of the FD, who are engaged in regular patrolling and monitoring. For direct sighting and indirect evidences for the tigers' presence both up- and down-stream surveys were carried out by using a mechanised boat along the rivers and creeks at a speed of 3-5 km per hour (kph) from 5:30 h to 17:00 h.

Tracks were defined by either single or pair of pug marks, or a set of continuous pug marks associated with the movement of a single animal. *Khals* (small creeks) were checked approximately three hours either side of low tide when tiger tracks below the high water mark could be detected in the soft mud of the *khal* banks. To ensure high track detection rate, the boat was kept between 1 m and 3 m from the *khal* bank. The morning and evening foraging schedules of the tigers and prey species were followed. A break was taken at noon. For surveys near the coast FD launches were used at a speed of 12-16 kph.

Observations were made from both the front and back of the vessels with a binocular and GPS. During the Phase I 'khal' (creek) survey, FD personnel conducted boat transects for prey base density estimation through direct sightings and signs of smaller cats, otters, estuarine crocodiles, monitor lizards, wild pigs, spotted deer etc. and human disturbance along with vegetation covariates were collected. GPS coordinates along with type of mangrove, slope of the bank and width of the upper and lower bank were noted for each sighting/sign encountered.

In addition, observations were made from the mangrove cage-trail and watchtowers with freshwater ponds, by mud and canopy walk as well as upper beaches within the forests on foot, accompanied by armed guards. Moreover, evening and night observations were made from 17.00 to 23.00 hrs using torches, spotlights and digital night vision. The watchtowers referred to above are located at Sajnekhali, Sudhanyakhali, Dobanki, Netidhopani, Bonnie camp, Burirdabri camp, Jhingakhali and Haldibari in SBR.

In 2010, National Tiger Conservation Authority (NTCA) launched the M-STRIPES (Monitoring System for Tigers- Intensive Protection and Ecological Status); a mobile monitoring system for forest guards for a significant check in anti-forest and anti-wildlife activities and the feedback was taken into account in this study.

Chapter 3
Mangroves of the world and Tiger Colonisation

"If we choose to walk into a forest where a tiger lives, we are taking a chance." – Peter Benchley

Tiger colonisation

The Bengal tigers in the Sundarbans are actually an isolated population, showing identical allele or gene pattern of the big cats. It is impossible to confirm exactly when tigers colonised or physically walked into the Sundarbans mangroves as there is hardly any evidence to prove this. The tiger was not very much here in historic times, say 5,000 years ago or during the pre- and proto-historic period around from 3,000 B.C. to 600 B.C. covering the Bronze Age and Chalcolithic culture from the beginning of Harappan civilisation up to the beginning of the historical period in 6th century B.C. which is the period of Buddha and Mahavira.

Hait and Behling (2019) collected sediment cores from two widely separated islands in STR- Chamta (CMT) and Sudhanyakhali (SDK). The pollen records indicate that mangrove existed at CMT from ~5,960 cal yr BP (calendar years before the present) and at SDK from ~1,520 cal yr BP. Changes in relative sea level, including the frequency and intensity of inundation as well as fluctuating precipitation, have been the major factors along with geomorphic processes that control the development and dynamics of the mangrove in the area during the Holocene. The mid-Holocene mangrove at CMT declined, to be progressively replaced by successive communities, and eventually reached climax stage, while the SDK site is transitional in nature. The mangrove responds rapidly to changes in environmental conditions at both locations. With the colonisation of these islands by the salt tolerant halophytic mangroves plants, the mangrove forests on the 'Sundarban surface' came into existence from 6000-3000 YBP (Vidyanadhan and Ghosh, 1993). Recent flood plain and channel deposits of the Sundarban delta system are formed from 560 BP+100 to present.

In fact, when the habitats in the Sundarbans were appropriate for the big cats, they diverged from the mainland and became distinct from other tiger populations. Baker (1887) reported status of the predator and prey animals during late 19th century:

"In the 'Soonderbuns' (Sundarbans) between the mouths of the Hooghly and Meghna rivers, tigers, deer, wild hog, rhinoceros, and buffalo are abundant, and probably are increasing, since comparatively few are ever shot, and the decrease in the forest area is so slow as to be unlikely to affect the numbers of its wild denizens. Spotted deer may there be seen in herds; but their pursuit by stalking is rendered extremely difficult by the nature of the thickets, and is more than ordinarily dangerous by the tigers, which also abound, having the reputation of being the most fearless and the most confirmed man-hunters and man-eaters of any in India. Nor is it possible to follow sport in the 'Soonderbuns' on the backs of elephants, which can neither penetrate the tree jungle, nor cross the endless streams which intersect it at every turn, and are as deep with salt water at flood time as they are with bottomless mud at ebb."

Mangroves

Mangroves, derived from the Portuguese word 'mangue' and the English word 'grove', are facultative salt-tolerant halophytic plants like woody trees and shrubs that grow in shallow coastal water and intertidal zones in the tropics and subtropics and even some temperate coastal areas. However, there is variation among mangrove species in the degree of tolerance. These mangals exclusively occur in the intertidal zones of coastlines with a diverse and vibrant ecosystem. The mangrove islands are known as hammocks and form teardrop-shaped patterns parallel to the flow of water and to the coast.

Mangroves in the world

In 2020, the total mangrove cover was estimated to be 1,47,900 km² globally in 113 countries, of which 55,500 km² occurred in Asia, followed by 32,400 km² in Africa, 25,700 km² in North and Central America, 21,300 km² in South

America and 13,000 in Oceania. India accounts for about 3.3% of the total mangrove cover of the world and 45.8% that of South Asia. None of the mangrove habitats in the world except the Sundarbans mangroves is inhabited by the tigers. During the past three decades between 1990 and 2020, 8,600 km² of mangrove forests had disappeared in south and southeast Asia, with Indonesia topping the countries, due to forest clearing and pollution originating from aquaculture and agriculture, climate change and associated sea-level rise.

Mangroves in India

India with a long coastline of about 7,516.6 km, including the island territories has a mangrove cover of about 6,749 km, the fourth largest mangrove area in the world. These mangrove habitats comprise three distinct zones-

(i) East coast habitats having a coastline of about 2,700 km, facing Bay of Bengal;

(ii) West coast habitats with a coastline of about 3,000 km, facing Arabian Sea; and

(iii) Island territories with about 1,816.6 km coastline.

In India, the states like West Bengal, Odisha, Andhra Pradesh, Tamil Nadu, Andaman and Nicobar Islands, Kerala, Goa, Maharashtra, and Gujarat occupy vast areas of mangroves. The area under mangroves in Gujarat is the second largest along the Indian coast, after Sundarbans in West Bengal. But they are largely distributed in the high energy tidal coast of two extreme conditions: (i) humid and wet in Sundarbans with rich biodiversity, and (ii) arid and dry in Gujarat with low biodiversity. Interestingly the Andaman and Nicobar islands have the third largest mangrove forest in India, occupying 13% of the total cover, located in low energy tidal coast with rich biodiversity.

The tidal mangrove forests grow by the side of the coast and on the edges of the fertile deltas like the Cauvery, Krishna, Mahanadi, Godavari, and Ganges. About 60% of the mangroves occur on the east coast along the Bay of Bengal, 27% on the west coast bordering the Arabian Sea, and 13% on Andaman and Nicobar Islands. In 2021, mangroves were reported to be

4,995 km² in India against 4,975 km² in 2019 (break up shown below), which is 0.15% of the country's total geographical area.

The differential distribution mentioned above may be attributed to two reasons:

(i) the east coast has long estuaries with larger deltas and runoffs due to the presence of mighty rivers, whereas the west coast has funnel-shaped estuaries with absence of delta formation; and

(ii) the east coast has a smooth slope providing larger areas for mangrove colonisation, whereas the west coast has a steep and vertical slope.

Mangrove forest in India (2021 Assessment in km²)

State/UT	Very Dense Mangrove	Moderately Dense Mangrove	Open Mangrove	Total	Change with respect to ISFR 2019
West Bengal	994	692	428	2,114	2
Gujarat	0	169	1,006	1,175	-2
Andaman & Nicobar Islands	399	168	49	616	0
Andhra Pradesh	0	213	192	405	1
Maharashtra	0	90	234	324	4
Odisha	81	94	84	259	8
Tamil Nadu	1	27	17	45	0
Goa	0	21	6	27	1
Karnataka	0	2	11	13	3
Kerala	0	5	4	9	0

State/UT	Very Dense Mangrove	Moderately Dense Mangrove	Open Mangrove	Total	Change with respect to ISFR 2019
Dadra & Nagar Haveli, Daman & Diu	0	0	3	3	0
Puducherry	0	0	2	2	0
Total	1,475	1,481	2,036	4,992	17

Mangroves in Sundarbans

Most spectacular mangroves are found in the Sundarbans. As mangrove habitat is naturally inaccessible, the region offers a suitable diverse habitat, such as beaches, mangrove swamps and tidal flats with the potential for long-term preservation of tigers and prey in adequate core areas and under low to moderate poaching pressure on the predator-prey species as a result of inhospitable terrain, remoteness and vigilant protection.

SBR

The State of Forests in India reports over the years show that between 1987 and 2021, Sundarbans mangrove cover varied between 2,076 km² and 2,155 km². However, in the 15 years between 2007 and 2021, Sundarbans' very dense mangrove cover (canopy density of more than 70%) has come down from 1,038 km² to 994 km², while moderately dense cover (canopy density of 70-40%) suffered a greater loss- down from 881 km² to 692 km². In contrast, open forest cover grew from 233 km² to 428 km². Open mangrove cover increased by 71 km² between 2003 and 2013, but it grew by only 17 km² between 2013 and 2021. About 110 km² of mangroves disappeared within the RF area due to erosion between 2000 and 2020 and 81 km² gained in SBR, corresponding to average losses and gains of 5.5 km²/year and 4.1 km²/year, respectively. The core area of the mangrove forest lost 58 km² (2.9 km²/year), the buffer area lost 52 km² (2.6 km²/year), while 81 km² (4.1 km²/year) were gained in the transition area outside the contiguous mangrove area (Samanta *et al.*, 2021). The rate of mangrove change was not uniform over time; the maximum loss rate was observed from 2000 to 2005 with losses of 4.5 km²/year in the core area and 3.4 km²/year in the buffer area, while maximum accretion was observed from 2015 to 2020 (7.7 km²/year). Shoreline changes on nine ocean-facing islands, where the erosion is concentrated, are- Land loss between 2000 and 2020 ranged from 1.3 km² in Jambudwip to 11.6 km² on Dalhousie Island. There was a net total loss of 49 km² from these sea-facing islands, all of which (except Jambudwip Island) provide tiger habitats.

SRF (Aziz and Paul, 2015)

From the 1970s to 1990s, forestland increased by 1.4%, but from the 1990s to 2000s, the area decreased by 2.5% and the net loss was 1.1%. The loss equals 110 km² for the total Sundarbans and about 66 km² for the Bangladesh Sundarbans. However, the forest is likely to incur more loss due to cyclones and severe floods. Of the total loss, approximately 50% was lost at the Bay of Bengal (the extreme southern edge of the forests where almost no compensating aggradations took place). Within the same tenure, the water bodies in the forests have been reported to increase by nearly 30 km² which, in other words, means a decrease in forestland. The decrease of sediment-laden water from the Ganges every year results in the further erosion of forestland.

Evolution of the mangrove tigerland

The transboundary Sundarban mangrove forest has evolved over the millennia through natural deposition of upstream sediments accompanied by intertidal segregation. From ca. 7,000 years BP sea level rose very slowly and was virtually stable by 2,000-3,000 years BP (Pirazzoli, 1996). The present age of Sundarbans mangroves can be dated as far as 3,000 years BP. For example, the westernmost part, comprising the whole of 24-Parganas Sundarban, was accreted during 5-2.5 ka BP (thousand years before present).

In the opinion of geologists, the sea once rolled up to the vicinity of Rajmahal Hills situated in present state of Jharkhand (erstwhile Santhal Pargana Division). During the Vedic age (c.1500-c.500 BCE), lower Bengal, where the present Sundarbans is located, did not exist. According to the sonnet 'Rajtarangini' written by Kalhana in 1148-1150, king Lalitaditya of Kashmir came to Gour (Malda) and saw the eastern sea (Bay of Bengal), beyond it. I quote below the status of tigers in the Malda district as per statement of W.W. Hunter (1876a: 34-35).

"Malda has always been celebrated for the unusual quantity of large game which it affords, and especially for its tiger-hunting. The ruins of Gaur and Panduah, each of which extends over several square miles, are the favourite haunt not only of tigers. Comparatively few are destroyed by the native *shikaris*; but the Collector thinks that from thirty to forty tigers are annually killed by sportsmen, who do not claim the reward. So long as these animals refrain from the habit of attacking men, their presence is desired rather than dreaded by the cultivators. Dr. Buchanan Hamilton expresses the opinion, when writing of this very region, that a few tigers in any part of the country that is overgrown with jungle or long grass are extremely useful, in keeping down the number of wild hogs and deer, which are infinitely more mischievous than themselves.

Mr. Pemberton, the Revenue Surveyor in 1848, also states that 'the inhabitants of Gaur are rather partial than otherwise to the tigers, and are unwilling to point out their lairs to sportsmen. They call them their *chaukidars*, as being useful to them in destroying the deer and wild hog, with which the place abounds, and which make sad havoc on their crops. The other side to the picture may be learned from the story of the notorious man-eater of 1863. This animal had its favourite haunt in the ruins of Pandua, but infested the whole of the high road between Malda and Dinajpur. It is reported on the authority of the gentleman who was at that time the Magistrate of Malda that this mischievous and cunning beast killed no less than 13 persons before it was finally shot."

During the Ramayana period, the river Bhagirathi met the sea at present Murshidabad district or Nabadwip (New isles, Nadia) (Das, 1896). The Revenue Surveyor (1857) gave the following information:

'Tigers are occasionally found in the hilly parts to the north-west, which have probably strayed from the Santal Parganas' (Hunter, 1876b).

Some idea of the numbers of tigers infesting Bengal less than a century ago, may be formed by a reference to that interesting old work, "Oriental Field

Sports," by Captain Thomas Williamson (1819), who served upwards of twenty years in Bengal, and describes vividly, not only the sports he shared in, about the end of the 18th century, but the state of the country. Writing of the river running under the field of Plassey (Pâlāshi), Williamson names it the Cossimbazar river, commonly called "Baugrutty" i.e., the Tigers' River (now known as Bhagirathi); it was formerly surrounded by large grass jungles teeming with tigers, buffaloes, etc. Williamson continues to inform that about Daudpoor, Plassey, Aughadup, and especially along the banks of the Jellinglue (Jalangi), which borders the Cossimbazar Island to the eastward, they (i.e. tigers) are known to cross and recross during the day, as well as by night; seeming to consider the stream as no impediment. From Aughadup in particular, they pass over to the extensive jungle of Patally, which has ever been famous for the number it contained. Williamson had in passing through it seen four tigers within the space of two hours. The country about Plassey and Rajmahal literally swarmed with game; while the road about Santipoor, Ranaghat, and Chogdaho (Chakdaha), were unsafe on account of tigers a hundred years ago, concluded Williamson.

Lower Bengal or the delta of the Ganges and the Brahmaputra was then a part of the sea, which gradually receded southwards in the course of time, a theory which receives corroboration from the Mahabharata, Vana Parva, Ch. 113, that Yudhisthira bathed at the *Sagar-Sangam*, which must have shifted southwards with the recess of the sea.

It is estimated that the separation of the mainland tigers began only about 1,000 years ago and was accelerated about 200 years from now due to habitat fragmentation, historical events, human pressure, and land-use patterns (Sharma *et al.*, 2013).

The Pāla dynasty ruled parts of Bengal during 750-1161 CE. It is known from the Pala inscriptions that there was a place called 'Byaghratatimandal' in southern Bengal, facing the sea. The literary meaning of the term, as Niharranjan Roy (*Bangalir Itihas* cit., p. 105) has pointed out, is "a forested seashore infested with tigers". Ralph Fitch, who visited the area in the 1580s, described south-eastern Bengal as a dense forest infested by ferocious wild animals such as tigers and buffaloes. The earliest concrete reference to the notoriety of the tigers of the Sundarbans can be found in the writing of Francois Bernier (1916), who visited the area in 1665: "Among these islands, it is in many places dangerous to land and great care must be had that the boat, which during the night is fastened to a tree, be kept at some distance from the shore, for it constantly happens that some person or another falls prey to tigers. These ferocious animals are very apt, it is said, to enter into the boat itself while people are asleep, and to carry away some victims who, if we are to believe the boatmen of the country, generally happen to be the stoutest and fattest of the party."

It appears that at some point in the 15th or 16th centuries many areas in the coastal tract of the 24-Parganas district were abandoned and overrun with forest and jungle owing to some disorder of a nature that is not clear to historians: it may have been either political or environmental (Chakrabarti, 2009). Pratapaditya, a Bhuian king of Jessore, Khulna, Barisal and Greater 24-Parganas at the end of 16th Century, used to hunt tigers in the Sundarban mangrove forest ('Jaśōra-Khulanāra Itihāsa', 1914). He used to take his companions for hunting. The capital of Jessore is on the edge of the mangrove forest. Bagmara Forest Block in STR contains the ruins of a city built by the Chand Saudagar merchant community in approximately AD 200-300. Much later, during the Mughal Empire, Raja Basant Rai and his nephew took refuge in the Sundarbans from the advancing armies of Mughal Emperor Akbar. The buildings they erected subsequently fell to Portuguese pirates, salt smugglers and dacoits in the 17th century.

There remains a lack of clarity about when, historically, Bonbibi, a legendary female deity of the mangrove forests, came to exist in the Sundarbans. Some folklorists are of the opinion that Bonbibi gained popularity with the emergence of Islam in the area in the early 15th century. Others argue that she was a Hindu goddess who was turned into a Muslim-Sufi figure over time.

European traveller François Bernier (1914), who visited the Sundarbans in 1665-1666, accounted for the tiger cult as "it constantly happens that some person or another falls prey to tigers". Krisnaram Das composed "Ray Mangal Kavya" (poetry of the auspiciousness of Dakshin Ray) in 1684, which was mostly dedicated to depicting the acts and glory of Dakshin Ray, the male deity of tiger-cult.

A news report (Gentleman's Magazine, June 1738-1739) stated that in 1737, during an autumnal super cyclonic surge with an estimated height of 10-13 metres and earthquake, many tigers (tygers), *inter alia,* were drowned near the Hooghly river-mouth.

The story of a man-eater in the Sundarbans, first widely published in 1793, became a classic in the genre. The announcement of the death of 17-year-old Hector Sutherland Munro, son of the British Major Hector Munro, then a cadet in the military service of the East India Company, who headed out on RF a tiger-hunting trip in the Saugor (Sagar) Island with his friends Downey and Lt. Pyefinch on 21st December, 1792. On 22nd December they disembarked from the ship on the Island. As the group sat eating lunch of cold meats by a fire, they were disturbed by a tiger hidden in the undergrowth. On Wednesday, 3 July 1793, the Times carried a short paragraph on this incident:

"A son of Sir Hector Munro was killed on passage to India. He went ashore with a party at an island where they had put into water; and reclining with his companions under some trees a tiger sprang from an adjoining thicket, and seizing him in his mouth, tore the unfortunate young man to pieces."

A more detailed report was published the next day in the *St. James's Chronicle* 2-4 July 1793, and again in the *Star* 5 July 1793, reproducing part of an eyewitness account in a letter written by one of Munro's companions, which had been published in the *Calcutta Gazette* on 1st January 1793, and had now reached London on a ship from India. The text of one of the letters from an eyewitness is reproduced below.

To describe the awful, horrid, and lamentable accident I have been an eye-witness of, is impossible. Yesterday morning Mr. Downey of the Company's troops, Lieut. Pyefinch, poor Mr. Munro (son of Sir Hector) and myself, went on shore on Saugur Island to shoot deer. We saw innumerable tracks of tigers and deer, but still we were induced to pursue our sport, and did the whole day. About half past three we sat down on the edge of the jungle, to eat some cold meat sent us from the ship, and had just commenced our meal, when Mr. Pyefinch and a black servant told us there was a fine deer within six yards of us. Mr. Downey and myself immediately jumped up to take our guns; mine was the nearest, and I had just laid hold of it when I heard a roar, like thunder, and saw an immense royal tiger spring on the unfortunate Munro, who was sitting down. In a moment his head was in the beast's mouth, and he rushed into the jungle with him, with as much ease as I could lift a kitten, tearing him through the thickest bushes and trees, every thing yielding to his monstrous strength.

The agonies of horror, regret, and, I must say, fear (for there were two tigers, male and female) rushed on me at once. The only effort I could make was to fire at him, though the poor youth was still in his mouth. I relied partly on providence, partly on my own aim, and fired a musket. I saw the tiger stagger and agitated, and cried out so immediately. Mr. Downey then fired two shots, and one more. We retired from the jungle, and, a few minutes after, Mr. Munro came up to us, all over blood, and fell. We took him on our backs to the boat, and got all medical assistance for him from the Valentine East India-man, which lay at anchor near the island, but in vain. He lived 24 hours in the extreme of torture; his head and skull were torn, and broke to pieces, and he was wounded by the claws all over his neck and shoulders; but it was better to take him away, though irrecoverable, than leave him to be devoured limb by limb. We have just read the funeral service over the body, and committed it to the deep. He was an amiable and promising youth. I must observe that there was a large fire blazing close to us, composed of ten or a dozen whole trees; I made it myself, on purpose to keep the tigers off, as I had always heard it would. There were eight or ten natives about us; many shots had been fired at the place, and much noise and laughing at the time; but this ferocious animal disregarded all. The human mind cannot form an idea of the scene; it turned my very soul within me. The beast was about four and a half feet high, and nine long. His head appeared as large as an ox's, his eyes darted fire, and his roar, when he first seized his prey, will never be out of my recollection. We had scarcely pushed our boats from that cursed shore when the tigress made her appearance, raging mad almost, and remained on the sand as long as the distance would allow me to see her.

Munro's friend reported that he 'lived 24 hours in the extreme of torture; his head and skull were torn, and broke in pieces, and he was wounded by the jaws all over his neck and shoulders.' On 23 December his body was 'committed to the deep'.

Pennant (1798:152-154) in his account of the tigers in Sundarbans (spelt Sunderbund) wrote,

"...tigers, the most voracious and destructive animal of the peninsula. This part was probably famed for the tremendous animal: *Seneca* distinguishes it in his *Oedipus* by the epithet *Gangetica Tigris.* Those which supplied the *Roman* amphitheatres with the objects of the sport, were procured from some part of this empire, which produced the largest and fiercest...In the Sunderbunds, the tigers are particularly fatal to the wood-cutters and salt-makers, who resort there in dry season; they will not only seize on them in the islands, but even swim to the boats at anchor and snatch the men from on board. The pietists, who annually visit one particular island for the sake of washing themselves in the sacred water, often fall victims to these terrible animals: they have such power as to carry off a man with the utmost facility; they will even go full speed with a buffalo, which they will seize out of the field or pasture".

He also cited specific instances of tiger's springing among a party of gentlemen and ladies recreating themselves on the islands of the Ganges, and carrying away one of the company. Such accidents were not uncommon. Another party in the beginning of the 18th century was seated under the shade of trees on the banks of a *Bengalese* river; a lady among them observed a tiger preparing to take his fatal spring, and with amazing presence of mind laid hold of an umbrella, and furling it full in the animal's face, terrified it so that it instantly retired. Pennant was told that the tigers are sometimes plagued with flies, which settle about their eyes, and frequently make them almost blind. These wander remote from their usual haunts, and give themselves up to destruction.

Pennant also referred to the practice of payment of large rewards for destroying tigers in general; the skins, the claws, and the teeth, were articles for exportation. Records of repeated man killing were required before the FD would declare a man-eater to be decimated. The last Bengal tiger declared as a man-eater was killed in 1989 at Kachikhali, Sundarbans East WLS.

Therefore, the Bengal tigers might have colonised the Sundarban mangroves much earlier, i.e. possibly during the early mediaeval period. Munro's death may have had some impact on the fate of tigers on Saugur Island. In 1819 the Saugor Island Society was formed to reclaim and develop the island, with rewards given for every tiger killed. There are now no tigers on Saugur Island.

Hugh Morrieson wrote in his Field Book in 1812:

"Having finished the observation, one of the sepoys said there was a tiger close alongside, that had been creeping up towards us, and for the last minute he and the animal had sat looking at each other; we now heard a slight noise in the jungle, the two sepoys fired, out sprang a tiger and ran off; he was only about 4 yards from us."

Transformation of ecological regions

The wetland region south of the contemporary Kolkata up to the Bay of Bengal was divided into two terrestrial ecoregions, spread over about 25,000 km^2 (14,600 km^2 freshwater swamp was mostly extinct and almost 50% of 20,400 km^2 mangrove forests vanished).

Under the colonial government, reclamation of the Sundarbans was initiated in 1770s. Whereas reclamation continued sporadically up to the 1980s in the western 24-Parganas between the rivers Hooghly and Raimangal, deforestation of the eastern Bakarganj (now Barisal) between the rivers Baleswar and Meghna was almost complete by 1910 and the intervening forests of Khulna, bounded by the rivers Raimangal and Baleswar, were subjected to minimum reclamation up to 1920. In this process, thousands of tigers were hunted in the eastern and western Sundarbans during the colonial period and even after independence. During the 18th and 19th centuries, the RFs and PFs were mostly hunting grounds of the British rulers and local elites. By the 1890s, the shift to preferential treatment for European hunters also coincided with the rapid decimation of game and the shift of the game frontier to remote areas like the Sundarbans, where today's wildlife reserves are located (Mani, 2012).

The effective protection efforts in both India and Bangladesh began only during the early 1970s.

Mangroves with tiger population

West Bengal

West Bengal has 2,112 km^2 mangrove forests or 42.45% of India's mangrove-cover. South 24-Parganas district alone accounts for 41.85% mangrove cover of the country. The State of Forests in India report over the years showed that between 1987 and 2021, Sundarbans mangrove cover varied between 2,076 km^2 and 2,155 km^2. However, in the 15 years between 2007 and 2021, Sundarbans' very dense mangrove cover (canopy density of more than 70%) has come down from 1,038 km^2 to 994 km^2, while moderately dense cover (canopy density of 70%-40%) suffered a greater loss – down from 881 km^2 to 692 km^2. In contrast, open forest cover grew from 233 km^2 to 428 km^2.

Scientists largely attribute the growth in open forest cover to afforestation initiatives. According to a study published in December 2021, about 110 km^2 of mangroves disappeared within the RF (STR) area due to erosion between 2000 and 2020. During the same period, the inhabited part of the SBR gained 81 km^2 through plantation and regeneration.

Extant tiger population

There are, no doubt, less demanding habitats to make a living, but the tiger is in its element in the Sundarbans, and looks like it. That makes the Sundarbans and the tigers living there unique today. The Sundarban forest is big and there is a potential for a viable population of tigers. The Sundarbans [designated as a Biosphere Reserve (9,630 km^2) including the reclaimed areas (5,367 km^2)] is the largest coastal mangrove forest region in India, harbouring a viable tiger population. However, there are no corridors that would allow immigration to the Sundarbans from adjacent areas; there are no tigers in the highly populated adjacent areas. Any tiger that ventures out of forests is death-prone. The tiger's survival hinges on those living there now and how well we can provide for its needs under ever-increasing population pressure.

Extinct tiger population

The islands in the coastal Midnapore had a tiger population during the 18th century. Due to conversion of the mangrove forests into human settlements the predator along with its prey species had gone extinct during the 19th century. Before the opening up of the district by railways and the destruction of the jungle which has accompanied extension of cultivation, tiger, leopard. pig and deer were to be found in the eastern alluvial portion of the district especially near the mouths of the Haldi and other rivers. O'Malley (1911) wrote:

"The annals of the old Calcutta Tent Club contain references to the sport obtained in Tamlūk, and old cultivators there mention the name of Lord Mayo as having visited the place for sport. Now the only tigers and leopards seen there are occasional visitors from the Sundarbans or from the western jungles. There were also many wild buffaloes in the south of the district in former years, but these have all disappeared with the extension of cultivation and growth of population".

Hunter (1858) gave an account of Hijli, then encircled by an encampment to defend it from inundation, which was an island swamp, separated by channels from the mainland, but half rescued from the sea; having a great store of wild hogs, deer, wild buffaloes and tigers." In the old European accounts Khedgree, mentioned under various names, was also an island just north of Hijli Island and separated from it by a narrow stream. This island appears in Valentijn's map (1664 AD), Bowrey's Chart (1688), and the Pilot Chart of 1703; while Streynsham Master referred to it as Khedgree in December 1876. On the 11th March 1683, W. Hedges, on arriving at "Khagaria", found the island with a great store of wild hogs, deer, wild buffalo, and tigers. Gradually the intervening belt of water was silted up, and Khejri, like Hijili, became united with the mainland and the other island, this junction taking place before the compilation of Rennell's Atlas (1779).

Mangrove states without tiger population

(i) Gujarat (1,177 km²) in Gulf of Kutch, Gulf of Khambhat, Dumas-Ubharat; (ii) Andaman and Nicobar Islands (616 km²) in North Andaman and Nicobar;

(iii) Andhra Pradesh (404 km²) with Coringa East Godavari Delta, Krishna Delta;

(iv) Maharashtra (320 km²) in Achra-Ratnagiri, Devgarh-Vijay Durg, Veldur, Kundalika-Revdanda, Mumbra-Diva, Vikhroli;

(v) Odisha (251 km²) with Bhaitarkanika, Mahanadi, Subarnarekha, Devi-Kauda, Dhamra, Chilka;

(vi) Tamil Nadu (45 km²) in Pichavaram, Muthupet, Ramnad, Pulicat, Kaznuveli;

(vii) Kerala (9 km²) in Vembanad, Kannur (North Kerala);

(viii) Goa (26 km²) in Terekhol, Chapora, Mandovi, Zuari, Cumbarjua, Sal, Talpona and Galgibag;

(ix) Karnataka (10 km²) in Coondapur, Dakshin Kannada/Hannavar, Karwar, Mangalore FD;

(x) Daman-Diu (3 km²) and

(xi) Puducherry (2 km²).

Rich mangrove or tidal forests

There are 54-75 species of true mangroves, which are taxonomically distinct from their terrestrial counterparts. The mangrove associates lack many of the characteristic adaptations of mangroves. Mangrove forests are among the planet's most complex ecosystems, flourishing in environmental conditions that would swiftly kill most other types of vegetation. Mangroves can survive within reach of the tides in extremely salty and oxygen-poor soils because they are highly adaptable to their surroundings and capable of excluding or expelling salt. The mangrove habitat is one of the most productive ecosystems and iconic last resort for the tiger with designated wilderness landscapes.

Mangrove forests are mainly confined to the supratidal and upper inter-tidal zones of the Sundarbans. Fringe mangroves border the tidal creek while basin mangrove forests are found more inland and the riverine mangroves are located in between these two categories. Fringe mangroves grow as a relatively thin fringe along the coast and are, therefore, directly exposed to the tides and sea waves as well as storm surges. Featured with prop roots, buttresses and pneumatophores they are tide-dominated, showing intermediate productivity and important primarily for shoreline protection. The riverine mangroves are river-dominated, most productive, commonly flooded by river water (and sometimes tides), with moderate salinity. The basin mangroves behind riverine and fringe mangroves occur in inland depressions or basins, often behind fringe mangrove forests and in drainage depressions, where water is stagnant or slowly flowing. The ground surface of such forests is often covered by pneumatophores from the dominant trees. Basin forests serve as nutrient sinks for both natural and anthropogenically enhanced ecosystem processes. Dwarf mangroves are forests that are limited in growth by soil conditions that may include low nutrients, high salinity, or excessive sulphide concentrations.

Habitat Suitability/Quality for tiger

Habitats are either suitable or unsuitable for a given species like the tiger because suitability depends on the potential of a habitat. Habitat suitability is defined as the potentiality of a habitat to support a viable population of a species over an ecological time-scale. Tigers are habitat specialists preferring the areas with dense vegetation cover, high prey density, and less human footprint. During the field exercise, camera trap stations were established in locations with varied topographic features such as the bank/slope of fresh water ponds, meanders, and inland vegetation as these are frequented by tigers and prey animals most often. This was not done by design except in case of ponds, but dictated by suitability of the location based on tiger/prey signs, unlikelihood of inundation of camera traps and pilferage/vandalism. Photo captures from all trap stations indicate the maximum use of freshwater ponds by tigers, followed by "inside vegetation", and then meander.

Habitats are normally evaluated by relating changes in habitat features to changes in species density, richness, and diversity. Evaluation of HS in purely demographic terms is not without flaws. Over the last century the ecological landscape has changed drastically resulting in isolation of tigers to small patches enclosed by agricultural land and human settlements (Dinerstein *et al.*, 2007). Key elements of tiger habitats include areas with rich ungulate population and dense and undisturbed vegetation along with water availability (Sunquist *et al.*, 1999; Karanth and Sunquist, 1995).

The tiger's population stability at an optimum level and survival depends on the genetically determined reproductive rate, ecologically influenced turn over-cum-recruitment rate and food supply, and the behavioural thresh-hold, wide enough to compensate for the imbalances in the reproductive and ecological factors. Kurup (1980) has calculated that a tiger population of 10 animals would increase by 1.75 young annually or about 17 young per hundred of the total population subject to adult mortality. If adult mortality is taken as 5%, the net annual increment of tiger population is obtained as about 12%. As regards the food demand and supply in relation to tiger requirements, he showed that a standing ungulate crop in a minimum area of 5.18 km² is needed to sustain one tiger for one year in a sample habitat. Available data on the population dynamics of three ungulate prey species are worked out, and the annual rate of net increment in population in numbers as well as in biomass for these species is shown. Annual biomass increment of prey species per 2.59 km² of habitat is shown to be about 336 kg in this example. Based on this, it is estimated that the tiger population would require about 24 km² of habitat per adult tiger to be stable and viable, using only the food increment and not the food capital stock. Although these figures are applicable only to the sample area, it is believed that similar figures would hold good for comparable reserves.

Kellner *et al.* (1992) recognised four problems, which are applicable in case of the changing mangrove landscape in the Sundarbans:

1. Populations occupying suitable habitats can undergo long-term declines associated with factors outside a habitat, e.g. destruction of the wintering habitat.
2. Habitat alteration may reduce carrying capacity but not eliminate a habitat's potential to support a viable population. Habitats in which populations are steadily declining are considered unsuitable.
3. Population persistence is problematic when populations are maintained through immigration, or when population size declines through high rates of emigration. On the other hand, birth and death rates are more important in evaluation of habitats.
4. In a highly fragmented landscape none of the habitat fragments may represent suitable habitat. In fragmented landscapes, the value of a particular patch depends on its contribution to species persistence within the landscape and not persistence in an individual patch.

Secretive and mysterious in many ways, adult tigers are inherently peripatetic. Occupying the top of the ecological pyramid with no predators of their own, and also highly territorial, these carnivores need large inviolate tracts to search for prey, rest, loaf, breed and rear and train their cubs. Such tracts tend to harbour quality habitats, which in turn add to the prey base for tigers. Large inviolate tracts and tranquillity are directly correlated to viable tiger populations. This rule of thumb suggests ensuring an inviolate core or critical tiger habitat (CTH) of around 1,000 km², with around 100 tigers of either sex and of different ages. This population should, however, contain around 20 breeding females to ensure viability in the foreseeable future.

Habitats are valued for their contribution to abundance or density of the target species, reproductive output of the species, their numbers, population, stability or representation of historical composition of the species. Anthropogenic disturbance is found to be an important factor influencing habitat quality and differential use of habitats by animals (Mathai, 1999). HS analysis of tigers is often calculated based on four factors: vegetation types, drainages, roads and settlements (Tiwari *et al.*, 2021).

Distance from settlements	Sustainability based on vegetation type	Rating
3,000-4,000 m	Water Bodies and wetlands; grasslands; forest plantations	Highly suitable
2,000-3,000 m	Pine and pine mixed	Suitable
1,000-2,000 m	Himalayan moist temperate	Moderately suitable
Up to 1,000 m	Agriculture/non-forest settlements	Least suitable

Comparatively habitat suitability is known recently from Bandhavgarh NP in Madhya Pradesh, where 80% area is suitable for tiger Habitat (Mishra, 2022):

Highly suitable 96.25 km² (4.30%);

High suitable 1,167.76 (52.18%);

Moderate suitable 531.22 (23.74);

Less suitable 393.40 (17.58) and

Not suitable 49.76 (2.22).

The presence of tigers is more likely at a distance away from the roads and settlements while they are more likely to be found near water bodies and drainage areas. Earlier tigers are found in dense forests, grasslands, wetlands etc. Therefore, these categories are considered most suitable for the species, while agriculture and non-forest areas are considered as least suitable for the species.

Sex-specific habitat suitability models for endangered species like the Bengal tiger are uncommon. An *et al.* (2021) developed sex-specific models for Bengal tigers (*Panthera tigris*) based on camera trapping data collected from 20 January to 22 March 2010 within Chitwan NP, Nepal, and its buffer zone. Differences in the geographic distribution of suitable habitats between sexes are likely the result of different reproductive strategies. Males do not aid in rearing or provisioning young, and so are less restricted to those areas with the highest concentrations of prey and water. Instead, they can cover a more extensive territory in search of mates and resources. Females, however, must stay near their young to feed and protect them. Consequently, female home ranges prefer those areas with the highest prey densities and most abundant water supplies, found primarily in the floodplain zone. Additionally, unlike males, whose home ranges overlap those of several females, female home ranges are smaller and generally do not overlap with other females.

Criteria

Five criteria fixed for tiger habitat suitability are discussed below.

(i) Distance (in metres) from drainage

Distance of 1,000 m is regarded as highly suitable because drainages are the primary source of water for wildlife including tigers and their presence in close proximity of drainages is highly likely. Distance between 1,000 m and 2,000 m is suitable, 2,000 m and 3,000 m is considered as moderately suitable, 3,000 m and 4,000 m is considered as least suitable.

(ii) **Distance (in metres) from roads and waterways**

Due to difficult physiography, there is limited road connectivity of the Sundarbans with the mainland. Moreover, no human settlements, trails or roads are available within the Sundarbans due to the thick mangrove vegetation. Thus, the passenger and cargo movement is solely dependent on waterways beyond certain connected centres. Besides causing water and air pollution, the effect of lights and noise from the ships affects tiger behaviour and that of their prey. Distance of 1,000 m is regarded as least suitable because it is highly unlikely that roadways would be ideal habitat for tigers. Distance between 1,000 m to 2,000 m is moderately suitable, 2,000 m to 3,000 m is considered as suitable and regions as far as 3,000 m to 4,000 m are regarded as highly suitable.

(iii) **Distance (in metres) from human settlements**

Human settlements are located at the northern north-western and north-eastern periphery of the Sundarbans. Human intrusions are reported to negatively influence tiger abundance, such that impacts are decreased with increasing distance from villages. Most of the incidents occurred close to villages bordering the forest. In terms of human-tiger conflict, distance of 1000 m is regarded as least suitable for them. Distance between 1000 m to 2000 m is moderately suitable, 2000 m to 3000 m is considered as suitable and regions as far as 3000 m to 4000 m are regarded as highly suitable.

(iv) **Vegetation types**

Indian Sundarbans

The Sundarbans tiger-land, the largest contiguous mangrove forest in the world, is classified by Champion and Seth (1968) as Type Group 4 Littoral and Swamp (Wetland) Forests, which are covered under the sub-group 4A/L1 Littoral Forest, 4B Tidal Swamp Forests with subdivisions of mangrove type (4B/TS1 Mangrove Scrub and 4B/TS2 Mangrove Forest), 4B/TS3 Salt Water Mixed Forest (*Heritiera*), 4B/TS4 Brackish Water Mixed Forest (*Heritiera*) and 4B/E1 Palm Swamp Forest.

The mangrove forests in reserved and PAs are classified into three types-Littoral forests, Tidal Swamp forests and Palm swamp. The vegetation in the core area is considered highly suitable and that of the buffer area is regarded as moderately suitable, whereas agriculture and non-forested vegetation are put in the least suitable category. The vegetation pattern of different ecological zones of the Sundarbans mangrove forest shows that the Polyhaline zone with salinity level (18.0 to 30.0 ppt) and also to a lesser extent Mesohaline zone (5.0-18.0 ppt) show formation of consociation of *Sonneratia, Rhizophora, Avicennia, Bruguiera* and *Acanthus* spp. in the former zone and *Kandelia, Avicennia, Rhizophora, Aegiceros* and *Sonneratia* spp. in the latter zone and the Oligohaline zone (0.5-5.0 ppt) has mixed plant community of *Sonneratia, Acrostichum* and *Cyperus* spp. A very large, dense and closed vegetational matrix enables the wild tiger and its prey to choose the habitat in spite of the least vegetation diversity (both generic and specific), possibly for other extraneous natural factors.

Similarly, in the Indian Sundarbans, tiger's presence was commonly observed in the mangrove forest having plant consociations of mainly *Rhizophora-Bruguiera, Avicennia-Oryza, Excoecaria-Phoenix, Excoecaria-Ceriops,* pure *Ceriops* and pure *Oryza* (Mallick, 2013). However, the latest exercise transpired the following habitat formations and captures of tiger and its prey base (Roy Choudhury *et al.,* 2018):

Sl.no	Habitat	Prey-base captures (%)	Tiger captures (%)
1	*Avicennia - Oryza*	7.62	12.73
2	*Avicennia - Sonneratia*	3.48	3.27
3	Pure *Cereops*	41.39	33.09
4	Pure *Excoecaria*	1.98	3.64
5	*Excoecaria - Cereops*	37.16	30.55
6	*Excoecaria - Phoenix*	0.56	8.73
7	Pure *Phoenix*	1.41	6.18
8	Beach Forest	4.61	1.45
9	*Rhizophora - Bruguiera*	1.79	0.36

The biomass productivity has been measured to be 212 mt/ha in case of *Avicennia-Sonneratia* association in the Sundarbans. Most of the mangrove leaves like *Sonneratia* or *Avicennia* are good fodder for the preferred prey animals of the tiger, i.e. deer and monkey, while the roots are available in plenty for the wild boar.

Bangladesh Sundarbans

In Bangladesh Sundarbans, the habitat use of spotted deer (*Axis axis*), the prime tiger's prey, was studied by Dey (2006) following the pellet group counting and transect method. The results are shown below.

Vegetation types	Area (km²)	Mean density (no/km²)	Total number of deer
Heritiera	749.92	3	2,250
Heritiera-Excoecaria	1059.73	10	10,597
Heritiera-Xylocarpus-Bruguiera	95.56	12	1,146
Excoecaria	215.20	12	2,582
Excoecaria-Heritiera	757.03	3	4,542
Excoecaria-Ceriops-Xylocarpus	346.04	15	5,190
Ceriops-Excoecaria-Sonneratia	648.07	43	27,867
Xylocarpus-Bruguiera-Avicennia	40.30	14	5,642
Sonneratia-Excoecaria-Open grassland	82.86	174	14,417

(v) Land use

Thakur *et al.* (2021) observed that there has been a marked reduction in the areas of plantation, mangrove swamp, mangrove forests and agricultural land since 2000. In contrast, an increase in sand beach, waterlogged areas, mudflat, river, and agriculture area was observed. There is also an increase in open patches/non-vegetated cover. The carrying capacity in Sundarbans is three to five tigers per 100 km² but in multiple blocks the density is more than that. When there are too many tigers in a particular forest, the "young adults" might be compelled to move out in search of new territory.

Blasco and Aizpru (2002) reported 6,840 km² of classified mangrove stands in the Gangetic delta as follows:

(a) India 1,920 km² distributed in four categories-

(i) Dense mangrove 1,330 km²
(ii) Unexplained top dying 220 km²
(iii) Degraded mangrove or young stands 140 km²
(iv) Very degraded mangrove or young stages 230 km²

(b) Bangladesh 4,920 km² distributed in four categories-

(i) Dense mangrove 3,570 km²
(ii) Unexplained top dying 510 km²
(iii) Degraded mangrove or young stands 50 km²
(iv) Mangrove afforestation areas 790 km²

Tall dense mangrove forests with *Heritiera fomes* (Bangladesh Sundarbans) or with *Avicennia officinalis* (Indian Sundarbans) are common. But the assemblage of species varies from one place to another. *Excoecaria agallocha* is abundant and heavily exploited. *Ceriops decandra* forms low dense forests and *Nypa fruticans* form narrow fringes at the edge of the mangroves close to the principal water courses. Due to heavy exploitation of *Heritiera fomes* trees over 1 m in girth are at present uncommon. The progressive degradation of the standing biomass of the Sundarbans is continuing. Some serious changes have occurred since the 1980s. There are indications of a tendency towards deciduousness or leaf-shedding which is often a signal of unhealthy situations and a probable ecological disturbance called '*Heritiera* top dying' which opens the canopies. This alteration has probably a deep impact on the ecosystem processes, in stand structure and ecological services. Blasco and Aizpuru (2002) anticipated that the Sundarbans are probably entering a transitional phase to a closed canopy mangrove of unknown components for which the species like *Excoecaria agallocha* show a natural tendency to spread all over the mangroves.

Density in the coastal afforestation areas sometimes does not exceed 30%. Two main species used are- *Sonneratia apetala* (80% of mangrove plantation, especially in islands where the salinity is lower, and about 20% in Rangabali and Kukri Islands) and *Avicennia officinalis,* which is planted in the eastern coastal belt where the salinity of the water is higher- about 25%.

So-called 'top dying' affected the populations of *Sundari* since the 1980s. Some replanted areas have disappeared due to cyclonic deluge or converted to new croplands after consolidation of the sediments by the mangrove trees.

Chapter 4

Migrations from Mainland to Islands: The Last Resort

"The most magnificent creature in the entire world is the tiger."
– Jack Hanna

Tiger landscape in India

In India, Bengal tiger populations constitute more than 70% of the global population of tigers and are mainly confined to four forested landscapes- Northeast, Central India, Western Ghats and the *Terai* with regional connectivity. These landscapes had contiguous forests and probably shared a common gene pool until recent times. These are divided into ecologically and biogeographically meaningful landscape clusters as follows:

(a) Central India and Eastern Ghats (3,38,378 km²)

(i) Odisha (Simlipal, Satkosia), (ii) Madhya Pradesh (Bandhavgarh, Kanha, Pench), (iii) Chhattisgarh (Achanakmar, Gurughasidas, Indravati, Udanti-Sitanadi), (iv) Jharkhand (Palamau), (v) Maharashtra (Tipeshwar, Umred), (vi) Andhra Pradesh (Nagarjunasagar-Srisailam), and (vii) Telangana (Amrabad).

Tigers disperse over large distances (up to 650 kms) in this landscape.

(b) Sundarbans (SBN) in southern West Bengal, interconnected with the Bangladesh counterpart;

(c) Northeast (NE)- Several TRs have been established in the region, including (i) Assam (Kaziranga, Manas, Nameri, Orang), (ii) Arunachal Pradesh (Kamlang, Namdapha, Pakke), (iii) Mizoram (Dampa) and (iv) northern West Bengal (Buxa), of which only Kaziranga and Manas have substantial tiger populations.

Namdapha, Dibang, Buxa, Manas, Kaziranga

The population of tigers in this region is the most distinct among Bengal tiger populations. The tiger-occupied habitat has increased in the landscape after recent photographic evidence of tigers from Buxa TR, Neora Valley NP, and Mahananda WLS in northern West Bengal. In 2023, a tiger was photo-captured in Namdapha TR, Arunachal Pradesh, and in 2021, in Dampa TR, Mizoram.

(d) *Terai* - Uttarakhand (i) Rajaji and (ii) Corbett; Uttar Pradesh (iii) Dudhwa, and Bihar (iv) Valmiki;

(e) Western India (WI) - Rajasthan (i) Ranthambore and (ii) Sariska;

(f) Northern Western Ghats (N-WG)- (i) Goa; Maharashtra (ii) Sahyadri, (iii) Dandeli-Anshi; Karnataka (iv) Bhadra, (v) Biligiri Ranganatha Temple, (vi) Bandipur; Tamil Nadu (vii) Mudumalai; and

(g) Southern Western Ghats (S-WG)- Tamil Nadu (i) Anamalai; (ii) Kalakkad-Mundanthurai; and Kerala (iii) Periyar (Kolipakam *et al.* 2019).

Among all the above tiger sub-populations in India, the meta-population (a group of separated populations of the same species which interact at some level) in the Central Indian Tiger Landscape with 23 TRs and about 46 other PAs with tiger's presence has the highest genetic diversity of wild tiger populations anywhere in the world (Pariwakam *et al.* 2018). Although 26 corridors have been identified in the Central Indian and Eastern Ghats tiger landscape, in practice, there are many more corridors. Tiger movement is negatively impacted by human settlements, road, rail, irrigation and other linear infrastructure developments within this landscape.

NTCA estimates that a tiger population of 80-100 tigers (with 20 females of breeding age) is required to be genetically viable on its own. Moreover, immigration and emigration of adult tigers from other sub-populations within the landscape is essential and the long-term survival of tigers is highly dependent on the connectivity of tiger populations through a network of intervening forest corridors. Several small tiger sub-populations in the landscape face numerous threats that make those populations even smaller. Even though there is habitat connectivity, many of the PAs and forested areas are lacking in tiger populations. Hence, maintaining a genetically viable population of tigers would require larger areas than the reserves and their contiguous forests provided.

All India Tiger census is conducted once every four years following double sampling- ground-based surveys (collection of tiger signs like scat and pugmarks) and camera-traps. According to the fifth cycle of All India Tiger Estimation 2022 in 20 states, a report released by NTCA and Wildlife Institute of India (WII), a total of 3,080 individual tigers (>one year of age) were photo-captured and a minimum tiger population was estimated to be 3,167.

Table: Unique tiger individual's photographs captured in each Indian landscape from 2006 to 2022 (Range: Minimum-Maximum)

Tiger landscape (States)	2006	2010	2014	2018	2022
Shivalik hills and Gangetic plains (Uttarakhand, Uttar Pradesh and Bihar)	297 (259-335)	353 (320-388)	485 (427-543)	646 (567-726)	804
Central Indian highlands and Eastern Ghats (Rajasthan, Madhya Pradesh, Chhattisgarh, Jharkhand, Maharashtra, Telangana, Andhra Pradesh and Odisha)	601 (486-718)	601 (518-685)	688 (596-780)	1,033 (885-1,193)	1,161

Tiger landscape (States)	2006	2010	2014	2018	2022
Western Ghats (Karnataka, Tamil Nadu, Goa, and Kerala)	402 (486-718)	534 (500-568)	776 (685-861)	981 (871-1,093)	824
North-eastern Hills and Brahmaputra Plains (Northern West Bengal, Assam, Arunachal Pradesh, Mizoram, Nagaland)	100 (84-118)	148 (118-178)	201 (174-212)	219 (194-244)	194
Sundarbans (West Bengal and extends into Bangladesh)	Not sampled	70 (62-96)	76 (62-96)	88 (86-90)	100
Grand total India	1,411 (1,165-1,657)	1,706 (1,507-1,896)	2,226 (1,945-2,491)	2,967 (2,603-3,346)	3080 *

*Ranipur (Uttar Pradesh) is added in Shivalik landscape because three tigers were common between Ranipur and Madhya Pradesh (Central Indian landscape); hence three tigers were subtracted from the total of All India landscapes.

The landscape of the Northeastern Hills and the Brahmaputra is secure, with 194 distinctive tigers captured on camera traps, compared to an estimated 219 (SE 194-244) tigers in 2018. Better protection and augmentation of prey, securing migration corridors to maintain the genetic diversity, regular ecological monitoring and mitigating the impact of development activities in PAs such as Nameri, Buxa, Namdapha, and Kamlang as well as outside PAs could help increase the tiger population.

Historic transboundary tiger landscape in Sundarbans

Kolkata (Calcutta) is situated on the east bank of the river Hooghly in latitude 22° 33 '47 N and longitude 88° 23' 34 E. During and towards the end of Pleistocene, the Ganges and its tributaries were building deltas to the west of Kolkata. The course of the Ganges gradually shifted towards the east and built a delta in the Jessore district, while extending it to the present Sundarbans location later (Karim, 1994). On the Indian side, a curious discovery was made in re-excavating the Esplanade tank on the north of Chowringhee in erstwhile central Kolkata. In the month of April, 1815 at a depth of four feet below the level of its dry bed, trunks of the sundari tree standing upright were found embedded at short intervals. The land which originally bore the trees on its surface had been covered over by subsequent fluvial deposits. In his Revenue Survey Report, Colonel Gastrell (1866) quotes interesting evidence on the subject of the depression and subsidence of the Sundarbans from time to time and incidentally mentions the Kolkata borings in Fort William to a depth of four hundred and sixty feet below the mean sea-level from December 1835 down to April 1840. James Fergusson (1891), the architectural historian, opined that hardly more than 4,000 years have elapsed since the tide was near Rajmahal. He believes there always existed- in historical times at least- a bar or barrier where the tides turned somewhere very near where the Sundarbans now are, and that between this and the apex of the delta there was a tidal swamp. The stones and red soil of the Chota Nagpur hills were found at a depth of 300 m at the Sagar Island and at several other places in the western Sundarbans delta.

It may be mentioned here that the tidal effect of the river Hooghly stretches to approximately 282 km length up to Nabadwip in Nadia district, where two non-perennial rivers Bhagirathi and Jalangi, meet. Ferguson referred to those 'dwipas' or islands as merely elevated spots in the tidal swamp that first attracted a population. These are -

(i) Navadwipa meaning a new island (*dwipa*) in the olden times when the neighbourhood was a sea or at least a tidal swamp.

(ii) Agradwipa, a little in front of the Navadwipa, in Purba Bardhaman district.

(iii) Further down is 'Sukhsagar' in Nadia district, indicating the site of a dried-up sea.

(iv) Lower down Chakdaha (Chakradwipa, the circular island) in Nadia district.

(v) Dumurdaha (Dumurdwipa, the island of the fig trees) in Hooghly district.

(vi) Khardaha (Khargadwipa, the spear-shaped island) in North 24-Parganas district.

(vii) Ariadaha (Aryadwipa, the island of the Aryans).

It appears that the extension of the delta was from north and west to south and east. Near Kolkata, an elevation of the area has alternately been followed by a subsidence. The extreme south-eastern portion, including the districts of Khulna, Jessore, the Sundarbans, and Kolkata, was not fully formed in the seventh century of the Christian era, when Eastern Bengal was sufficiently inhabited to form the nucleus of a kingdom of the Sen dynasty about the year 994 CE.

Molecular data also supports migration of the mainland (peninsular) tiger population to the Sundarbans mangroves during ±1,000 years ago due to fragmented tiger-lands, absence of any viable corridor and complete isolation during all these years from the nearby Bengal tiger populations by 200-300 km of arable land and human settlements; there was little scope for migration out of the Sundarbans population, particularly during last 600 to 800 years due to anthropogenic activities. Human colonisation of this region happened relatively late due to the inhospitable conditions though some people did occupy the area even in the 6th century. The present day district of the 24-Parganas was ceded to East India Company as part of the treaty of 1737 and thereafter became the 'jagir' of Lord Clive. However, it was only in 1770 that serious efforts were made to reclaim land for agriculture by Claude Russell, the then Collector General of the district.

In West Midnapore, Hijili emerged as an island from the estuarine surroundings around 1400-1500 AD and later covered with natural mangroves to harbour tigers. At the time of emergence of this land, it was intersected by Cowcolly River, a tidal creek, and the two islands namely

Khejuri and Hijili were formed. After the decaying of Cowcolly River, the said islands were merged and the total area is called Khejuri. To the Muslim historians of the eighteenth century, the coastal strip from Hijli (Midnapore) to the Meghna in the east was known as 'Bhati', meaning lowland.

During the 18th century, the Nawabs of Chitpur used to hunt the tigers at Baranagar a little way above the temples of Dakshineswar. In 1756, when Siraj-Ud-daulah recaptured Calcutta (now Kolkata) from the East India Company, beyond the Lower South Circular Road on the south, now known as Chowringhee, were the Sundarbans forests. Before the East India Company settled in Kolkata in August, 1690, on the Hooghly riverside to the south of the settlement was the village of Gobindapur, founded by the Setts and Basaks from Satgaon (Saptagram) around the end of 16th century, which was surrounded by a thick tiger-haunted jungle where expanded the grassy level of the maidan. Sir William Jones, who was appointed puisne judge to the Supreme Court of Judicature at Fort William on 4th March, 1783, lived in Garden Reach and walked to the Court and back everyday, when people did not travel beyond Chowringhee Road in the evening, for fear of tigers! Kashipur (Cossipore), where Chief Justice Sir Robert Chambers lived during the late 18th century, was also quite dangerous at night from the visits of tigers. Opposite Baithakkhana, in the southern corner of Sealdah, there were, in those days, the houses which formed the Jockey Club and the Restaurant for the sportsmen of Calcutta, who enjoyed their holiday by visiting the neighbourhood of Dum-Dum for shooting tigers and boars. Further south, Belvedere in the suburb of Alipore became the official residence of the Lieutenant Governor of Bengal, in 1854 and could easily be recognised from the figure of the Bengal tiger surmounted on the central gateway.

In fact, reports of tiger phobia among the local citizens were elaborately published in journals and newspapers of the era. In 1819, a news report in 'Samachar Darpan' appeared that described how a woman was killed by a tiger in the Gouripur area. After killing the victim, the tiger forcefully entered a neighbour's house. The residents of the house somehow managed to trap the tiger in a room and locked the door from outside. The police were informed, who, with the help of local administrative staff, managed to kill the tiger. Present-day Bagmari was once the hunting resort of the sahibs of Calcutta and a tiger was captured and killed here.

In 1837, the Autumn issue of Bengali journal 'Samachar Darpan' reported on an adventure by a tiger-hunting party. There had been rumours that a tiger was on the prowl in the northern parts of Calcutta. But when Babu Dinanath Dutt, along with some Europeans, went to Shyampukur with a double-barrel gun and his hunting dogs, he didn't find any tiger there. The team, however, saw a full-grown leopard jumping out of the undergrowth and vanishing into a bush. The forest cover in that area of the city, according to the journal, had increased significantly and the police force had been instructed to cut down the bushes. Three years later, in 1840, 'The Asiatic Journal' published an interesting piece of news. Quoting a report published by 'Prabhakar', it wrote that a tiger would be sacrificed in front of the goddess in a temple in Shobhabazar. A few locals had reportedly bought a small tiger for the purpose (God knows from where). The paper also reported that although there were mentions of tiger sacrifice in some scriptures, nobody had ever heard of anything like that before.

Stereoscopic photograph of a tiger behind bars in the Zoological Gardens at Calcutta, taken by James Ricalton in c. 1903, is described by him in 'India Through the Stereoscope' (1907), "The tablet on top of this animal's cage states that he has devoured two hundred human victims...Higher records than this are claimed for some man-eaters, but it is always difficult to obtain absolute certainty in such matters...There are several reasons given as to why the tiger acquires the habit of killing human beings. Sometimes a mother tigress with cubs finds it difficult to supply her young with food, and this urgent family requirement compels her to make indiscriminate attacks. In other cases decrepitude or physical disability renders it necessary to capture something less agile and fleet than common game."

From the fifteenth century, men carried out the work of reclamation in the Sundarbans, fighting with the jungle, the tiger, the wild buffalo, the wild boar and the crocodile, until at the present day, nearly half of what was fortunately an impenetrable forest has been converted into gardens of graceful palm and fields of waving rice. The region has been focussed on as being largely a forest, but human settlements have been grafted onto forests, which were systematically cleared to make way for human habitation.

James Rennell's map of Bengal (1779) is considered the first most reliable geographical document of lower Bengal. Rennell's map showed Bakerganj district (east of the Beeskhallee River in the eastern Sundarbans) as "completely depopulated by the Maghs".

Between 1784 and 1786 Tilman Henckell tried to demarcate a boundary between the *zamindari* lands and the Sundarban forest. He is known to have defined the boundary of Sundarbans as Bay of Bengal in the South, Haringhata river in the East, Raimangal river in the West and the village of Dulyanpur, Kagrighat, Chingrikhali, Dhaki creek, Serpatlya, Kachua and the rivers Kalinga, Jamuna, Kabadak, Marjatar, Pandora, Dankhali and Baleswar in the North.

Hugh Morrieson and W.E. Morrieson, the surveyor brothers of the Sundarbans adjoining the Jessore district, wrote many accounts in their field books (1812-1818) referring to the depredations caused by the tigers. In one such account they found that "just as the theodolite was rectified and we were about to take the first angle, a tiger made a great spring from somewhere into a bush, about six yards from us, and there we lost sight of him". Moreover, these surveyors had no tiger charmers in their retinue as their field books record numerous attacks made by the tiger on the attendants. In the most dangerous tigerish parts of the Sundarbans, at the mouths of the rivers Malancha and Raimangal, they found many rhinoceros and deer.

In 1831 Captain Hodges published a map of the Sundarbans he divided all the forest as far as the River Pasur into blocks, and revising the numbering, reduced the whole of his and Princep's blocks into a series numbered from 1 to 236. The aggregate area of these 236 "Sundarban lots" was computed at 1,702,420 acres or 2,660 square miles. Since 1830, 3,737 km² of the region to the South of Dampier-Hodges line had been cleared for cultivation and settlement. This line denoting the then forest limit extended from Kulpi to Basirhat and comprised of (a) deforested colonised area and (b) the residual forested land.

When the first volume of Hunter's statistical account of Bengal (1875) was reprinted in 1998, the section on the Sundarbans was reprinted separately.

O'Malley's gazetteer (1914) was reprinted in the same year with the picture of tigers on the cover page indicating the importance of the tiger when talking about the District of 24-Parganas. When the West Bengal Government's Information and Cultural Department published its volume on the district of South 24-Parganas a tiger not only appeared on the cover page, but on virtually the first page of each article.

"The Southern portion of the Sundarbans, which comprises the jungle tract along the seashore, is entirely uninhabited, with the exception of a few wandering gangs of woodcutters and fishermen". The whole population is insignificant according to Hunter (1998 [1875]:35). Therefore, the humans were all 'immigrants' and the tigers and crocodiles were the only 'aboriginals'. When Hunter talked about nature the first thing that caught his attention was the tiger.

"Tigers are very numerous, and their ravages form one of the obstacles to the extension of cultivation... The depredations of a single fierce tiger have frequently forced an advanced colony of clearers to abandon their land. and allow it to relapse into jungle (Hunter 1998 [1875]: 33)."

Hunting the tiger, mostly man-eaters, was essentially the prerogatives of the white hunters who sometimes in collaboration with the indigenous elites were keen and passionate about tiger hunting. As far as the hunting of the tigers of mangroves of Sundarbans was concerned, the colonial government thought it safe to engage the indigenous elite shikaris in the destruction of man-eaters of the mangroves and also adopted a policy of endowing monetary reward to encourage indigenous shikaris in the mission destruction of man eaters (Chatterjee, 2019).

When Tilman Henckell was appointed as Judge and Magistrate of Jessore in 1781, his agent, who was clearing parts of the Sundarbans, was very much disturbed by tigers which attacked his people. In 1852 special measures were taken up for the destruction of tigers in the 24-Parganas and the reward was raised from Rs. 5, which afforded very little profit, to Rs. 10. Government sanctioned the increase in 1853 and doubled the reward in 1860, for every tiger killed near the Matla in order to aid the work of reclamation there. A government notification dated 16th November 1883 and published in the Calcutta Gazette authorised the rangers and foresters in charge of the eight chief revenue stations in the Sundarbans RFs to pay rewards for the killing of tigers. In 1883 the amount of the reward was Rs. 50 for each full-grown tiger and Rs. 10 for each cub. To receive their reward, the *shikaris* (native professional hunters recruited by British officers because of their local knowledge of the jungle and hunting skills) were required to produce the skin and skull of the animal for the forest official. The reward was gradually raised over time, each increase following fresh depredations of tigers in the jungle. In 1906 the reward was raised to Rs. 100 per full-grown tiger and Rs. 20 per cub. In 1909, the amount of reward for a full-grown animal was further raised to Rs. 200. This last raise was prompted by the loss of 500 lives to tigers between 1906 and 1909. Reward policies were adopted by the government to induce hunters into killing tigers that disrupted land reclamation. This led to a large-scale slaughter of tigers. More than 2,400 adult tigers were killed between 1881 and 1912. The Divisional Forest Officer of the Sundarbans Division used to recruit professional hunters for the posts of Boatman and Forest Guards (Gani, 2002). Their services were used to kill tigers, especially man eaters or tigers who were disturbing people near the timber coupe during harvesting operations. The poachers commonly used shotguns for killing Bengal tigers. Hunters also use muzzleloaders and rifles to shoot tigers. Poisoned baits are also used by poachers against tigers. Efforts were also made to destroy the tigers and prey animals by setting plain traps or traps with spring-loaded bows and poisoned arrows. Such traps could be successful only in the winter, as tidal waters flooded them at other times of the year. By 1947, the forests in the Sundarbans were only 50% of their pre-colonial size.

Human-tiger conflict

Haque *et al.* (2015) opined that the climate change effects in the Sundarbans mangrove forest through changing its biodiversity composition in terms of loss of wildlife habitats which is responsible for accelerating tiger-human conflicts and suggested a social and cultural revolution for sustainable alternative livelihood of forest-dependent population i.e. Alternative Income Generation (AIG), modification of the formal legal system, institutional development and in depth research can minimise these issues towards the sustainability of Sundarbans mangrove forest.

The north-western side of the STR is surrounded by many revenue villages (0.22 million people are living in 66 *maujās* within only 2 km of the buffer zone), thus making the reserve ever more vulnerable to ever increasing biotic interference in the form of livelihood forest explorations, illegal fishing, timber smuggling, and poaching. Reports of tiger attacks continued with usual regularity. Much of the tiger strikes occur in the northern and north-western mangrove jungle because most of this area falls within the buffer zone of the TR, where permit holders are allowed to fish and collect forest produce. Usually thousands of men used to enter the deep forests every day without permit, driven largely by lack of alternative sources of income.

Usually, the tiger attacks the fishermen and others when it is low tide. While they get off their boats and walk through the dense thickets in search of trees to chop or honeycombs and for fishing in the prime tiger habitat, killings are unavoidable, just like the mainland tigers in fringe forests with more cattle than wild prey live mostly by cattle lifting just because of easy availability. Tigers deep inside the mangroves do not differentiate the targeted prey roaming inside the forest, be it human beings or deer. In the dark forest, tigers find it easy to stalk and attack men absorbed in their work. Even fishermen in small boats have been attacked due to tigers' strong swimming abilities. In all, 67.2% of all tiger attacks occurred as a result of illegal forest entry. Over three dozen tiger-related deaths have been recorded in the Sundarbans in West Bengal since April 2020. In almost all cases, attacks occurred while fishing in the narrow creeks or on riverbanks or catching crabs on mudflats adjacent to RFs defying prohibitions. A majority of the deaths in the past two years have happened around the Jhila forest area, which is part of the buffer area, but the tiger density is very high in this forest block. Tigers in STR (25%), from the Vidarbha region of Maharashtra (25%) and from around Pilibhit-Dudhwa TR (20%) accounted for 70% of all human deaths in India (Jhala *et al.,* 2021).

Only three straying cases took place in 1970s (2nd August 1974, 14th August, 1974 and 13th July, 1977).The tiger straying incidents during the late 20th century increased from the 1980s to the 1990s.

1986- 14 (Male or M 12+Female or F 2);
1987- 20 (M16+F4);

1988- 4 (M);

1989- 3 (M3+F1);

1990- 11 (M10+F1);

1991- 1 (M);

1992- 3(M);

1993- 1(F);

1994- 8(M6+F2), four sent back to the forests, three killed by villagers; 1 tiger tranquillised and released in STR area;

1995- 25(M19+F6), 13 sent back to forests; one killed by villagers; one sent to Calcutta zoo after tranquillisation;

1996- 21, 20 sent back to forests; one sent to Calcutta zoo after tranquillisation;

1997- 3, all sent back to forests;

1998- 5, all sent back to forests except one killed by villagers;

1999- 10 (one suspected to be killed by villagers);

2000- 6;

2001-2002- 8;

2002-2003- 24;

2003-2004- 23;

2004-2005- 17;

2005-2006- 2;

2006-2007- 11;

2007-2008- 11;

2008-2009- 15;

2009-2010- 24; and

2010-2011- 3.

The problem of human-tiger conflict in the Sundarbans is classified into two types:

i.   Conflict outside the forest area: This is caused when the tiger stray out of the forests and enter the villages on the opposite side by crossing the channels. In the past, these fringe villages were established on the reclaimed forestland, where the interface is mostly demarcated by a small creek, e.g. Kalitala and Samsernagar on the boundary of Arbesi-1.

ii.   Conflict within the forest areas: This is caused due to intrusion (with or without legal permits) of people in the tiger habitat. Mostly the honey collectors and fishermen fall prey to the tigers inside the forest. The honey collectors penetrate deep inside the forest in search of beehives and, in the process, mostly get isolated from the other group members because of the thick and almost impenetrable forest. As a result, they become easy prey for the tigers. Sometimes these groups accidentally and unknowingly get close to the tiger habitat and the disturbed tiger attack them. The honey collectors also burn the beehives by using the Golpata (*Nypa fruticans*) and Hental (*Phoenix paludosa*) leaves, which creates heavy smoke and causes irritation and disturbance for the tiger and they are attacked by the big cat.

It is a good indication that although many tigers strayed, the villagers help the forest staff to rescue them safely with the people's cooperation in STR as follows: 2002-2003- 11, 2003-2004- 4, 2004-2005- 4, 2005-2006- 1, 2006-2007- 1, 2007-2008- 2, 2008-2009- 3, 2009-2010- 14, 2010-2011- 11.

Between the World Environment Days of 2009 and 2010, no less than 25 tigers strayed into Sundarbans villages and left around a dozen people injured. Never before in recorded history has this tide country seen so many tigers venturing beyond the forest limits. The wildlife experts believe this unique habitat of the Royal Bengal Tiger is undergoing a change and there are enough indications to suggest that the terrain is turning increasingly hostile not only for the big cat but also several other species of flora and fauna. A chain of ecological mutations followed by the rise in water level and increasing salinity in the rivers and creeks has resulted in the flight of tigers from the southern end of the forest to the northern part, leading to a higher density of big cats in the area, which is hemmed by human habitation, and frequent straying has been the inevitable impact of these natural threats.

The Sundarbans tigers stray into the neighbouring villages because they are situated in the reclaimed forest lands and in the same places the boundary between the forest and agricultural land is also not distinct. Some villages have small patches of mangrove forests and the tiger enters into these forests by losing direction. Sometimes tigers swim and cross the small creeks to capture easy prey of cows and goats.

Still in the second half of 2020, the whole of 2021 and the early months of 2022 there were increasing instances of tiger straying in localities because the protective fences remained collapsed or damaged at several places in the aftermath of cyclones, allowing tigers to stray into human inhabited islands. Sometimes fishermen also removed fences to enter RFs for catching crabs or collecting honey, etc.

It is well known that Bengal Tiger, the apex predator, represents the only population in the mangrove forest, which is adapted to living and hunting in both the terrestrial and aquatic ecosystems. Mallick (2013) recorded that almost 35% of the water regime in the Indian Sundarbans forms a normal part of the tiger's movement zone. The straying tigers were observed to cross the creeks or rivers between 50 m and 150 m in width to enter into the villages in Basirhat Range (Hingalganj block), but 300-900 m in cases of the villages of Sajnekhali Range (Gosaba block) and cross more than one km width of the River Raimangal (Hingalganj block). The most vulnerable rivers related to the tiger straying incidents are listed below.

Most Vulnerable Rivers in the Forest-Village Interface in SBR

River/creek	Community Development block	Width (m)	Percentage of tiger crossings
Kurekhali/Shakun khali	Hingalganj	25	36.3
Pitchkhali	Gosaba	150	36.3
Gomdi	Gosaba	150	7.5
Korankhali/Bagna	Gosaba	100	6.5
Rangabelia	Gosaba	600	6.5
Raimangal	Hingalganj	800	2.2
Kapura	Hingalganj	50	1.8

River/creek	Community Development block	Width (m)	Percentage of tiger crossings
Mokri	Kultali	40	1.7
Dakuran	Kultali	75	1.0
Others	-	-	2.9

Out of the above twelve rivers or creeks, the Kurekhali or Shakunkhali is the most vulnerable river in Hingalganj block as far as tiger crossing is concerned (36.3%), followed by the Rivers Pirkhali, Gomdi and Rangabelia (33.6%). The creek Kamalakhali is the most convenient for the tiger to cross because, at places, it is very narrow, i.e. a width of 15 m, which separates the Samsernagar village of Hingalganj from the Arbesi forest block. Hence, this is one of mostly affected villages visited by the straying tigers.

Long march of a straying tigress in STR

De (1991) recorded a historical movement of a straying tigress during 1985 and 1986. On 10th November 1985, the inhabitants of the Puinjali village near Gosaba discovered that a tiger had carried off a sheep at dawn. The tell-tale blood-trail and pug marks were proof of the attack but remnants of the kill could not be recovered. Puinjali is 8 km away from the Jhilla block of STR with two rivers- Korankhali and Puinjali in between. Apart from a few thorn bushes around homesteads, it has little tree-cover. At that time the paddy fields were to be harvested soon. The assistant FD of STR arrived at the village the next day. He detected the pugmarks of an adult tigress. He first sighted the animal on the 12th. It was believed that the tigress had strayed from the Jhilla block of the Sundarbans and crossed the two rivers on her way. She took cover in the paddy fields. The forest officials were besieged by angry and excited villagers and could not implement their plan of driving the animal back to the Jhilla forest. On the 13th, FD armed with tranquillisation equipment arrived at Puinjali. The trap was laid with the bait of a live goat. In order to prevent the tigress from entering other villages, the forest officials and villagers organised patrols on the northern outskirts of the village. The patrols used crackers and high-power torches. But the tigress sneaked through the cordon at night and moved to

Moukhali, 17 km inside human habitation. The residents had not come across any tiger in the village in their living memory. Very soon a large crowd gathered to have a glimpse of the unwelcome guest. Tension ran high and some men pelted stones at the animal. The unnerved tigress could have mauled anyone in self-defence. All this made the forest official's task really daunting. Plans for tranquilisation of the animal had to be abandoned as the crowd could hardly be controlled. On November 16 the tigress moved on a south-western course and reached Mollakhali. Alarmed on being chased by a screaming mob, she mauled a person. Next day, the tigress prowled about Kalidaspur, a hamlet hardly a few km to the north of the forests. With the arrival of a police contingent the problem of crowd-control became somewhat easy. In order to arrest the animal's movement, watercrafts were placed on a creek on the northern fringe of the village. Shouts of the men on board foiled an attempt by the tigress to swim across the creek. Unfortunately, the tigress had no intention of getting into the nearest forest. She continued wandering on a south-western course and slunk by Satjelia and Dayapur villages. On November 18, many had a glimpse of the majestic animal as she stalked along the embankment at Pakhiralay. Thereafter, she swam across Pichekhali river to enter the Pirkhali forest close to the Sajnekhali tourist lodge. The forest officials trailed the tigress all along her 60-km journey. Except for the loss of one sheep at Puinjali and minor injuries to a man around Moukhali, there were no other casualties. The tigress again came to the notice of forest officials on 19th March 1986 at Lahiripur. Some women were fishing near the Duttar forest checkpost, when the tigress suddenly appeared and mauled one of them. The woman subsequently died in the Gosaba hospital. The tigress next appeared at Sudhanyakhali on April 3, and carried off a fisherman. His body could not be recovered. The tigress was identified by her pugmarks. On several occasions from the beginning of March 1986, the tigress killed cattle, goats and pigs at Jamespur and Dayapur villages on the bank of Pichekhali river opposite Sajnekhali. On her way she traversed the forest around the Sajnekhali tourist lodge and visitors had the thrill of seeing her on a number of occasions. The villagers were panic-stricken and their rage against the forest officials mounted.

Towards the end of March, the forest officials attempted to trap the animal with the bait of a live goat at Jamespur but failed. The villagers were at the end of their tether and sent a deputation to the FD in Kolkata. The department directed that the animal be trapped and released deep inside the forests. In view of the recent death of a tiger after being tranquilised at Hiranmoypur, this process was not recommended.

On 20th April 1986, a trap with a wild pig as the bait was laid in the forest about 200 m to the north-east of the Sajnekhali tourist lodge. Forest officials stationed in a nearby launch kept vigil. Simultaneously, an electric fence was erected along the bank of Pichekali to prevent the tigress from entering the villages. The tigress was perhaps hungry and walked into the trap in the early hours of April 22. The forest headquarters in Kolkata instructed the local officials to release the tigress around Haldi block in the core area of STR.

The tigress was transferred to a lighter trap to facilitate handling. In the morning of April 23 the launch Monorama with her Royal Bengal passenger cruised towards Haldi. Meanwhile, the animal had finished the bait and also drunk some water. No veterinary aid was readily available but the tigress looked cheerful. She enjoyed another meal of wild pig that was offered. The Monorama reached Narayantala in the vicinity of the Haldi watch tower at around 4 pm. The trap was placed on the mud-flat under a tall keora tree from the branches of which men could operate the trap-door. The forest officials were ready for release of the animal, but under instruction from the headquarters, they awaited the arrival of FD, who had rushed from Kolkata.

With the arrival of the flood-tide, the water started rising. The mud-flat was inundated. The tigress was in a pitiable condition, and the water was almost touching her neck when FD's launch finally arrived. Dusk had set in and visibility was poor. Two forest officials, Maity and Gopal Tanti, well-known for their courage and dedication, volunteered to operate the trap-door. They swam through the swirling waters, braving the menacing crocodiles and a pair of tigers said to be on the prowl in the area. As they lifted the trap-door from their perch on the tree, the drowning tigress jumped out and bounded beyond the edge of the water. Before slinking away into the forest, she had a lingering look at her benefactors on board. She was hardly aware that some of her saviours were on the tree. For many weeks, the tigress's movements were monitored by the officials of the Haldi watchtower and her pug marks were detected at regular intervals on the banks of the fresh tank adjoining the tower. It was a happy end to a traumatic chapter of her life, wrote Rathindranath De (1991).

Over the years the site management has several actions taken to contain human-tiger conflicts by -

(i) stoppage of forestry operations;
(ii) closure of issuance of permit for *Phoenix* and *Nypa*;
(iii) digging of fresh water ponds, 3.5 m deep and monsoon-fed inside the forest to accustom the tigers to a permanent source of freshwater and re-excavation annually; and
(iv) introduction of human face masks and clay models wrapped with energisers, which are charged to 230 Volts and 12 Volt battery source.

As a result of all this, man-killing incidents have decreased from an average of 25 annually during the 1980s and 1990s to only 7 and 5 in 1994 and 1995 respectively. But none of these measures have conclusively proved to be effective. The incidents of tigers trying into the fringe villages of the STR and other areas of SBR seem to have increased in the next two decades for killing livestock. During their stay in the villages of the Sundarbans, they generally lift cattle, goats or pigs from the village. It shows that the tigers generally visit the villages during the monsoon and early winter months. The heavily affected villages, located on the north-eastern part of STR, are Shamshernagar, Kalitala and Kumirmari in Bagna range and Rajat Jubilee and Jamespur in Sajnekhali range. All the straying incidents have occurred at night, with male tigers being responsible for the majority of the incidents as against minimum incidents by females. The tigers had crossed rivers between 25 m and 150 m in width (Kalitala 25 m, Kapura 50 m, Gomdi 150 m and Pitchkhali 150 m) to enter into the villages in Basirhat range but 300-900 m in width (Rangabelia 600 m and Raimangal 800 m) to enter into the villages in Sajnekhali range. In most of the cases the tigers had resorted to cattle-lifting or poultry feeding. Only in four incidents men were attacked or killed.

Fencing the boundaries of the vulnerable forest and village areas by vegetative cover i.e. *Ceriops-Excoecaria* species and mechanical methods by nylon-net fencing using *Avicennia* posts along the forest fringe has been found to be very effective. Both these fencings generally last for about three years. However, *Ceriops-Excoecaria* fencing is not encouraged now because it leads to a heavy toll of vegetation. Net fencing is not foolproof. Nets serve mainly as a psychological deterrent for tigers, but they can easily bring them down. Also, receding tides bring dead leaves and branches that get stuck in nets and pull them down, which need to be checked and fixed regularly, but lack in manpower makes it difficult to cover the extensive forest boundaries regularly. Field observations indicate that the tiger is able to negotiate the eight feet high fence by jumping over the same. Use of RCC posts and Bamboo poles can also help erect the fence at a height of 10 - 12 ft. Changing the vegetative fencing with Nylon-net fences, having mesh size 4" X 4" (to avoid any strangulation of wild animal) with a height of 8-10 feet and erected with *Avicennia*/bamboo/Reinforced Cement Concrete posts at a 2 m distance with the metal runner both at the top as well as bottom, has been found to play an important role in preventing the straying out of tigers into village's areas from forest. In STR, such fencing has been erected over a stretch of 96 km including 54 km of forest fringes with the following distribution:

Range	Length including *khal* guards	Location
NP (West)	21 km	Belegudam *khal* to Nawbanki *khal*
Sajnekhali WLS	35 km	Pirkhali *khal* to Pakhir *khal*: c. 5 km Pakhir *khal* to Lalitkhali: c. 9 km Lalitkhali to Ranjit's *khal*: c. 5 km Ranjit's khal to Duttar Beat: c. 6 km Dattar beat to Kamari *khal*: c. 10 km
Basirhat	40 km	Shamshernagar (Kalindi River side) to Jhingekhali Beat: c. 13 km Raimangal river to Jhilla river (Arbesi-2 comp): c. 7 km Jhilla Compound to Kakmari *khal*: c. 20 km

A tiger strays due to age, injury or pregnancy which impairs its ability to hunt. Straying may be classified on the basis of duration. Many times it becomes possible for the staff and local villagers to drive the tiger back to the forest by using traditional methods of drums, crackers, fire etc. Sometimes the tiger also goes back to the forest on its own after spending some duration in these villages. These straying incidents are termed as 'Temporary Strayings' and an iron cage with live bait is sometimes used to trap the tiger and then relocate in the forests. In 'Longer Strayings', the tiger takes refuge in a cattle shed or inside any village hut, and chemical capture and tranquilisation becomes the only resort to rescue the animal.

Tiger bite is a major public health problem in the greater Sundarban areas. Tigers have the ability to cause significant trauma and hidden injuries. The most common location for these injuries is the nape of the neck- tigers can realign their jaws so that they can bite down between a victim's vertebrae and into the spinal cord, resulting in serious craniofacial and cervical spinal injuries. Proprioreceptors, found in the mouth and teeth of tigers, enable them to align their bite parallel to the cervical vertebra and during subsequent cervical extension they purposively sever the cervical spinal cord. The mechanism of injury mirrored the typical nature of tiger attacks where, post biting into the victim's neck, and subsequent lateral extension, an open cervical fracture occurred complicated by a vascular injury. Post violently pulling him to the ground the animal's massive forebody weight was further exerted onto his left shoulder resulting in a closed 4-part proximal humeral fracture (Kelly *et al.*, 2020).

Predatory behaviour is also triggered by movement, making humans particularly stimulating as "prey" for big cats. For example, tigers can frequently be seen stalking the running human beings. Bite wounds can also result in significant bacterial infections. In fact, the deep nature of the wounds inflicted, that are difficult to clean, results in them incurring a high rate of infection. *Pasteurella multocida*, a frequent commensal found in the mouths of tigers, is one of the most frequently offending organisms. Although *Pasteurella multocida* is often involved, actual bite infections are commonly polymicrobial.

Rahman *et al.* (2009) reported that six patients were admitted at the Department of Surgery, Khulna Medical College Hospital, Bangladesh, situated 30 miles from the forest area, from 2003 to 2006. Injury involved different parts of the body, head and neck region; predominantly, the upper part. All patients needed emergency surgery including wound toileting, debridement and even reconstructive surgery. One patient died due to infective complications.

Case studies in Indian Sundarbans

Case 1: A 17-year-old boy, along with his brother, went to the deep jungle for fishing. They were sleeping on a country boat in a narrow tunnel inside the jungle. He woke up to find a big Royal Bengal tiger trying to drag him from the boat and biting his face. They fought with wooden sticks and ultimately escaped from the claws of the tiger but the boy received an injury on his face. After that they managed to come home, after an 8-9 hours journey by country boat, the wound was dressed locally by village doctors. On the next day, they managed to get admission in the emergency department of the hospital. On examination the boy had a thready pulse, was dehydrated and examination of the face and neck showed that a major part of the face including nose, cheek, left mandible and eye ball were absent with bleeding from the wound. He was immediately resuscitated with parental fluids, blood transfusion and sedation. The wounds were cleaned and dressed in the operation theatre. He was fed through a nasogastric tube. After controlling the infection in two months, reconstructive surgery was attempted, offering a deltopectoral flap for replacement of the soft tissues of the cheek and lower lip and applying a forehead flap to the nose. Seventy percent of the tissues survived and he got some configuration of the face as well. Oral feeding was possible at last. He was discharged after five months of hospital stay.

Case 2: A 31-year-old woodcutter was suddenly attacked on his head during the daytime, as he was busy cutting a tree. He fought back with his axe and managed to come to the hospital after 24 hours of the injury. On examination he was pale and dehydrated. Skull examination revealed a huge loss of scalp on the back of the head with a bleeding surface. After resuscitation, the wound was cleaned in the operation theatre and bleeding secured. He also developed postoperative infections and took nearly two months to recover. Later, the scalp was reconstructed with skin cover.

Case 3: A 42-year-man, a part time woodcutter, was attacked by a tiger at night while he slept on a boat. He was attacked on his neck and back, but managed to escape. When he was examined after 20 hours of the incident, there was evidence of claw injuries on the neck and loin and subcutaneous emphysema in the neck region. The apex of the lung was also injured. There was no major sepsis and wound complications and the patient was discharged within 15 days of admission.

Case 4: A 33-year-old man was admitted with scalp injury, through the emergency department within 10 hours of injury. He was fishing in the outer area of Sundarbans. On examination, he was anxious, pale and bleeding from his scalp. He was taken to the operation theatre, bleeding was stopped followed by debridement and stitching of the wound, as there was no tissue loss. In the postoperative period he developed infection but daily dressing of the wound led to wound healing. He was discharged within 20 days of the admission.

Case 5: A 40-year-old woodcutter was admitted after 16 hours of tiger bite on his abdomen and right loin with evisceration of the gut. A chunk of the abdominal muscles were taken away with part of the iliac bones. He was given first aid and debridement of the wound on different parts of the body. Daily cleaning and dressing was carried out. A pedicle flap replacement on the big defect of the loin was planned. Unexpectedly, he developed uncontrollable sepsis and died on the seventh day.

Case 6: A 46-year-fisherman was attacked by the tiger. His scalp was partially avulsed giving a profuse bleeding surface. He was admitted on the same day to a hospital. The wound was toileted and stitched primarily. The patient could be sent home within seven days.

Although some experts believe tigers strike only from the rear of their prey, it was also observed to attack forest entrants from the front, grabbing his shoulders with its paws and trying to bite his neck. Two of his companions drove off the tiger by hitting it with sticks. Moments later, as the men desperately pulled the victim toward their boat by his feet; the tiger returned, grabbed him by the neck and dragged him away as the others fled to their fringe village. Elsewhere, tigers seldom kill people unless they are defending their food or cubs or recovering from wounds inflicted by a hunter. The swamp's unparalleled big tigers are completely at home in the channels that separate the many islands. They have been known to rocket out of the water onto the deck of an open boat and pull a sleeping crewman off by the back of his head. The tigers often consider man an easier catch than their main prey: wild pigs, spotted deer and even fish. The deer are faster. Both male and female boars have a pair of razor-sharp tusks, which are elongated canine teeth, to defend themselves. In males, the tusks can grow to 25 cm long. Even the fish negotiate the swamps and waterways better.

Ecological and social isolation

All the above mentioned factors probably caused complete isolation of the Sundarbans tigers, genetically and ecologically distinct from the mainland tigers. The estimated migration rates from the Sundarbans to the northern and Peninsular India were lower (0.005% and 0.007% respectively) and the lower limit of 95% credible intervals of both estimates was 0, suggesting no migration (Singh *et al.* 2014).

Vivekanandan (2021) critically examines the conservation politics in a transboundary protected area (TBPA) in South Asia, the Sundarbans mangrove forests in Bangladesh and India. She explores the reasons why, despite collaborative measures by the two states, conservation has largely tended to conform to sovereignty practices, making it top-down and exclusionary. She examines the cultural politics of conservation since contestations to state power have often entailed the articulation of popular sovereignty in the Sundarbans and argues that the social sustainability of conservation will critically hinge on how issues of resource access and governance are framed, negotiated, and addressed.

"...the portrayal of the tiger is significant in understanding how the Sundarbans came to be regarded as a land of migrants from both, across the border and within the state. Local beliefs hold that the tiger migrated from Java and Bali to the Sundarbans to escape extensive hunting (not scientifically proven so far). The tiger's search for refuge that led it to eventually settle in the islands is an articulation of the ordeals that the migrants endured before making the Sundarbans their home (Jalais, 2010, p. 147). Cohabitating a shared mythical space led each to empathise with the other. This understanding, borne out of a shared sense of dispossession, the settlers believe, defines the dynamic balance in the mangrove forests. However, for the locals, the entire tiger conservation campaign amounted to fetishising the animal and its subsequent alienation from the people (Jalais, 2010, p. 172). Tiger tourism has further transformed and packaged this fetishised animal into a commodity. The official recognition extended to the tiger by both India and Bangladesh was to them an unequivocal indication that, on either side of the border, the Sundarbans mattered solely as tiger habitat".

At present, the breeding tiger populations in the mangrove Sundarbans must have a possibility to share genetic material and exist in a meta-population framework, thereby enhancing the possibility of their survival. As such, low genetic variation in the Sundarbans tigers [He (expected Heterozygosity or Gene diversity) = 0.58] is indicated by the microsatellite markers as compared to other mainland populations, such as northern and Peninsular (He = 0.67-0.70), whereas average observed heterozygosity (Ho) ranged from 0.40 in northern to 0.49 in the Sundarbans and Peninsular India tigers.

Chapter 5
Impact of Biotic and Abiotic Environment

"The tiger's world, by contrast, is not only amoral but peculiarly consequence-free, and this-the atavistic certainty that there is nothing more lethal than itself-is the apex predator's greatest weakness."
– John Vaillant

Together, biotic and abiotic factors make up an ecosystem like mangroves. Mangroves are a critical forest ecosystem, dominating coastlines of the Bay of Bengal. They straddle the connection between land and sea as well as nature and humans; nurture the estuaries; provide habitat and refuge to a wide array of wildlife such as birds, fish, invertebrates, mammals and plants and fuel the nature-based economies. On the contrary, when mangrove forests are cleared and destroyed, they release massive amounts of carbon dioxide into the atmosphere, contributing to climate change and stop providing ecosystem services for sustainable production and consumption.

A biotic factor is a living thing that has an impact on another population of living things or on the environment. These are both organisms and the food the organisms eat, which are classified into autotrophs (producers like green plants and algae), heterotrophs (consumers of plants and/or animals like bacteria, protists, fungi, herbivores, carnivores, and omnivores), and detritivores (either eating dead organisms directly or breaking down dead things to get energy like earthworms, fungi, dung beetles, millipedes, sea stars, and fiddler crabs).

Abiotic factors are nonliving, categorised into climatic conditions such as water, waves, tides, wind, sunlight exposure, humidity, climate, temperature, and pH, edaphic (terrain, soil composition, texture, structure, and density), and social (human activities) having profound impacts on other abiotic factors, biotic factors, entire ecosystem, and even entire biome.

Origin

The Sundarbans is the largest delta in the irregular estuarine areas of low lying, meso-macrotidal, humid and tropical belt, crisscrossed by several tributary rivers, creeks and waterways which flow into the Bay of Bengal. The entire land mass of Sundarban is of recent origin. The Himalayan orogenesis, carriage of world's largest sediment load to the sea by the rivers Ganges and Brahmaputra and the tectonic movement of the continental plates have built the unique geological characteristics of Sundarban. Neotectonic movements induced an easterly tilt in the Bengal Basin between the 12th and 15th AD and in the 16th century an earthquake tilted the Bengal basin eastward. These constant shifts due to tectonic movements leading to changing sedimentation patterns have greatly influenced the hydrology of the Sundarbans.

Describing the region, Hunter writes that the general aspect of the Sundarbans gradually changes as one travels from west to east, from the river Hooghly towards the river Meghna. Thus "the superficial aspect of the three divisions is what might be expected from their physical character". The three divisions of Sundarbans were the belt of cultivated land from the Hooghly to the Jamuna or the western division of the Sundarbans, "the marshy tract of the middle Sundarbans, between the Jamuna and Baleswar" or the middle section of Sundarbans and "the third division or the Bakarganj Sundarbans, between the Baleswar or Haringhata and the Meghna rivers". The Bakarganj Sundarbans was a pleasant change from the hot and dry lands of the 24-Parganas, and the depressing and swampy atmosphere of the Jessore Sundarbans. Here the river water is comparatively sweet and the soil richer than in the first two sections of Sundarbans.

Other than Hunter's accounts we have a number of gazetteers and handbooks written on the Sundarbans which covered three districts of the Bengal province- Jessore, 24-Parganas and Bakarganj. These included H. Beveridge's "The District of Backerganj- Its History and Statistics" (1876); L.S.S. O'Malley's "Bengal District Gazetteer: 24-Parganas" (1914) and J. Westland's "A Report of the District of Jessore, Its Antiquities" (1871). Apart from these gazetteers we also have revenue and geographical reports compiled on the Sundarbans. These included F.E. Pargiter's "A Revenue

History of the Sundarbans 1765-1870" (1885); F.D. Ascoli's "A Revenue History of the Sundarbans 1870-1920" (1921); Ralph Smith's "Statistical and Geographical Report of the 24-Parganas District" (1857) and G.P. Gastrell's "Revenue Survey Report" (1864). These records explained the process of reclamation, the history of the region as well as the physical terrain of the area.

J. Westland, the 19th century Magistrate and Collector of Jessore, described the tiger's depredation during the reclamation works in his 'A report on the district of Jessore' (1874):

"...tigers are not infrequently met with and they occasionally attack the defenceless forest-clearers, if the latter approach their lair too closely. Sometimes a tiger takes possession of a tract of land and commits fearful havoc there. The depredations of some unusually fierce tiger or tigers, again, often compel an advanced colony of cleaners to abandon, through fear, their land which they reclaimed after years of labour. A deforested land springs back into the jungle to become as bad as ever unless the greatest care be taken of the land so cleared."

In the General Report on the Census of India for 1871, 1881 and 1891, "the deltaic wastelands of the Sundarbans, which fall within the province of Bengal, were apparently excluded". In 1911, Sundarbans was described as a tract of waste country which had never been surveyed, nor had the census been extended to it. It then stretched for about 266 km from the mouth of the Hooghly to the mouth of the Meghna and was bordered inland by the three settled districts of 24-Parganas, Khulna and Backergunje. The total area (including water) was estimated at 16,902 km².

Terrain

Innumerable small creeks (locally called *khals*) caused by the scour when the water drains off the swamps at each successive ebb tide, flow into the rivers. Except where the new *chars* are forming, the banks of the rivers and *khals* are generally the highest grounds; the levels become gradually lower as one proceeds towards the interior of the island. Some *khals* join up one steam with another. These are called *bharani*, as opposed to *mara khal*, which gradually split up into smaller *khals* until they are lost in the forests. *Bharani khals* are important particularly when they run east-west to join two rivers or estuaries. The inhospitable terrain (mud, tide, salinity, innumerable protruding mangrove roots, cyclonic storms, crocodile in water and poisonous snakes on land) has moulded the Sundarbans tiger to acquire unique characteristics that have branded the big cat as most cunning, strong and having supernatural powers. This terrain not only makes it difficult for people to eke out a living but also forces people and tigers to tread into each other's territories for meeting their biological needs.

Climate

The Sundarbans experiences tropical humid coastal climate characterised by moderate rainfall, temperature and high humidity with fierce cyclonic storms and depressions during pre-monsoon, monsoon as well as post-monsoon months- dry weather remains in pre-monsoon from March to May; monsoon from June to September and post-monsoon from October-November excluding winter from December to February with an average wind speed of 11.5 km, 11.1 km and 60-65 km per hour respectively. The climate is subtropical- temperature varying from 20°C (December-January) to 33°C (June-July), average annual rainfall 1,920 mm and average relative humidity between 70% and 82%. During the winter, relative humidity remains below 60% and cloud cover is the lowest (about 10% country-wide) due to the cold dry winds from the north-western part of India. In contrast, during the monsoon, humidity is more than 80% with 80-90% cloud cover. There was a decrease of 40% in sunshine during the winter and post-monsoon seasons (decrease of 0.6 hours a day per decade in winter and 0.4 hours a day per decade in post monsoon).

Seasonality

Seasonality plays a very important role in the Sundarban, which may be seen from the comparative data collected in 2006 and 2010 respectively. There were a total 2471 number of tiger pugmarks observed in the year 2006 during river transects as compared to only 818 of tiger pugmark observations made in 2010 during river transects. There is a sharp decline of 66.89 % in tiger pugmarks in 2010 as compared to 2006. Similarly, the numbers of signs/evidences of other wild animals (spotted deer and wild boar) were 18,293 in 2006 as compared to only 4,267 in 2010 exercise showing a decline of 76.67%. There may be many reasons for the same but one of the important reasons is that in 2006 the exercise was done in the first week of January when temperature was low (23°-18°) whereas the exercise in 2010 was done in March, when temperature was high (22°-33°).

Impact of climatic changes

Recent observations by big cat experts like Dr. Pranabesh Sanyal have revealed serious physiological changes in this endangered cat, considered Bengal's pride because of climatic change and increased salinity in the water. These include weight loss, shrinking size and fading lustre. While the average weight of an adult Royal Bengal tiger in other parts of India is over 180 kg, in the Sundarban region it has fallen to around 110 kg. And it hasn't just lost weight. The length of a Sundarban tiger has come down from over 2.7432 m to about 2.4384 m. The lustre of its coat is also fading because of the increased salinity in the river water. Similarly, the tiger's weight loss is being attributed by some experts to the food shortage and spending more time trying to catch its prey, even migrating from the core area to the buffer zone because of the unavailability of food in addition to their consumption of saline water as the number of freshwater ponds inside the core area of Sundarbans has gone down drastically because of rising salinity. Currently, the average salinity in Sundarbans is >21 parts per thousand (ppt), which touches 25 ppt during winter. Experts claim that the salinity level in the mangrove forest has increased 15% in the last two decades.

Deltaic zones

The Sundarbans tidal delta plain can be divided into four specific zones: the inactive delta (once an active part of delta, but with currently reduced or no fluvial activity), mature delta (moribund deltas that have ceased), tidally active delta (a still active part with water channelling through it carrying sediment) and sub-aqueous delta (part of the delta that is below the low-tide mark).Although there have been some micro-topographical exceptions due to rivers, streams and tidal flow, this deltaic wetland of the Sundarbans is almost flat. The land slope in this region is 0.03 m vertically per km of horizontal distance from north to south. The elevation of the forest area varies between 0.9 and 2.1 m above mean sea level.

During the period May to October tidal inundation regulates the hydrology of the area. The Sundarban can be divided into four hydrological zones:

Inundated area normal high tides: It is the more significant part of the Sundarban.

Inundated area only by spring high tides: This zone includes the northern part of the Sundarban. Inundated area by the high monsoon tides: It includes levees located in the north east of the Sundarban.

Inundated area by all tides: It is located in the mouth of big rivers and along the sea coast.

Major river system of the Sundarbans

Eastern (Bangladesh) side	Western (Indian) side
Mathabhanga	Hooghly
Gorai	Muriganga
Bhairab	Saptamukhi
Madhumati	Bidyadhari
Kapotaksha	Thakuran
Raimangal	Matla
Arpangaisa	Gosaba
Sibsa	Harinbhanga
Pasur	-
Baleshwar	-

The Sundarbans region is criss-crossed by numerous south-flowing rivers. The Ganges along with the inter-connected distributaries in India and Bangladesh play a vital role in the supply of freshwater into the Sundarbans. The river Hooghly in the extreme west is the only river carrying freshwater from upstream of the lower Gangetic delta in the Indian Sundarbans. Due to upstream diversion and narrowing of the distributaries of the Bhagirathi-Hooghly in the Sundarbans, presently freshwater discharge in these rivers of Sundarbans are negligible or mostly absent except during monsoon months when downpours in the southern parts of Sundarbans occur. Within the Bangladesh part, these distributaries fall into two zones: (a) upstream, dynamic, fluvial, freshwater channels, and (b) downstream tidal rivers. The upstream fluvial channels are dominantly freshwater with dynamic processes (i.e. active bank erosion and changing courses), e.g. the Gorai-Madhumati, Gorai-Nabaganga and Mathabhanga-Bhairab-Kobadak river systems. The off-take of the Mathabhanga River has become closed and currently there is no supply of fresh water flow to the Sundarbans through this system. On the other hand, the Gorai River dries up during low-flow dry season periods but it receives freshwater from Ganges River during the monsoon period and provides the only supply of freshwater flow in the region to the Sundarbans. Tide- and wave-dominated rivers are found closer to the Bay of Bengal, e.g. Baleshwar, Pasur, Shibsa and Arpangaisa rivers. Compared to the upstream rivers, the tidal rivers carry much less sediment and are relatively stable with far less short-term bank erosion and course shift or river widening. The embankments restricted the natural river flow, cut off numerous interconnected branches and creeks, prevented sediment being deposited on adjacent land and caused it to build up the river beds.

The average width of the Baleshwar River increased from 3,730 m in 1990 to 4,065 m in 2019. The main increase occurred during 1990-2003. Over this same period, the average width of the River Pasur increased from 1,700 m to 2,100 m whilst the Shibsa River increased from 2,110 to 2,320 m.

Estuaries

Due to extensive reclamation during the colonial period, the Hooghly-Muriganga-Saptamukhi estuaries in India and Meghna estuary in Bangladesh are human-dominated landscape and devoid of any big cat population. All major rivers flow from north to south, which are interconnected by the east-west narrow channels of rivers and creeks. Three mighty rivers in the Bangladesh Sundarbans, namely the Arpangaisa (width: 1.2-2.8 km), the Sibsa (width: 1.3- 3.4 km) and the Pasur (width: 1.4-3.1 km) divide the forest into three larger isolated regions. In addition, the rivers Raimangal (width: 1.5-2.8 km) and Harinbhanga (width: 2.9-4.4 km) mark the international boundary and bisect the entire forest into Bangladesh and Indian Sundarbans. Most of these major rivers have been used as cargo channels as well as other human activities for centuries.

India

Among the seven estuaries in the Indian Sundarbans- Hooghly, Muriganga, Saptamukhi, Thakuran, Matla, Bidyadhari, Gosaba and Harinbhanga, Bidyadhari is a dying one. Mitra *et al.* (2022) inferred that the freshwater supply in the main channel of Hooghly estuary has increased since the commissioning of the Farakka Barrage; however, the same has decreased substantially in the estuaries and tidal creeks situated in the north-eastern part and south-western part of the Indian Sundarbans over the last two to three decades. The mangrove plants are at present thriving under potential salt-stress, which is manifested by stunted growth, poor health, improper physiological functioning and degradation of forest cover in several places within this ecosystem. The freshwater loving mangrove species are diminishing in large numbers and these are getting replaced by more salt-tolerant species, resulting in a net loss of species biodiversity.

Bangladesh

Five estuaries in the Bangladesh Sundarbans namely Baleswar, Bangra, Kunga, Malancha and Raimangal are highly affected by saline water intrusion and tidal inundation which is a new threat for estuarine ecosystems (Islam *et al.,* 2011).

Tidal range

Tidal range varies spatially and is semidiurnal with tidal amplitude between 2.5 (neap) to 4.8 m (spring), with waves reaching 7 m during violent cyclonic storm surges from mid-May to mid-June due to north coastal winds followed by floods, which more or less damage the vegetation cover and have negative impact on the tigers and other wildlife. These storms have registered an increase in frequency and intensity over the last decades.

Problem of sweet water

The Farakka Barrage on the Ganges, operative since 1975, has been blamed for reducing water flow, causing salinity ingress and drying up of the Sundarbans delta. Since the late twentieth century, the Hooghly-Matla has become known as an estuary of vanishing islands- subsidence and eroded coastline. While parts of the delta's coastal edge are shrinking, especially in the west, areas to the east are stable or growing.

Dr. Pranabesh Sanyal, the then FD, STR during early 1980s and Director, SBR during the 21st century, once observed that the southern islands were full of tigers; but now we are not finding as many because the tigers are moving northwards, resulting in a higher density of the tiger population in the northern Sundarbans, which, again, are closer to the human habitations. During the 1990s, most sightings were in the central Sundarbans, i.e. at the Netidhopani watchtower. Now, maximum sightings are near the northernmost watchtowers- Sajnekhali and Sudhanyakhali.

Changing Tidal Islands Landscape (Bandyopadhyay *et al.*, 2023)

The partly-reclaimed Sundarban Mangrove Wetlands occupy the fluvially abandoned western part of the macro-mesotidal Lower Ganga-Brahmaputra-Meghna Delta (GBMD) in India and Bangladesh. To ascertain the evolution of the planform of this 15,500-km² region over a century, maps and images pertaining to 1904-24 (Survey of India topographical maps), 1967 (Corona space photographs), 2001 (IRS-1D LISS-3 + Pan merged images), and 2015-16 (Resourcesat-2 LISS-4 fmx images) were digitally compared for documenting the changes in the areas of ~250 mangrove-covered (but not all tiger-occupied) and reclaimed tidal islands above the spring high water level. Area change of the individual islands was also studied based on their relative north–south and west–east positions.

The results indicated that while erosion of the estuary margins and the sea facing coastline- up to 40 m/yr- was continuing for decades in the southern islands, intervening channels between the northern islands were getting silted up, especially in the western sector, resulting in land gain. Area change in the central sector mostly tended to be small and erosional. Overall, the total island area and counts changed from 11,903 km² and 253

(1904–24), respectively, through 11,663 km² and 250 (1967), 11,506 km² and 244 (2001), and 11,455 km² and 251 (2015–16). The reduction rate of area, at –4.46 km²/yr, remained noticeably similar across all intervals of mapping/imaging years and projects that the region will lose 3.4 % of its present extent by 2100 if the observed tendencies continue. The trend of area reduction was 2.55-times higher in the western (Indian) segment of the region, than the eastern (Bangladeshi) section. At the level of individual islands, the trends of area change were classified into nine types involving linear and non-linear changes, with dominance of continuous (post-2001) erosion in 48.3% (68.9 %) of the islands. Conversely, continuous (post-2001) accretion dominates in 15.8% (31.1%) of the islands.

Haliday Island WLS has shrunk to less than 1% (from 5.95 km² to 0.04 km²) of its original size in the last more than 40 years.

Forest block (Compartments) and status	Area (km²)	Names of Islands (Number)
STR		
Buffer Area		
Arbesi (1-5) RF	149	Jhingekhali, Kalindi, Raimangal, Arbesi, Burirdabri (5)
Jhilla (1-6) RF	121	Jhilla (3)
Harinbhanga (1-3) RF	115	Kalu *khal*, Boro Bholakhali, Hari *khal*, Chhagkapura (4)
Khatuajhuri (1-3) RF	130	Boro Tushkhali, Khatuajhuri, Chhota char, Uttar char, Dakshin Char, Burirdaha (6)
Panchamukhani (1-5) WLS	174.6	Duttar Island, Chhenra Dutta Island, Dutta-Pasar dwip, Panchamukhani dwip, Deulbharani dwip (5)

Forest block (Compartments) and status	Area (km²)	Names of Islands (Number)
Pirkhali (1-7) WLS	183.6	Sajnekhali, Sudhanyakhali, Sundarkhali, Pirkhali, Gajikhali, Nabanki, Dobanki (7)
Core Area (NP/CTH)		
Chandkhali (1-4)	154	Khirodkhali, Chhoto Chandkhali, Garankhali, Khejurtala, Chandkhali (5)
Chamta (1-8)	218	Chhamta, Chhoto Chamta, Boro Chamta, Durania, Chhoto Durania, Boro Durania, Dhutar (7)
Bagmara (1-8)	290	Bharkunda, Bharkunda badwip, Bagmara Banka, Hangsaraj, Bakultala, Mechhua (7)
Gona (1-3)	137.4	Lodhiduani, Khejurtala (2)
Mayadwip (1-5)	270	Chaimari, Mayadwip, Bhangaduani, Kendo (4)
Chhotohardi (1-3)	174	Choubanka, Bego, Kachikhali (3)
Gosaba (1-4)	169	Marabuni, Maya *khal*, Chhotohardi, Gosaba island (4)
Matla (1-4)	1,742	Lakshmi, Surajmukhi, Habatidoni, Chhoto Habatidoni, Khalsibuni, Narayantala (6)
Netidhopani (1-3)	91.9	Netidhopani, Bhagaban Bharani island, Storekhali,-1, Storekhali-2, Chandramukhi (5)
24-Parganas (South) FD		
Thakuran RF	110	Dhanchi (1)

Forest block (Compartments) and status	Area (km²)	Names of Islands (Number)
Saptamukhi RF, WLS	212	Dia, Jemson, Lothian, Prentice (4)
Chulkati (1-8) WLS	248.96	Kalas, Bijayara, Bhaijora, Pashemari (4)
Dulibhasani WLS (1-8) WLS	313	Kalibera dwip, Kamli, Chikua, Jhinjira, Chilkamari, Haliday, Matla (7)
Ajmalmari RF	216	Gura, Olion, Baniban, Kankalmari, Benupheli, Sundrikati, Lakshmikhali (7)
Herobhanga RF	133	Herobhanga, Surajmani, Hamalbere, Baenchampi (4)

Table: *Mangrove forest islands in SBR (Rudra and Lahiri, 2021)*

Soils

Soils of the Sundarbans mangrove forest differ from other inland soils in that they are subjected to the effects of salinity and waterlogging, which naturally affect the vegetation (Ray *et al.*, 2014). The salient features are-

1. In places soils are semi-solid and poorly consolidated.

2. pH ranges widely from 5.3 to 8.0.

3. Texture: soil is in general medium textured; sandy loam, silt loam or clay loam. The grain size distribution is highly variable. Silt loam is the dominant textural class. 72-87% silt content in 0-60 cm depth is reported from the Indian Sundarbans.

4. Porosity: 0.7.

5. Sodium and calcium contents of the soil vary from 5.7 to 29.8 meq/100g dry soil, respectively, and are generally low in the eastern region and higher towards the west.

6. Available potassium (K) content of the soil is low, 0.3-1.3 meq/100g dry soil.

7. Organic matter content varies between 4% and 10% in dry soil.

8. Soil salinity increases from 5 ppt in east (slight to moderate) to 30 ppt in west (highly saline), but the salinity is not uniform from north to south throughout the forest.

9. In spite of being situated under inter-tidal zones receiving highly saline estuarine water at frequent intervals, the average Electrical Conductivity (EC) values of the mangrove soils are marginally lower, average 22.6 (dsm^{-1}) than the non-mangrove soils 25.9 (dsm^{-1}).

The non-mangrove soils show slightly higher levels of salinity as compared to the mangrove soils. The non-mangrove soils occur mostly in slightly higher elevations and under more dry conditions. This dryness of non-mangrove soils is due to regular evaporation of the soil-water from the surface area leading to increased capillary movement of the highly saline groundwater from lower levels to surface soils. The bulk density of the mangrove soil increases with depth. Deep muddy and sticky soil makes it very difficult for the tigers to chase and hunt a fast moving ungulate near the shoreline after each tidal submersion. When the ground is slippery or marshy, the animal has to expend more energy just to keep from sliding or sinking, which also slows it down. A young, healthy animal will be able to run faster than an older one that is not in good shape. If an animal has weak or injured muscles, it will not be able to run as fast as one with strong, healthy muscles. However, the tiger's skeletal structure limits its top speed because it affects its acceleration; heavier ones take longer to reach their top speed than lighter individuals.

Most of the physical parameters of the water in the Sundarbans are governed by seasonal variability, with high salinity, pH and nutrient loads during the pre-monsoon period and the converse pattern during the monsoon.

pH

The observed water pH (January) is generally between 7.38 and 8.31 in the two ranges of 24-Parganas (South) FD. Water salinity is not uniform across the study area. Ajmalmari forest compartments bordering Kultali community development blocks are less saline relative to the Ajmalmari forest compartments lying alongside river Matla and Dhulibasani forest compartments. Chulkati forest compartments show a high salinity range relative to Ajmalmari and Dhulibasani forest compartments. The salinity during the study period was found to be in the range between 26.4 mg/kg and 31.2 mg/kg.

Nitrogen, phosphorus and potassium

Nitrogen, phosphorus and potassium levels are not uniform. Dhulibasani and Chulkati forest compartments are rich in nitrogen as compared to the Ajmalmari forest compartments bordering the Kultali community development block. The nitrogen levels, observed in the study area, range between 110 mg/kg and 247 mg/kg. The southern part of the study area is rich in potassium relative to the northern part with levels between 351 mg/kg and 1346 mg/kg. From east to west, the level of phosphorus has been noticed to be in the range between 7 mg/kg and 218 mg/kg.

Nutrient concentration

Seasonal and tidal variations in nutrient concentration and water quality were investigated by Rahaman *et al.* (2013) in the western Sundarbans of Bangladesh during the post-monsoon, winter and monsoon seasons during 2010-2011. Mean nutrient levels during post-monsoon, winter and monsoon seasons, respectively, were in the following ranges: nitrate (0.06–0.40, 0.06–0.46 and 0.08–0.46 mg/L); phosphate (0.09–0.18, 0.05–0.42 and 0.10–0.16 mg/L); sulphate (58.71–86.14, 68.68–119.01 and 78.15–136.47 mg/L) and ammonia (0.02–0.08, 0.02–0.04 and 0.26–0.38 mg/L). Increased levels of PO_4–P, SO_4 and NH_3–N and lower DO and salinity were recorded during the monsoon period. Most of the experimental sites showed higher NO_3–N content during monsoon, whereas few elevated concentrations were observed during post-monsoon and winter periods. High and low tidal waters contained mean nutrient levels in the following ranges: nitrate (0.05–0.46 and 0.04–0.40 mg/L); phosphate (0.05–0.42 and 0.07–0.18 mg/L); sulphate (63.63–125.36 and 58.71–136.47 mg/L) and ammonia (0.02–0.38 and 0.02–0.37 mg/L) without following any distinct fluctuation patterns. The western part of the Sundarbans receives less freshwater input during the monsoon season than other areas of the ecosystem, which reduces the variability of nutrient levels and water quality components.

Surface water salinity

Trivedi *et al.* (2016) have recorded that during 1984-2013 surface water salinity has decreased in Indian Sundarbans by 0.63 and 0.86 psu per year in the western and eastern sectors respectively, whereas in the central sector, it has increased 1.09 psu per year. Climate change-induced frequent cyclones are pumping saline seawater into the Sundarbans. Fani, Amphan, Bulbul, and Yaas were the major cyclones that hit the region during 2019–2021. Increasing tidal water salinity (parts per thousand) trends in both pre-monsoon (21 to 33) and post-monsoon (14 to 19) seasons have been observed between 2017 and 2021 (Chowdhury *et al.*, 2023). A 46% reduction in the soil organic blue carbon pool is observed due to a 31% increase in soil salinity.

Soil organic blue carbon

Soil organic blue carbon has been calculated by both wet digestion and the elemental analyser method, which are linearly correlated with each other. A reduction in the available nitrogen (30%) and available phosphorus (33%) in the mangrove soil has also been observed. Salinity-sensitive mangroves, such as *Xylocarpus granatum, Xylocarpus moluccensis, Rhizophora mucronata, Bruguiera gymnorrhiza,* and *Bruguiera cylindrica*, have seen local extinction in the sampled population. An increasing trend in relative density of salinity resilient, *Avicennia marina, Suaeda maritima, Aegiceras corniculatum* and a decreasing trend of true mangrove (*Ceriops decandra*) has been observed, in response to salinity rise in surface water as well as soil.

Erosion-accretion (Jayanthi *et al.*, 2023)

The shoreline change by endpoint rate indicated the erosion rate by 5.81 m yr−1 in 2000–2020 compared to −0.90 m yr−1 in 1980–2000. The weighted linear regression indicated the average erosion and accretion rate of −5.96 m-yr −1 and 4.92 m yr−1, respectively, with erosion in 76% of transects and accretion in the remaining 24% between 1980 and 2020.

Land Use change in SBR vis-a-vis tiger-human conflict

From the land use change analysis between 2005 and 2020 land use and land cover map, it is understood that the area under human settlement has increased significantly (26.58%), whereas the area dedicated to agriculture, and forest has decreased (Dhar and Mondal, 2023).

Elements	2005	2020	Percentage of change
Water	24,99,755	24,65,747	-1.36
Forest	23,78,896	23,09,276	-2.93
Settlement	14,54,434	18,41,760	+26.63
Arable land	18,55,086	13,61,939	-26.58
Vacant land	6,31,056	8,50,073	+34.71

Table: *Land-cover and land-use change data*

Between 2005 and 2020, the buffer mangrove has increased from 2.15% to 4.32%; the outer core mangrove decreased from 48.18% to 45.49% and the core jungle has increased from 22.33% to 23.51%. In 2021 the area of mangrove forest was 2,114 km². Considering the fluctuating nature of mangrove forest area in the present study area from 1987 to 2021 shows that, the area of mangrove forest was much higher from 2005 to 2011 but has been declining sharply in 2013. Subsequently, it increased slightly and in 2019 it declined marginally. It shows a negative change in forest cover for the years 1999, 2011 and 2017. For the rest of the years the figures are positive.

Comparison of built-up area in the study area shows that non-built-up area has decreased from 78.14% to 73.37% between 2005 and 2020, whereas built-up area has increased from 21.85% to 26.62% during the same time span, whereas the area dedicated to agriculture, and forest has decreased. This signifies the pressure of increased human population on the forest. At the same time, vacant or barren land has increased a lot. It is quite clear that such conflict areas are found near the forest boundary in most cases. Such an outer buffer location of the forest is a favourable place where both the tigers come sometimes in search of a few cattle residing in the human settlement and people also visit those places quite regularly in search of forest products.

The tiger census is done quadrennially (every four years) by NTCA with technical help from the WII. The tiger population in Indian Sundarban does not remain stable at all times. The figures for the tiger population in the successive years of 2010, 2014, 2018, and 2020 were 70, 76, 88, and 96 respectively (Jhala *et al.*, 2020). With this increasing nature of tiger population, it is obvious that its visit probability in or around human habitation has increased gradually. Considering a 5-km circular buffer area around the tiger attack spots, it is revealed that 62% of the surrounding area of tiger attack consists of the outer core mangrove forest. 18% area is under the core mangrove jungle, whereas 17% is covered by water bodies and only 3% is under buffer mangrove forest. This distribution of forest categories around the tiger attack spots indicates that people frequently reach the outer core and core area of the Sundarban frequently, which is a natural habitat of tigers. This practice of entering the core forest without proper training and protection is increasing the risk of casualties of people due to tiger attacks.

Chapter 6
Habitats: Plants-Tiger Interactions

"Tigers, except when wounded or when man-eaters, are on the whole very good-tempered." – Jim Corbett

Animals and plants are intricately connected through a complex ecological network that includes pollination, seed dispersal, and herbivory. Plants provide habitats and shelter for animals and serve as a source of food. Without plants, the herbivores die. The soil becomes unstable, erosion is high and the nutrient-base is stripped from the land. This scenario is exactly why it's crucial we protect the tigers. It's never about a single animal- it's about an entire ecosystem. The living, growing, or thriving area of the tiger is called a tiger habitat. It provides the condition for a species to grow, adapt, reproduce and flourish. The inland forest areas near the coastlines in the subtropical and tropical regions are known as the mangrove forest.

Tigers are found in amazingly diverse and healthy natural habitats. Habitat is everything the tigers need to survive in the wild. Three such fundamental elements are food, water and shelter. Tigers, like most animals, are dependent on plants for their survival in a number of ways as detailed below.

Firstly, tigers are obligate carnivores, also called hypercarnivores, whose diet consists of at least 70% meat. The tigers cannot properly digest vegetation. Despite unsuitable morphological and physiological traits for plant consumption, the presence of plants in scat or stomach contents has been reported in various carnivorous species (Mallick, 2013, 2015). But the frequency of plant occurrence has not always been reported and a significant negative relationship is found between the frequency of plant occurrence and body mass. The plant contents in their scats are estimated to vary from 0.132 to 0.147 (Yoshimura *et al.*, 2021). This may be because plant-eating behaviour reduces the energy loss caused by parasites and increases the efficiency of energy intake.

Secondly, plants also provide shelter and cover for tigers, which they use to hide from predators and prey.

Thirdly, plants play a crucial role in maintaining the ecosystem and biodiversity of the tiger's habitat, which in turn supports the survival of the tiger population.

Additionally, plants also help regulate the climate and water cycle of the tiger's habitat, which is essential for the tiger's survival.

Moreover, tigers are a lynchpin in the watersheds throughout their habitats. Watersheds are the area surrounding and leading to bodies of water. Inasmuch, the health of the entire area directly impacts the purity of the water sources animals and humans rely on.

Overall, plants are an essential component of the tiger's ecosystem and play a critical role in ensuring their survival.

The Sundarbans stand in contrast to the Asian region as a whole in terms of the following criteria:

(i) the world's largest area of mangrove forest in India and Bangladesh, where the tiger population can potentially share their gene pool,
(ii) a famous biodiversity hotspot protecting threatened and endangered species,
(iii) the only area known to have tigers that are ecologically adapted to mangrove habitats, and
(iv) while climate change, salinity, pollution, heavy biotic pressure for forest resources, fisheries, and non-timber forest produce (NTFP) collection, aquaculture and agriculture all pose threats to this ecosystem, conservation efforts have largely minimised forest loss in the 21st century.

Habitat classification

Marine Intertidal - Mangrove Submerged Roots

Marine/Freshwater/Terrestrial- 1. Brackish tidal and 2. Intertidal forests and shrublands

Biogeographic Realm- Indomalayan

Key biodiversity hotspot- Home to rare and threatened mammals, birds, reptiles, fishes and crustaceans.

Although the tigers have an enormous habitat range across subspecies, from temperate Eastern Himalayas to equatorial- hot and wet year round, resulting in a humid climate, one of the most unique tiger habitats are mangrove swamps, lacking freshwater sources apart from some temporary pools during monsoon. The Sundarbans mangrove ecosystem is considered to be unique because of its species richness due to its diverse mangrove flora, which constitutes the mangrove associated flora, back mangrove species and several endemic typical mangrove associated fauna of different groups. The tigers do not live among the tangle of mangroves, but are elusive and often hide in the thick Hental (*Phoenix paludosa*) bushes for stalking prey and, unlike their mainland counterparts, lead a nearly amphibious life. As apex predators, Bengal tigers balance their mangrove ecosystem and help keep a significant carbon sink.

Today, Bengal tigers (*Panthera tigris tigris*) are found in India, Bangladesh, Nepal, Bhutan, and south-western China. They inhabit tropical moist evergreen forests, tropical dry forests, tropical and subtropical moist deciduous forests, mangroves, subtropical and temperate Himalayan foothills and in montane forests, flood plains, and alluvial grasslands. In India, most of the tiger population (about 70%) is primarily restricted to the TRs and, as of 2020 nearly 30% of tiger population is present outside the TRs. A TR consists of an undisturbed core area wherein human settlements, livestock grazing, and resource utilisation are prohibited under the Wildlife Protection Act, 1972. The core zone is further supplemented by the surrounding multi-use buffer zones, which act as population sinks.

TR	Habitat types
Achanakmar, Chhattisgarh	Northern Tropical Moist Deciduous and Southern Dry Mixed Deciduous Forest
Amrabad, Telangana	Dry deciduous scrub, woodland and grassland
Anamalai, Tamil Nadu	Wet evergreen forests, semi evergreen forests, moist deciduous, dry deciduous, dry thorn and shola forests

TR	Habitat types
Bandhavgarh, Madhya Pradesh	Moist Peninsular Low-level Sal Forest, West Gangetic Moist Mixed Deciduous Forest, Northern Dry Mixed Deciduous Forest, Dry Deciduous Scrub Forest and grasslands
Bandipur, Karnataka	Dry Deciduous Scrub, Tropical Dry Deciduous Forest, Tropical Moist Mixed Deciduous Forest
Bhadra, Karnataka	Southern Moist Mixed Deciduous Forests, Southern Dry Mixed Deciduous Forests, and Semi-evergreen Forests
Biligiri Ranganatha Temple, Karnataka	Scrub, dry deciduous, moist deciduous, evergreen, semi evergreen and shola forests
Bor, Maharashtra	South Deccan Plateau dry deciduous forests
Buxa, West Bengal	Northern dry deciduous, Eastern Bhabar and Terai sal, East Himalayan moist mixed deciduous forest, Sub-Himalayan secondary wet mixed forest, Eastern sub-montane semi-evergreen forest, Northern tropical evergreen forest, East Himalayan subtropical wet hill forest, Moist sal savannah
Corbett, Uttarakhand	Sal and mixed forests, interspersed with grasslands and riparian vegetation
Dampa, Mizoram	Tropical evergreen to Semi-evergreen forests
Dandeli-Anshi (Kali), Karnataka	South Indian moist deciduous teak forest, southern moist mixed deciduous forests, west coast semi-evergreen forests, moist bamboo brakes and cane brakes
Dudhwa, Uttar Pradesh	North Indian Moist Deciduous forest
Indravati, Chhattisgarh	Tropical dry deciduous forests

TR	Habitat types
Kalakad-Mundanthurai, Tamil Nadu	Dry thorn forest, dry deciduous, moist deciduous and a patch of West coast wet evergreen forests
Kamlang, Arunachal Pradesh	Tropical temperate, Alpine forests, Tropical wet evergreen forests
Kanha, Madhya Pradesh	Moist Deciduous Forest, Dry Deciduous Forest, valley meadow and plateau meadow
Kawal, Telangana	Southern Tropical Dry Deciduous Forest, Dry Teak and Southern Dry Mixed Deciduous Forests
Kaziranga, Assam	Alluvial inundated grasslands, alluvial savanna woodlands, tropical moist mixed deciduous forests, and tropical semi-evergreen forests
Manas, Assam	Sub-Himalayan High alluvial Semi evergreen forests, Eastern Bhabar Sal type Forests, East Himalayan Moist mixed deciduous forests, Eastern wet alluvial grassland, low alluvial savannah woodlands, Riparian fringing forest and Khair-Sisoo forests
Melghat, Maharashtra	Tropical dry deciduous forests, dominated by teak *Tectona grandis*.
Mudumalai, Tamil Nadu	Southern Tropical dry thorn forest, Southern Tropical dry deciduous forest, Southern Tropical moist deciduous forest, Southern Tropical semi-evergreen, moist bamboo brakes and Riparian fringing forest
Mukundara Hills, Rajasthan	Dry deciduous forests and grasslands in between.
Nagarahole, Karnataka	North Western Ghats moist deciduous forests with teak (*Tectona grandis*) and Roseta rosewood (*Dalbergia latifolia*) predominating in the southern parts

TR	Habitat types
Nagarjunsagar-Srisailam, Andhra Pradesh-Telangana	Southern tropical dry deciduous miscellaneous type 5A/C3
Namdapha, Arunachal Pradesh	Assam Valley Tropical Wet Evergreen Forests (IB-C-1), Upper Assam Valley Tropical Evergreen Forests (IB-C2a), Upper Assam Valley Tropical Evergreen Forests (lB-C2b), Assam Valley Tropical Semi Evergreen Forests (2B/C-1), Sub-Himalayan light alluvial semi evergreen Forests (2B/C1/S1), Secondary moist bamboo tracts (2/251), Eastern Hemlock Forests [3/1S2(b)], East Himalayan moist temperate Forests (12/C -3), Moist Alpine scrub Forest (15/C-1)
Nameri, Assam	Tropical evergreen, semi-evergreen, moist deciduous forests with cane brakes and narrow strips of open grassland along river
Navegaon-Nagzira, Maharashtra	5 A/C3 Southern Tropical Dry Deciduous forests, vegetation ranging from dry mixed forest to moist forest
Orang, Assam	Eastern Himalayan moist deciduous forest, Eastern seasonal swamp forest, Eastern wet alluvial grassland, Savannah grassland, Degraded grassland, Bodies of water, Moist sandy area, Dry sandy area
Pakke, Arunachal Pradesh	Lowland semi-evergreen, evergreen forest and Eastern Himalayan broadleaf forests
Palamau, Jharkhand	Northern Tropical Dry Deciduous Sal (*Shorea robusta*) forest and its associates
Panna, Madhya Pradesh	Southern dry deciduous teak forest, northern dry deciduous mixed forest, dry deciduous scrub forest, Salai forest, dry bamboo forest, and Kardhai forest

TR	Habitat types
Parambikulam, Kerala	West Coast Tropical Evergreen Forests, Moist Deciduous Forests, Dry deciduous forests, Teak plantations, Shola forests, Vayals
Pench, Madhya Pradesh	Tropical Moist Deciduous Forests, Tropical Dry Deciduous Forests
Pench, Maharashtra	South Indian tropical Moist Deciduous Forest 3B/C, Southern Tropical Dry Deciduous Forest 5A/C, Southern Dry Mixed Deciduous Forest 5A/C3
Periyar, Kerala	Tropical evergreen forests, tropical semi-evergreen forests, moist deciduous forests, grasslands and eucalyptus plantations
Pilibhit, Uttar Pradesh	North Indian moist deciduous type, with sal Shorea robusta forests
Rajaji, Uttarakhand	Moist Shivalik Sal Forest, Moist Mixed Deciduous Forest, Northern Dry Mixed Deciduous and Khair-Sissoo forests in the Southern Slopes, Low Alluvial Savannah woodlands
Ramgarh Vishdhari, Rajasthan	Dry Deciduous Forest, dominated by Dhok (*Anogeissus pendula*) trees
Ranipur, Uttar Pradesh	Northern tropical dry deciduous forests
Ranthambore, Rajasthan	Dry deciduous forests and open grassy meadow
Sahyadri, Maharashtra	North Western Ghats moist deciduous forests, North Western Ghats montane rain forests
Sanjay-Dubri, Madhya Pradesh	Evergreen Sal, Bamboo and mixed forest
Sariska, Rajasthan	Tropical dry deciduous and tropical thorn forest

TR	Habitat types
Sathyamangalam, Tamil Nadu	Tropical evergreen (Shola), semi-evergreen, mixed-deciduous, dry deciduous and thorn forests
Satkosia, Odisha	North Indian tropical moist deciduous Forests and Moist peninsular low level sal
Satpura, Madhya Pradesh	Moist deciduous forests, dry deciduous forests, bamboo forests, and grasslands
Similipal, Odisha	Northern Tropical Moist Deciduous Forest, Dry Deciduous Hill Forests, High Level Sal Forest, Grassland and Savanna
Srivilliputhur Megamalai, Tamil Nadu	Tropical evergreen and semi-evergreen forests, dry deciduous and moist mixed deciduous woods, and grassland
Sundarbans, West Bengal	Mangrove scrub, littoral forest, saltwater mixed forest, brackish water mixed forest and swamp forest
Tadoba-Andhari, Maharashtra	Southern tropical dry deciduous forest
Udanti-Sitanadi, Chhattisgarh	Dry Teak Forest 5A/c 1b(iii), Dry peninsular Sal Forest 5B/c 1c(iv), Moist peninsular Sal Forest 3C/c 2 e (ii), Northern Dry mixed deciduous forest 5B/c 2(ii)
Valmiki, Bihar	Moist mixed deciduous, Open-land vegetation, Sub-mountainous semi-evergreen formation, Freshwater swamps, Riparian fringes, Alluvial grasslands, high hill savannah and wetlands

Table: *Habitats of Bengal tigers in various TRs in India*

Transboundary Sundarbans

The Sundarbans is the only mangrove tiger habitat not only in India or the Indian subcontinent, but also in the world. In fact, the tiger-centric transboundary (India and Bangladesh) Sundarbans, having the status of

Level I Tiger Conservation Unit (TCU), is the only exclusive mangrove habitat to harbour a viable tiger population in a unique ecological setting, despite its reduction to almost half the size of the area that existed in the late 1800s due to large-scale reclamation. The mangrove forest is also referred to as tidal forest because this forest can thrive both in salt and freshwater.

STR, situated in the coastal districts of West Bengal, i.e. South and North 24-Parganas (Basirhat Range), was launched on 23rd December, 1973 over 2,584.89 km² RF area (land area 1,600 km² and water body rounded to 985 km²) and got the distinction of being one of the first nine TRs to be declared in the country.

Within STR, 1,699.62 km² has been designated as the Critical Tiger Habitat (CTH) vide Notification No. 6028-Forest, dated 18.12.2007. Comparatively other TRs also have CTH like Corbett (Uttarakhand): 821.99 km², Ranthambore (Rajasthan): 2595.72 km² and Nagarjunsagar- Srisailam (part of Andhra Pradesh): 1330.12 km². Within the STR's CTH lies the Sundarbans NP having an area of 1330.12 km² area notified vide No. 2867-Forest, dated 04.05.1984. An area of 124.40 km² (Chamta 4-8) within the core area is preserved as a primitive zone to act as gene pool.

The area outside the CTH over RFs of 885.27 km² is known as Buffer Zone notified vide No. 615-For/11M-28/07, dated 17.02.2009. Sajnekhali WLS, covering an area of 362.42 km² vide latest Notification No. 5396-Forest, dated 24.06.1976, is located on the confluence of the rivers Matla and Gomdi within the buffer area.

The buffer zone has twin functions, viz.

To provide habitat supplement to the spillover population of wild animals from the core area, conserved with the active cooperation of stakeholder communities, and …

To provide site-specific, need-based, participatory eco-development inputs to local stakeholders for reducing their resource dependency on the core zone and for eliciting their support towards conservation initiatives in the area. The habitat conserved in the buffer zone also serves as a corridor for wild animals.

Outside the buffer area the multiple use zone is used by the local population to fulfil their bona fide needs. The importance of this area was recognized by UNESCO and the area was accorded the status of a World Heritage Site in the year 1987 (now renamed as World Heritage Property), being an outstanding example of ecological and biological processes in the evolution and development of coastal communities of plants and animals and for the importance of this region for biodiversity conservation. A decade later the Bangladesh part of Sundarbans was also added to the same list.

Tigers and people's belief system

The transboundary Sundarbans is home to a large population (minimum 214) of Bengal tigers now. But they are elusive. There are many who have only witnessed pug marks the animal left behind on soft clay. Bittu Sahgal, the editor of Sanctuary Asia magazine, once told: "The first time I saw a Bengal tiger in the Sundarbans, it was some 45 years after I had first been there. So you only see them when they decide that you're good enough to be given a vision of orange and black." There are also superstitions around seeing a tiger. Culture, traditions and gods of the Sundarbans are inseparable and intertwined with the forest and its most charismatic creature, the swamp tiger.

Although the people's reverence towards tigers may be rooted in religious superstition, there is a practical element too. Some of them simply expressed their belief, "The Sundarbans needs its tigers otherwise the forest will not survive." They further expressed, "Without tigers [and the government's enforced protection of their habitat], woodcutters will come and cut down the forest." This is why, despite seeing friends and relatives maimed before their eyes, the people of the Sundarbans do not hate tigers. They know their fate is ultimately tied to the killer cats: the tigers protect the forest and the forest protects the people. The tigers, the people, the vegetation and all animals that rely on the area will suffer significantly if the ecosystem is permanently damaged. The Sundarbans tigers, though hazardous, ultimately protect the people's homeland.

Tiger's habitat

SBR

There are two distinct ecological units of mangrove vegetation in SBR.

1. The western portion lies west of river Thakuran where a trickle of sweet water reaches from the river Hooghly, i.e. mostly in the South 24-Parganas FD.

2. The central mangrove patch, which is practically cut off from the upstream flow and is fed by backwaters of Bay of Bengal lying between rivers Harinbhanga and Thakuran. STR falls under this area. The salinity regime is very high here. However, on the eastern side bordering Bangladesh River, the River Kalindi receives some sweet-water flow; as a result, the salinity levels are comparatively low here. Understanding the habitat preferences of, and ecological niche inhabited by, the tiger is fundamental to informing conservation and management effort. The habitat preference of the tiger is both pure and mixed vegetation. There is adequate mangrove forest cover, which is used by the tiger for hunting, hiding and procreation. However, it is generally observed that the *Heritiera fomes* (Sundari) and *Phoenix paludosa (Hental)* formations, which are not regularly inundated or inundated for a short period, form the prime tiger habitat.

The tiger's presence is commonly recorded (direct sighting and foot-prints) in the mangrove forest having the following plant consociations:

(1) *Avicennia* spp.-*Sonneratia apetala* (1.43%), on the river banks, below natural tidal level;

(2) *Avicennia* spp. with foreshore grasslands of *Oryza coarctata* (14.29%), on the river banks, below natural tidal level;

(3) *Phoenix paludosa-Excoecaria agallocha* (12.86%), above natural tidal level;

(4) Pure *Ceriops decandra* (3.58%), above natural tidal level;

(5) Pure *Excoecaria agallocha* (2.15%), above natural tidal level;

(6) Pure *Phoenix paludosa* (14.29%), above natural tidal level;

(7) *Excoecaria agallocha-Ceriopes decandra* (10.71%), above natural tidal level;

(8) *Rhizophora-Bruguiera* (17.85%), above natural tidal level;

(9) *Heritiera fomes-Phoenix paludosa* (8.57%), above natural tidal level and submerged only a number of times during very high tides;

(10) *Heritiera fomes-Excoecaria agallocha* (1.42%), above natural tidal level;

(11) *Heritiera fomes-Excoecaria agallocha-Ceriops decandra* (2.14%), above natural tidal level; and

(12) Pure *Oryza coarctata* (10.71%).on the river banks, below natural tidal level.

Naha (2015) also showed that tigers preferred *Avicennia* patches whereas *Phoenix* and *Ceriops* habitats were used according to their availability. Though *Avicennia* was preferred, 50% of the locations were in *Phoenix*, followed by 20% within *Ceriops* dominated patches signifying their importance as critical habitat for tigers in mangroves. There was an avoidance observed for water in the overall preferred habitat use.

The abundance of tiger and the habitat used for specific purposes depends on composition, characteristics and quality of the pure or mixed forest types present under different tidal zones. The Sundarbans tidal delta plain can be divided into four specific zones: the inactive delta (once an active part of delta, but with currently reduced or no fluvial activity), mature delta (moribund deltas that have ceased), tidally active delta (a still active part with water channelling through it carrying sediment) and sub-aqueous delta (part of the delta that is below the low-tide mark).

The ideal tiger habitat, where the tiger is commonly found, is recorded above the natural tidal limits in the pure forests of *Phoenix* and mixed forests of *Heritiera-Phoenix, Excoecaria-Phoenix* and *Excoecaria-Ceriops*. However, tiger depredation is also not common in all habitat formations, but is mostly concentrated in pure *Ceriops, Excoecaria-Ceriops* and *Heritiera-Excoecaria-Cereops* association. Moreover, the tiger forages in the open forests with sparse vegetation during the low tide and swims in the open and inland water-bodies of variable-widths to shift from one island to another. They are also seen to stay long in the flowing water during a summer noon.

But an alteration in the composition of the mangrove forest has made it difficult for tigers to hunt and could be a reason behind their dwindling number. Till the late 1980s, Baen (*Avicennia* spp.), which has the highest pneumatophore density among mangroves, comprising less than 1% of all mangroves in the forest, now accounts for around 10% which is alarming. Its pneumatophores can spread across a radius of 6-7 m and make movement extremely difficult for tigers. Proliferation of Baen has been making it even tougher. As it is, tigers get little room to run and catch prey in the mangrove forests. They have to rely on short bursts to make a kill, catching their prey by surprise. More Baen trees means it has now become more difficult in certain baen-dominated hunting grounds of the tiger.

Tigers need extensive areas for meeting their biological needs. During the study, the tigers were observed to use different habitats for definite purposes like shelter, breeding, hunting, feeding, drinking, defecating, scent-marking, drinking, scratching trees, etc. Out of twelve forest types/associations, nine (88%) are used as habitats (shelter- 72%, breeding= 7%, and hunting- 9%) and four types (12%) are not normally preferred, i.e. rarely used by the tigers. Due to frequent shifting or island-migration and occupation of new habitat by the tigers, the habitat occupation and range of movement varies considerably from individual to individual and from time to time, mainly depending on availability of the prey species- be it wild, domestic or human beings entering forests, some with legal permits and mostly illegally. For example, between 1985 and 2010, 666 persons were attacked by tigers of which 410 were recorded by the Forest Directorate indicating that over 250 victims were illegal entrants. However, tiger attacks in the restricted and prohibited areas remain unrecorded and are not compensated.

The noted special features of the tiger habitat are that there is complete absence of the ground flora. Humus formation is also poor, the leaves and branches that fall to the ground are washed away in the high tide, leaving no chance to form a suitable soft cushion for their rest. There is no thick crowned bush or branchy trees to afford shade and shelter. The tigers were seen to occupy the coastal forests with *Casuarina* sp and others, bordered by a high retaining sand-wall for protection from the high tides, for breeding. The *Ceriops* and *Aegialitis* afford side shade and the tunnels in *Phoenix* bushes provide the tiger with best shade and shelter. However,

Netidhopani block with an open crop of low height and patches of *Phoenix* formation, which are above the daily tidal inundation level, recorded a high concentration of tiger and there were four old adult resident tigers. It was also observed that the salinity and tiger activity might have some positive correlations.

Alien species threatening tiger habitats (Mallick, 2023b; Islam *et al.*, 2019; Biswas *et al.*, 2007)

Invasive/alien/exotic species are any non-native species that significantly modify or disrupt the ecosystem it colonises. "Alien invasive" is a term used to describe plants that are moved out of their natural habitats and end up exploiting the local biota in their new environment. For example, introduced in India in 1807, *Lantana camara* is originally from South and Central America and has the dubious distinction of being one of the world's most invasive species. Several alien invasive plants growing together can have a detrimental effect on the biodiversities in tiger habitats.

India's biodiverse ecosystems are threatened by a variety of alien plants like *Lantana camara, Parthenium hysterophorus, Prosopis juliflora*, etc, introduced during British colonisation. Lantana alone has pervasively invaded 44% of India's forests. Many naturalist observations would rightly suggest that alien plants like *Lantana camara* are associated with rich sightings of birds and butterflies and provide 'green cover' for many other wildlife. At the same time, many others would count its harmful impacts on native plants and other ecosystem components. Native forests are packed with biodiversity, particularly with rare species and interactions. Contrastingly, invaded areas only sustain a few commonly found species. Co-occurring invasive plants like *Lantana, Ageratum conyzoides, Pogostemon benghalensis*, etc, have a magnified cumulative impact than their individual impacts, causing ecological homogenisation in invaded regions. Multiple alien species together affected soil nutrients, which may have depleted the richness of diverse plants. The abundance of rich grasses and herbs, the signature component of these ecosystems, was the most affected. There was hardly any regeneration of important plants or even the most common trees and these might lead to diseases in the herbivores. Invasive species flourish in non-native regions too due to soil microbes and fungi.

The invasive plants can put pressure on native forage plants and drive away wild herbivores- the food source for the big cats. Reduced forage availability for herbivores like chital, which is major prey for tigers in the mangrove landscape, threaten the sustenance of these carnivores in invaded regions. It is indicative of an 'invasion-centric forest ecosystem' historically unmatched, defaunated and functionally downgraded with homogenised biodiversity. Native wild herbivores do not prefer the commonly found plants in invaded areas. They prefer rare forage plants, which are already depleted in infested areas. Although wild herbivores consume a limited portion of alien plants, with monotonous invasion stands, their dependence on native forage increases. These aggressive weeds may even be triggering the spurt in the number of tiger strayings and human attacks.

Altogether 23 plant species of two broad types, viz. aquatic weeds and climbers were identified as invasive. Of the identified 23 invasive species, 19 are native or naturalised to Sundarbans mangrove. Invasives' abundance, diversity and rate of invasion (RI) were highest at the riverbanks and gradually decreased with increased proximity to the forests. Based on the severity of damage, species were classified as highly invasive, invasive and potentially invasive. The three highly invasive plants were *Derris trifoliata* (climber), *Eichhonia crassipes* (aquatic shrub) and *Eupetorium odoratum*, respectively. Among the identified invasive plants, *Derris trifoliata* and *Eichhonia crassipes* showed highest density followed by *Acrosticum aureum* and *Micania scandans*.

Alien plants have invaded the majority of natural areas, while invasion management still operates at a truncated scale. In the Bangladesh Sundarbans, 4 out of 23 invasive species are alien. Almost 88% of the area is affected by invasive species in which 55% severely damaged.

It is necessary to prioritise restoration investments in the least invaded regions to retain native biodiversity and slowly upscale such restored habitats.

Tiger Density and carrying capacity of tigerlands

Carrying capacity is the maximum number, density, or biomass of a population that a specific area can support sustainably. This likely varies over time and depends on environmental factors, resources, and the presence of predators, disease agents, and competitors over time. Limiting factors determine carrying capacity. The availability of abiotic factors (such as water, oxygen, and space) and biotic factors (such as food) dictates how many organisms can live in an ecosystem.

In an ecosystem, the population of a species will increase until it reaches the carrying capacity. Then the population size remains relatively the same. If abiotic or biotic factors change, the carrying capacity changes as well. Natural disasters can destroy resources in an ecosystem. If resources are destroyed, the ecosystem will not be able to support a large population. This causes the carrying capacity to decrease. Humans can also alter carrying capacity. Our activities can decrease or increase carrying capacity. We alter carrying capacity when we manipulate resources in a natural environment.

If a population exceeds carrying capacity, the ecosystem may become unsuitable for the species to survive. If the population exceeds the carrying capacity for a long period of time, resources may be completely depleted. Populations may die off if all of the resources are exhausted. Artificially boosting the prey base in a reserve is often an intuitive solution but it can be counter-productive. To harness the umbrella effect of tigers for biodiversity conservation, it is more beneficial to increase areas occupied by tigers. Create safe connectivity among forests and allow tigers to disperse safely to new areas.

To understand various factors that correlate and possibly determine tiger density in the Sundarbans, exploratory data analysis as well as formal hypothesis tests with a priori expectations were conducted as follows:

(1) Tiger density would increase with increase in encounter rate of tiger signs

The rate of tiger signs like pugmark, scat, rake-marks, and scrape marks are strongly and significantly related to tiger density.

(2) Tiger density would increase with increase in primary prey (in case of the Sundarbans, it is chital)

A study by WII suggests that the density of tigers in the Indian Sundarbans may have reached the carrying capacity of the mangrove forests, leading to frequent dispersals and a surge in human-wildlife conflict. The correlation between prey availability and tiger density in undisturbed areas is fairly established. A comparative chart is shown below.

Sl.no	Tiger conservation areas in different mainlands and mangrove islands	Estimated range of the carrying capacity
1	Terai and Shivalik hills habitat, e.g. Corbett TR	10-16 tigers per 100 km^2
2	TRs of north-central Western Ghats such as Bandipur	7-11 tigers per 100 km^2
3	Dry deciduous forests of central India, such as Kanha	6-10 tigers per 100 km^2
4	Sundarbans mangroves	Around 4 tigers per 100 km^2

Tiger densities increase as primary prey encounter rates as well as their pellet density increase. It may be mentioned here that the faecal pellets of chital are cylindrical with the bottom end rounded and the apical region tapering to a point. Adult chital pellets range in size from 1.2 to 1.8 cm.

(3) Tiger density would increase with increase in tiger habitat and its quality.

Amongst habitat indices the extent of forest in a grid, especially forest away from edge effects (core forest area) is significantly and positively related with tiger density.

(4) Tiger density should decline with increasing human impacts and decrease in protection regime.

Security in the form of PAs harboured higher tiger densities. Human impact indices in the form of night light area, distance to night lights, livestock presence, signs of wood cutting, signs of lopping, human and livestock trails are negatively and strongly related with tiger density.

Usually, physical (space) and biological (forest productivity) factors have an obvious influence on a reserve's carrying capacity of tigers. More so when different land uses overlap and a good number of people depend on forest resources for livelihood. However, not all dispersing tigers will chance upon corridors simply because many will find territories of other tigers between them and such openings. The corridors may not lead to viable forests in reserves such as Sundarbans, bounded by the sea and villages.

Out of 324 reserves embedded within all Tiger Conservation Landscapes covering > 380,000 km², Wikramanayake *et al.* (2011) intersected the reserve layer with biome layers to determine the extent of each biome that is included within each reserve to calculate the potential tiger densities (tigers/100 km²) assigned to biomes, or major tiger habitat types based on Miquelle *et al.* (2010) and Karanth *et al.* (2009) as described below.

Biome/Habitat type	Potential population density	Potential breeding female density
Rainforest/tropical evergreen forests	3	1
Dry deciduous forests	10	3.3
Subtropical pine forests	1	0.3
Broadleaf temperate forests	1	0.3
Alluvial savanna/grassland	15	5
Mongolian steppe/open woodland	0.6	0.2
Amur steppe	0.6	0.2
Thorn scrub/woodland	10	3.3
Mangrove	3	1

Table: *Potential Tiger Densities (tigers/100km²) assigned to biomes or major tiger habitat types*

It may be seen in the above table that the potential tiger population densities (tigers/100 km²) and potential breeding female density assigned to the mangrove tiger habitat by Wikramanayake *et al.* (2011) based on Karanth *et al.* (2009) was 3 and 1 respectively, which is considered to be similar to densities in rainforest/tropical evergreen forests and much better than four other habitats, viz. subtropical pine forests, broadleaf temperate forests, Mongolian steppe/open woodland and Amur steppe.

Dendup *et al.* (2023) reported that during a camera trap exercise from October 2021 to January 2022 in Jigme Dorji NP, Bhutan, a total of 56 camera stations (a pair in each station) was installed in a 5 × 5 km² grid along the latitudinal gradient of 1,200-4,300 m. With the sampling effort of 1,528 trap nights, 478 tiger images were captured in 12 stations (21.4 % of the camera stations). Their unique stripe patterns identified six tigers (4 male and 2 female). The tiger abundance was estimated at 15.16 ± SE 5.41 (9.13–32.79) and a density of 0.263 ± SE 0.115 (0.116–0.601 individuals) per 100 km². The home range analysis at 50% kernel density estimators indicated 151 km² and 90 km² as the home range for the male and the female tiger, respectively. Six main prey species were observed in this tigerland, namely barking deer (*Muntiacus muntjak*), Himalayan serow (*Capricornis thar*), Sambar (*Rusa unicolor*), wild pig (*Sus scrofa*), takin (*Budorcas taxicolor whitei*) and musk deer (*Moschus chrysogaster* and *M. leucogaster*).

A study using GPS collars recorded home ranges of two female tigers in the south-east of the Sundarbans. The two tigers were living in relatively good habitat with respect to other areas in the Sundarbans, but their small home ranges (<20 km²) are probably indicative of a very high density compared to other tiger habitats. Even if tiger home ranges are double this size in other areas of the forest, the Bangladesh side of the Sundarbans could still support 100-150 breeding females or 300-500 tigers overall (Barlow 2009).

During 2016, the enumeration exercise in the Indian Sundarbans was for the first time extended to five sectors- one in South 24-Parganas FD 546.62 km² with tiger density/100 km² 5.46 and four in STR- NP (East) Range 523.67 km² with tiger density/100 km² 2.09, NP (West) Range 580.34 km² with tiger density/100 km² 3.07, Sajnekhali WLS 413.42 km² with tiger density/100 km² 1.79 and Basirhat Range 403.83 km² with tiger density/100 km² 3.56. In spite of being RFs (with greater human use), both South 24-

Parganas FD and Basirhat Range of TR support a higher density of tigers than the adjacent NP. The overall female:male ratio is female-biassed, i.e. 2.73:1. Spatial overlap is minimal among the males whereas between females as well as males and females it is comparatively higher, which is consistent with the territorial behaviour of tigers.

Whereas the carrying capacity in the mainland landscapes is more than double in some TRs, for example, 10-16 tigers per 100 km^2 in Corbett within the Terai and Shivalik hills habitat, 7-11 in Bandipur within the north-central Western Ghats, 6-10 in Kanha within the central India, low density of tigers in the Sundarbans [for example, 3.7 tigers/100 km^2 in Bangladesh (Khan, 2007)] is an inherent attribute of the hostile mangrove habitat that supports low tiger prey densities. Compared to the Indian maximum of 120-130, Bangladesh's carrying capacity could be as high as 200 tigers. The higher level of human-wildlife conflict in the mangrove peripheries is due to frequent dispersal of tigers. The tigers are more frequently on the move as a result of hypertonicity, variations in prey availability, different hunting techniques or due to other factors, including human disturbance level. The increase in the number of tigers might be one major reason for the rise in straying incidents.

Diverse animal communities are associated with mangrove habitats, where tigers are both a Flagship and Umbrella species. As a flagship species the tiger acts as an icon or symbol for the mangrove habitat and thereby supports biodiversity conservation in the Sundarbans and as an umbrella species its conservation indirectly conserves many other species in the ecosystem. The tigers are also considered a landscape species since they are widely distributed, highly mobile or dispersed in different habitat types and best recovered by managing threats associated with habitat loss or degradation at a landscape scale.

Habitat types

Good quality habitat (i.e. the habitat that supports sufficient populations of large prey species, some vegetation cover so that tigers can ambush, and availability of drinking water) is important for tigers. The habitat types in the tigerland are described below as per field investigations in different sectors of the Sundarban mangroves.

24-Parganas (South) FD

(1) Confirmed Tiger Habitats with good conservation value (396 km² or 23.59% of the total area)

   a) Ramganga range: Population 8.0±0.2 individuals and density 4.3 individuals/100 km²; total landmass 204 km²; effective trapping area 184.5 km², i.e. 18.77%;

   b) Raidighi range: Population 13±3.5 individuals and density 7.08 individuals/100 km²; total landmass 192 km²; effective trapping area 141.3 km², i.e. 14.38%.

The conservation value of these habitats is considered good because of the protected status of the forests combined with evidences of the presence of tigers like the animal-kill, pugmark, claw-mark on the trees, urine-mark on the bush, scat, mating call, normal roaring, and the alarm call by the prey animals like the monkey or chital.

(2) Expected Tiger Habitats (586.56 km² or 34.94% of total area) with fair conservation value

These are forest blocks that are physically connected to confirmed tiger habitats. Tigers are expected to occur in these habitats because of their physical connectivity. The conservation value of these areas can be raised once tiger presence is confirmed.

(3) Possible Tiger Habitats (696.44 km² or 41.47% of total land area) with marginal conservation value.

These areas include forests in the tiger region that are isolated from confirmed tiger habitats. It also includes areas with natural vegetation not defined as "forests" (e.g. scrublands and abandoned agricultural fields), but where tigers are not recorded. Because the future of these lands is uncertain, their conservation value is marginal, except for areas considered as potential corridors connecting confirmed/expected tiger habitats.

STR (Naha et al., 2016)

(1) Water/Channels

This category consisted of open water and inland water channels exceeding 20 m in width. This habitat comprised 1,161.44 km² (41.48%) of the total area.

(2) *Phoenix* dominated

The habitat was predominantly of *Phoenix paludosa* along with *Excoecaria agallocha* forming dense thickets on high lands with compact soil and minimum inundation during high tides. This habitat therefore represents shady but thorny palms on drier ground and covered 1,033.76 km² (36.92%) of the total area.

(3) *Ceriops* dominated

Ceriops decandra was the dominant mangrove species here. It formed dense shrub vegetation of moderate height growing on comparatively higher ground. The habitat covered 333.48 km² (11.91%) of the total site.

(4) Barren dry areas

These areas covered 145.6 km² (5.2%) including the open areas that occasionally get tidal influx. This results in hard salt encrustation of the ground and low water availability. Vegetation cover was sparse.

(5) *Avicennia-Sonneratia* habitats

These habitats covered 126.56 km² (4.52%) and were primarily low lying areas with fresh deposition of silt that are characterised by *Avicennia* species and flanked by the foreshore grasslands of *Oryza coarctata*. *Sonneratia apetala* also grows along with *Avicennia* along banks of these fertile, moist and mostly flat lands.

Bangladesh

Khan and Chivers (2007) identified four major habitat types in the Sundarbans East WLS (312 km²):

(1) Mangrove woodlands dominated by trees such as *Heritiera fomes*, *Excoecaria agallocha* and *Sonneratia apetala*, comprising c. 70% of the Sanctuary, and including narrow creeks.

(2) Grasslands with species such as *Imperata cylindrica, Acrostichum aureum* and *Myriostachya wightiana*, comprising c. 10% of the area; this habitat type also includes some bare areas and sand dunes.

(3) Sea beaches, which are relatively open but narrow sandy strips with sparse reeds and other stunted vegetation, comprising c. 6% of the area.

(4) Transitional areas between mangrove woodlands and grasslands, characterised by having few trees and sometimes sungrass and reeds, comprising c. 14% of the area.

The mean density of all tiger signs combined was highest in mangrove woodlands and lowest on sea beaches but tigers did not exhibit an overall preference for any of the four habitat types. Means of feeding, resting, defecation and interaction signs were significantly different between habitat types, indicating that tigers probably have habitat preferences for these activities. The means of movement, scratch-scent-urinal, and other signs were not significantly different between habitat types, indicating that tigers probably have no significant preference for any one habitat type for these activities.

Most of the tiger signs (69.1%) were of movement, which is consistent with tigers' need to move frequently for hunting and territory patrolling. Although there were hunting signs in the grasslands (probably because the *Imperata* grasses, long enough to provide stalking cover for tigers), the density was lower than in transitional areas and mangrove woodlands. Prey densities are similar in these three habitat types. There were no hunting signs on the sea beaches, probably because it is almost entirely open and because prey density is low in comparison to other habitat types.

Density of movement and feeding signs were highest in mangrove woodlands and transitional areas, resting signs in grasslands and transitional areas probably due to the combination of the drier ground, presence of air flow and less human disturbance, and defecation signs in grasslands because tigers often defecate where they rest. Scats were commonly found in small dry sand dunes and besides footpaths in the grasslands. The density of interaction signs was highest on sea beaches, probably a nocturnal activity because of human disturbance on the beaches in the daytime, and scratch-scent-urinal signs in grasslands and mangrove woodlands. The density of other signs was highest in transitional areas.

Scratches were found on three tree species, two of which (*Syzygium* sp. and *Lannea* sp.) are relatively soft-barked and one (*Zizyphus* sp.) hard-barked; 13 of the 16 scratches found were on the softer-barked species. In general, relatively hard-barked trees (e.g. *Heritiera fomes*, *Sonneratia apetala*) are more available than soft-barked trees in the Sundarbans. Tigers often repeated scratches on the same individual tree at different times. Scratches were 0.0-2.0 m off the ground and all were on tree trunks with girths of c. 100 cm. Scratched trees were 0.5-7.0 km apart.

Karsch *et al.* (2023) reported that between 2014 and 2020 no major changes in the vegetation cover throughout the Indian Sundarbans, the coastal regions, especially the southern end, showed significant changes. There were both mangrove vegetation gains due to land accretion, for example on the east coast and mangrove loss due to erosion and cyclone damage. A large portion of the Sundarbans, especially in the northeast and central areas, also showed little change.

Chapter 7
Enumeration Dilemmas: State-of-the-Art

"The one certainty in tiger tracks is: follow them long enough and you will eventually arrive at a tiger, unless the tiger arrives at you first."
– John Vaillant

Global status

The IUCN's latest assessment (2022) estimates remaining wild tigers in Asia between 3,726 and 5,578, with an average of 4,500 individuals. Some 3,140 of the 4,500 are estimated to be adult tigers. In North-east Asia, numbers are relatively stable in Russia and likely increasing along the border with China. Of all regions, however, Southeast Asia's tigers are faring the worst, with the tigers having been lost from Cambodia, Lao PDR, and Vietnam since the turn of the century. In 2010, at the first Global Tiger Summit, the world committed to a highly aspirational goal of doubling global tiger numbers or at least reaching 6,000 individuals by 2022 but this target has not been reached.

The wild tiger population, first recognised as endangered back in 1975, was estimated at 5,000 to 7,000 (Seidensticker *et al.* 1999) as compared with 100,000 a century ago and a decline of about 50% (declining to approximately 3,500 in 2014) is anticipated, in which the population of mature individuals may be fewer than 2,500 individuals (Goodrich *et al.* 2015). In 2006, it was believed that breeding populations of this amazing species existed in 13 countries, now it continues to thrive only in eight [Bangladesh (114 in 2018), Bhutan (90 in 2015), India (3,167 in 2022), Indonesia (618 ± 290 in 2017), Malaysia (150 in 2022), Nepal (355 in 2022), Thailand (100 in 2022) and Russia (480-540 in 2015)]. The number of tigers held in captivity is more than double the number left in the wild- 7,000 in the United States, 1,600 in Europe, and 8,000 more in China, Laos, Thailand and Vietnam in private ownership, tiger farms, zoos, circuses and self-proclaimed 'sanctuaries', which are traded (loosely monitored), bred and exploited.

India

The then Prime Minister Mrs. Indira Gandhi spearheaded a fight against the growing tiger crisis, outlawing the export of skins in 1969 and appointing a Tiger Task Force two years later. By 1971, about 1,800 tigers were left alive and the Tiger Task Force predicted they would be extinct by the end of the century. That year, the Delhi High Court banned tiger killing, despite opposition from the trophy hunting industry. Then in 1973, Mrs. Gandhi launched "Project Tiger," and nine TRs were established, which still stands as the world's most comprehensive tiger conservation initiative and as of 2023, NTCA has declared 53 TRs. But the embarrassment over extinction of the tigers in Sariska (Rajasthan) prompted the creation of a new entity, NTCA. A camera trap survey by WII reported (2008) just 1,411 adult tigers- after a $400 million investment over 34 years to save the tigers under Project Tiger. Two years later, a wider national census raised the tiger estimates to 1,706.

The wild tigers thrive at optimum carrying capacities in the protected tiger landscapes so as to perform their ecological role, and which continue to provide essential ecological services to mankind. For the development of any tigerland, there is a need for information about the number of the tiger population, their behaviour, and aspect.

Tracking the untraceable Sundarbans tiger

In the old days there were confusions about the tiger's walking as there used to be differences in the distance between the track marks of fore and hind legs and at times the tracks were overlapping. Dunbar Brander writes: "If the tiger progresses more slowly, or with purposeful caution, the gap decreases, and in following up pug marks along a *nala* one comes on places where the tiger has been obviously 'interested' or has rounded a corner with great caution. In these circumstances one comes on places where the hind leg on the same side has been placed exactly on over the spot vacated by the forefoot. Of course, this action reduced the speed and noise. It is possible that a tiger, hungry and hunting, might continue this action over some distance, but it is not the tiger's carefree, habitual action, it is controlled and purposeful action".

Most of the time in Sundarbans you cannot see the tiger. Sy Montgomery talks about why it is hard to track the mysterious, mythical ruler of the Sundarbans mangrove forest in the book 'Spell of the Tiger' (2008). Why is it hard to track the mysterious, mythical ruler of the Sundarbans mangrove forest? How can you study an animal you cannot see? How can you manage an unseen population?

The FD has been attempting frequently as a daily routine work or periodical (two to four years) to trace the outlines of the tiger's mystery and all categories of staff along with the specialists, NGO members and volunteers, all well trained for the purpose, looking for the tiger impressions in the mud or direct sighting. Hundreds of water vehicles including the big launches, houseboats, small motor boats and dinghies are involved in such operations.

The first countrywide tiger population assessment was done in 2006 and it estimated the number to be 1,411. The Project Tiger authorities developed a new methodology for monitoring tigers, co-predators, prey and habitat. The new techniques include radio-collar/GPS collar, remote photography (camera trapping), genotyping of scats and hair, tracing of pugmark and other indirect signs like scats, scrapes, scent marks, rake marks, vocalisation (roaring) and kills. The tigers use scat, scrape, and rakes to advertise their presence. Scats are also collected for studying the big cat's diet.

Pugmark technique

Before technology took its due turn during the early 21st century, the traditional 'pugmark census' methods were relied on. It has been one of the foremost popular ways of counting tigers. Each tiger is understood to leave a definite pugmark on the bottom and these are different from the others within the cat family. Photographs or plaster casts of those pugmarks are then analysed to assess the tiger numbers. The study of pugmarks can provide subsequent information reliably if analysed skillfully:

- Presence of various species within the area of study.
- Identification of individual animals.
- Sex ratio and age of huge cats.

The gender of tigers was differentiated on the basis of the shape of the hind pugmark. In "Forty Years Among the Wild Animals of India, from Mysore to the Himalayas" by the British forester FC Hicks, published in 1910, described the male tiger's pugmark as being more circular than that of a female. He added that the pugmarks of a tigress were misshapen and ugly and her forepaw resembled the hind pugmark of a male. Sommerville (1933) suggested that the pugmarks of a male tiger were larger than the female's. He also noted the male's toes were square while those of the females were more rounded and slender. Sankhla (1978) described the front pugmark of a male tiger as regular and that of a female as irregular or zygomorphic. He also observed captive tigers and found that at 3 months of age the male's pad size was double that of the female. This difference was maintained throughout life. The male tiger's pugmark was also said to be more square, less angular, and relatively wider in relation to its length (McDougal 1977, 1999). Panwar (1979) suggested that the whole hind pugmark of a male tiger fit into a square frame, whereas that of a female fit into a relatively rectangular frame. He also suggested that the female's toes were slender and elongated compared to a male's toes, which were oval and more circular. This criterion is the most adopted and widely used field technique to differentiate the gender of a tiger based on its pugmark.

The first attempt to enumerate tigers from their pugmarks was made by WJ Nicholson of Imperial Forest Service in Palamau district, Bihar in 1934, which gave him a figure of 32 tigers for an area of 299 km². A systematic methodological approach for recording pugmarks for individual tiger identification and their census was formally conceptualised and advocated by a forest officer from Odisha, named Saroj Raj Choudhury. He invented the 'tiger tracer' and developed the methodology for a census of pugmarks in 1966. This method was again fine-tuned during the 1980s and 1990s. Features in a pugmark, e.g. the shape and relative size of the right, left or bottom lobe of the pad, the top edge of the pad, the relative sizes and placings of the toes with respect to the pad and several other features vary from tiger to tiger. An individual can be identified from a study of a combination of these features unique to itself, from frequent tracings of pugmarks recorded in the field (Sharma *et al.,* 2005).

Nine pugmark variables were measured from pugmark: (1) area of toe 3, (2) length of minor axis of toe 3, (3) distance between toe 2 and toe 3, (4) length of minor axis of toe 2, (5) distance between main pad top to toe base-line, (6) angle between toe 2 and toe 3, (7) heel to lead toe length, (8) distance between notch 1 and notch 2, (9) width of the pugmark (Sharma *et al.,* 2003).

It was difficult to obtain good pugmark sets with enough replicates in the field because tigers are elusive, primarily nocturnal, have large home ranges, and are sparsely distributed. It was more difficult to obtain male pugmarks than female pugmarks in the wild because the sex-ratio is biassed toward females and males have larger home ranges. In such conditions one has to rely on signs left by tigers to determine their presence.

Indian Sundarbans

Studies on Bengal tigers in the Indian Sundarbans date back to 1970, when Amal Bhushan Chaudhuri, the then Divisional Forest Officer (DFO) of undivided 24-Parganas FD, carried out ecological studies and on the basis of pug marks estimated 112-120 tigers in the Indian Sundarbans (Chaudhuri, 2007). The initial census in 1973 was limited to the Sajnekhali WLS and the rest was left out due to infrastructural problems. As a result only 50 tigers were calculated. No official census report is available for erstwhile 24-Parganas FD till 1984.

STR

After creation of STR, efforts were made by the FD to estimate tiger numbers based on a single approach called pugmark (plaster cast) method for individual identification of footprints, attacks on humans, and interviews with local communities. Up to 2004, the tiger population estimate (census or monitoring) was based on this method, in which the fresh left hind pugmark impressions were collected from the field and analysed on the basis of eighteen parameters. The first tiger census (conducted partly) estimate of 135 was available in 1972. Year wise tiger population in STR since its creation during the 20th century is placed below.

Year	Male	Female	Cub	Total	Remarks
1973	-	-	-	50	Incomplete census; census was done only for Sajnekhali WLS area; sex not determined.
1976	66	72	43	181	Entire STR area was covered for the first time.
1977	-	-	-	205	Sex not determined.
1983	137	115	12	264	Plaster casts of pugmarks were taken from the field, computer programming was applied to count the no. of tiger, determine sex and no. of cubs. During this census one staff member was killed, so the census was stopped and was repeated after two months for the remaining 30% area.
1989	126	109	34	269	Female and cub populations have increased.
1992	92	132	27	251	Computer programming introduced. Population reduced.
1996	95	126	21	242	
1997	99	137	27	263	NGO participation, use of nylon net and tiger guard first started during this census.
1999	96	131	27	254	Three piece tiger guard was used for the first time during this census.
2001	93	129	23	245	Population marginally reduced.
2002	-	-	-	271	Detailed break up not available.

Comparative block-wise results of tiger enumerations during the subsequent periods are shown below.

Block	Year	Male	Female	Male:Female (Ideal 1:3)	Cub	Total
Pirkhali	1977	8	10	1:1.25	1	19
	1983	10	9	1:0.9	2	21
	2004	8	14	1:1.75	5	27
Panchamukhani	1977	4	5	1:1.25	5	14
	1983	10	12	1:1.2	-	22
	2004	6	13	1:2.16	4	23
Netidhopani	1977	4	5	1:1.25	3	12
	1983	1	2	1:2	-	3
	2004	3	5	1:1.6	1	9
Jhilla	1977	3	5	1:1.6	8	16
	1983	8	7	1 : 0.875	2	17
	2004	3	5	1:1.6	-	8
Arbeshi	1977	6	9	1:1.5	0	15
	1984	12	9	1:0.75	4	25
	2004	8	12	1:1.5	1	21
Khatuajhuri	1977	4	5	1:1.25	2	11
	1983	9	7	1:077	-	16
	2004	3	6	1:2	3	12

Block	Year	Male	Female	Male: Female (Ideal 1:3)	Cub	Total
Chandkhali	1977	5	5	1:1	2	12
	1983	6	10	1:1.66	-	16
	2004	5	9	1:1.8	-	14
Chamta	1977	8	8	1:1	-	16
	1983	19	12	1:0.63	-	31
	2004	11	12	1:1.09	3	26
Harinbhanga	1977	4	4	1:1	-	8
	1983	4	4	1:1	-	8
	2004	4	6	1:1.5	3	13
Matla	1977	4	5	1:1.25	2	11
	1983	7	7	1:1	1	15
	2004	6	8	1:1.33	2	16
Chhotohardi	1977	4	4	1:1	3	11
	1983	6	5	1:0.83	1	12
	2004	3	10	1:3.33	2	15
Gosaba	1977	5	7	1:1.4	1	13
	1983	12	10	1:0.83	1	23
	2004	4	8	1:2	2	14
Mayadwip	1977	6	7	1:1.16	3	16
	1983	14	9	1:0.64	-	23

Block	Year	Male	Female	Male: Female (Ideal 1:3)	Cub	Total
	2004	4	6	1:1.5	2	12
Bagmara	1977	9	10	1.1.11	3	22
	1983	14	6	1:0.42	1	21
	2004	11	13	1:1.18	3	27
Gona	1977	3	3	1:1	1	7
	1983	5	6	1:1.2	-	11
	2004	4	6	1:1.5	2	12
Total	1977	77	92	1:1.19	36	205
	1983	137	115	1:0.83	12	264
	2004	83	133	1:1.6	33	249*

*Joint census with Bangladesh was attempted for the first time.

Since Sundarbans mangrove forest comprises highly difficult and muddy terrain, and many parts of forest are inaccessible, enumeration of exact number of tigers in such a forest is physically and theoretically not possible. It is possible that some parts of the forest, containing pug mark impressions, might have been left out. It is also possible that because of clay and muddy texture of soil, there may be some error in clear identification of the individual pug marks. It is, therefore, more scientific to state the estimated population of Sundarban tiger within the statistical arrived Range, instead of quoting the absolute figure. Therefore, the estimated population of Sundarban tiger, as per January 2004 tiger census will be as follows Minimum Population 237, Mid-value 249 and Maximum Population 261. Hence, the population of tigers in STR lies in the range between 237 and 261 as in January, 2004.

The block wise census figures of STR in 2004, as shown in the comparative table above, reveal that all the compartments hold a number of tigers, but if it is categorised as high (20+), medium (10+) and low (below 10) grades, Pirkhali, Bagmara, Chamta, Panchamukhani and Arbesi were high zone, whereas Matla, Chhotohardi, Chandkhali, Gosaba, Harinbhanga, Khatuajhuri, Gona and Mayadwip formed the medium zone and Jhilla and Netidhopani low zone. All the previous censuses (1976-2004) in STR recorded a population of ±250 tigers on the basis of pugmark analysis. Accordingly, tiger concentration was calculated to be <10 km²/tiger. The adult male female ratio during this year was 1:1.60, which should ideally be 1:3.

Data from different clusters of census blocks in STR during 1995, 1997, 1999, 2001

Cluster	Block-Compartment	Category	1995	1997	1999	2001
I	Arbesi -1	Male	-	-	1	1
		Female	3	5	1	2
		Cub	-	-	-	-
II	Arbesi- 2,3,4,5; Harinbhanga-1	Male	5	5	7	5
		Female	6	8	11	9
		Cub	2	1	2	1
III	Khatuajhuri-1&2; Harinbhanga - 2&3	Male	12	9	5	7
		Female	4	5	9	9
		Cub	1	1	3	1
IV	Khatuajhuri-3	Male	2	3	1	2
		Female	1	2	2	1
		Cub	-	1	-	-

Cluster	Block-Compartment	Category	1995	1997	1999	2001
V	Jhilla-1,2,3,4,5&6	Male	6	5	4	4
		Female	5	8	6	4
		Cub	1	1	-	1
VI	Panchamukhani - 1&2; Pirkhali - 1,2&3	Male	1	5	6	6
		Female	12	10	8	7
		Cub	1	3	2	6
VII	Pirkhali - 4,5,6&7	Male	4	4	4	4
		Female	8	4	4	5
		Cub	-	-	-	-
VIII	Panchamukhani- 3,4&5	Male	6	6	7	5
		Female	10	14	8	7
		Cub	1	5	4	2
IX	Netidhopani-1,2,3; Matla-1; Gosaba-1	Male	2	7	9	8
		Female	5	8	11	11
		Cub	5	1	2	1
X	Matla- 2,3&4	Male	5	3	5	5
		Female	8	5	6	6
		Cub	2	2	1	1
XI	Chottohardi-1&2	Male	4	4	4	4
		Female	3	7	6	6

Cluster	Block-Compartment	Category	1995	1997	1999	2001
		Cub	2	-	2	-
XII	Gosaba- 2,3&4	Male	3	6	4	5
		Female	6	4	7	7
		Cub	3	5	-	1
XIII	Chottohardi- 3	Male	2	2	1	-
		Female	-	2	2	3
		Cub	-	2	1	-
XIV	Mayadwip- 1,2&3	Male	3	2	3	3
		Female	9	7	3	4
		Cub	-	1	1	3
XV	Mayadwip- 4&5	Male	3	2	3	3
		Female	9	7	3	4
		Cub	-	1	1	3
XVI	Gona- 1&3	Male	2	2	3	1
		Female	3	4	6	4
		Cub	2	1	-	2
XVII	Gona - 2; Chandkhali- 4; Bagmara- 1	Male	8	4	4	5
		Female	8	6	7	5
		Cub	-	-	-	1
XVIII	Bagmara- 2,3,4&5	Male	4	9	6	6

Cluster	Block-Compartment	Category	1995	1997	1999	2001
		Female	10	5	7	8
		Cub	-	1	2	-
XIX	Bagmara- 6,7&8	Male	3	3	5	4
		Female	7	2	7	8
		Cub	1	1	2	2
XX	Chandkhali- 1&3	Male	6	3	5	5
		Female	6	4	4	3
		Cub	-	-	-	-
XXI	Chandkhali- 2; Chamta-5,6,7&8	Male	5	6	6	7
		Female	8	8	10	8
		Cub	-	-	-	1
XXII	Chamta-1,2,3&4	Male	10	7	5	4
		Female	5	10	7	9
		Cub	-	-	-	-
Total		Male	95	97	96	93
		Female	126	132	131	129
		Cub	21	27	27	23
Grand total			242	256	254	245

It may be seen from the following table that the sweet water ponds are not usually frequented by the tigers during the rainy season. The tigers only visited the ponds mostly during the dry season.

Tiger Sighting Data at Netidhopani, Sudhanyakhali, Sajnekhali and Haldi in STR during 1998-2001 (Mukherjee, 2004)

Month	Netidhopani				Sudhanya khali		Sajnekhali			Haldi		
	1998	19-99	20-00	20-01	20-00	20-01	19-98	19-99	20-00	19-99	20-00	20-01
Jan	5	-	2	3	3	2	-	-	-	-	2	3
Feb	3	5	2	4	4	1	-	-	-	5	1	4
March	2	4	2	4	4	2	-	-	1	4	1	2
April	3	2	1	3	2	2	-	-	1	4	-	3
May	4	2	-	4	4	-	-	-	-	2	1	4
June	-	1	2	1	3	2	-	-	-	1	-	1
July	-	1	2	1	2	2	-	-	-	1	3	1
Aug	-	2	1	2	2	3	-	-	-	1	-	1
Sep	-	3	3	3	9	5	-	-	-	1	1	3
Oct	-	4	1	4	2	2	-	-	-	1	-	3
Nov	-	1	4	1	2	5	-	-	-	-	3	-
Dec	-	5	2	5	4	-	-	-	-	1	4	-

It is found that the number of tigers visiting the ponds per year varies from 0 to 6 in numbers. The tiger uses the sweet rain water which is collected in small puddles during the rainy season. As rains are distributed beyond the rainy season in the Sundarbans, the tigers fulfil their need of sweet water from these puddles.

Mallick (2011) also recorded tiger sightings in STR during the periods 2007-2008, 2008-2009 and 2009-2010.

Month-wise tiger monitoring data (number) in STR during 2007-2008

Month	Direct tiger sighting		Roars heard	Sighting of fresh pugmarks	
	Adult	Cub		Adult	Cub
April	25	-	10	586	3
May	14	-	8	629	-
June	11	4	6	411	-
July	13	-	5	380	-
August	21	-	19	433	5
Sep	24	-	19	332	1
Oct	18	-	12	466	1
Nov	26	-	17	455	-
Dec	27	1	15	747	-
Jan	28	1	22	849	-
Feb	25	-	17	711	-
March	35	-	6	540	-
Total	267	6	156	6,539	10

Month-wise tiger sighting data (number) in STR during 2008-2009

Month	Direct tiger sighting	
	Adult	Cub
April	20	.
May	16	.
June	23	.
July	20	4
August	31	.
September	17	.
October	53	.
November	37	.
December	39	.
January	51	.
February	37	2
March	50	.
Total	394	6

Forest Block wise and month-wise tiger sighting data in STR during 2010

Block	M1	M2	M3	M4	M5	M6	M7	M8	M9	M10	M11	M12	Total
Panchamu khani	1	1	1	4	1	2	1	2	-	3	2	6	24
Pirkhali	33	31	23	27	9	10	16	4	3	3	5	16	180
Matla	1	2	2	1	1	2	1	-	-	-	-	-	10
Chamta	-	-	2	1	1	-	2	-	-	-	-	1	7
Chhoto-hardi	3	1	1	-	-	1	2	-	1	-	-	-	9
Gosaba	1	-	3	1	1	-	5	1	1	1	-	2	16
Gona	-	-	-	-	-	-	1	-	-	1	-	-	2
Bagmara	-	2	-	1	-	-	-	-	-	-	-	-	3
Mayadwip	-	-	-	-	-	-	-	-	-	-	-	-	0
Arbesi	2	2	6	1	4	4	9	3	6	5	-	13	55
Jhilla	2	1	3	2	1	-	3	-	-	1	-	-	13
Khatuajhuri	2	2	3	-	2	-	1	1	-	-	1	1	13
Harinbha nga	1	-	1	1	-	-	-	1	-	-	-	1	5
Netidhopani	10	4	11	11	4	11	3	5	2	10	1	5	77
Chandkhali	-	1	1	-	1	-	-	-	-	-	-	-	3
Grand total	56	47	57	50	25	30	44	17	13	24	9	45	417

The above table is prepared on the basis of direct sighting and indirect evidence along the river/creek banks, bank-side mangroves and at the watchtower locations, which has revealed that the tigers were present throughout the landscape with varying abundance. In all, the tiger was sighted 417 times (mean 13 in September and maximum 56 in January) in fourteen forest blocks during 2010. Such sightings were frequent during the winter and summer months. Maximum sighting was recorded on the north-western forest blocks like Pirkhali (43.16%), Netidhopani (18.46%), Arbesi (13.18%) and Panchamukhani (5.75%). Although tiger was not sighted in Mayadwip block, the pugmarks of a mother along with a cub and (separately) those of an adult male were traced in this island a number of times in May.

Four cubs were also sighted, two each at Pirkhali and Chhotohardi in November. However, the cub-sighting ratio is very low in the study area. In all, only five cubs were found in three out of fifteen blocks (20%). Earlier, the author found the signs of hunting a deer by the tiger at Burirdabri (Khatuajhuri block), opposite Bangladesh. There was a resident tiger family including a cub on this small island.

24-Parganas (South) FD

De (1991) first recorded that a tiger was recently found in Haliday Island in the Matla river. In order to reach the island, which is generally free from tigers, the tiger must have swum 8 km at a stretch. It may be mentioned that Haliday Island was the last resort of the barking deer which was earlier distributed in many areas of the eastern Sundarbans. The tiger might have also come to the island in search of food.

The block-wise results of tiger enumerations in 1984 are shown below.

Block	Male	Female	Cub	Total
Herobhanga	2	3	-	5
Aimalmari	7	3	1	11
Dulibhasani	2	-	-	2
Chulkati	3	2	-	5
Total	14	8	1	23

Comparative estimates of tiger population in 24-Parganas (South) FD on the basis of plaster cast method during 1997-2004

Year	Male	Female	Male: Female	Cub	Total
1997	13	16	1:1.23	6	35
1999	9	16	1:1.77	5	30
2002	7	13	1:1.85	6	26
2004	7	14	1:2	4	25

The erstwhile tiger estimation by pugmark method was field-friendly and cost effective, but criticised as deficient, prone to human error and runs the risk of overestimation. Critics of the technique believed that an individual tiger's pugmark changes in shape and size over different substrates (soil texture, moisture and depth). Another source of variability is the variation between different tracers' abilities to trace the features of the pugmark on the tracing sheet. There are logistical constraints in the Sundarbans.

In the year 1999 the pugmarks that were collected during the census period were analysed directly by using a Flatbed Scanner, further the contrast of the scanned pugmarks were increased so as to mark the actual contour lines. Then software was used with 18 standardised parameters of the pugmarks and the data generated was analysed. But this method was inexactitude due to scanning errors and a misrepresented census showed a huge number of tiger population, as such, the method has been abandoned.

The situation has become more complicated in Sundarbans because the fluctuations of the pug marks of different individuals occur not only because of usual variability, an important part of these fluctuations occur because of the difference in soil types like hard soil, muddy soil and mud in Sundarbans. In order to study the effect of splayed pugmarks on different soil conditions, pug-mark casts were taken of identified individuals in three soil conditions: hard soil, mud and muddy soil.

These data were taken for 13 tigers during the study period; although no data were collected from sandy soil with the assumption that a sandy soil does not have any effect on splaying. On analysing these data in ungrouped with disclosing the identities of the tigers, by different individuals, it was found that the number varied from 8-16. This led to standardising clustering techniques, which would readily work in this context and which can negate to a large extent human error. It was realised that it is necessary to use a formal model such as the two ways classification model with one random and one fixed effect, to explore the suitability of classification schemes based on the normal likelihood, as well as, some non parametric/exploratory methods. In order to reduce the possibility of error, it is suggested to look for features, which are quite stable across various pugmarks of a particular tiger, and yet have potential to discriminate against different tigers. As such, The Indian Statistical Institute, Baranagore, Kolkata were approached for analysis of the data, to devise more efficient computer software for pug mark analysis.

After preliminary analysis it was found on the basis of 17 extracted pugmark features of 10 different tigers in three types of soil showed that there is at least one linear combination of these features which discriminate the 10 different animals quite well. Whether the advantage of high dimensionality continues to be there when the sample size is much larger was a question at large. So it was realised that repeated pugmarks of a particular tiger in the same type of soil is also needed. This could serve as a reference for "within individual" (as for the tiger and also the soil condition) fluctuations. If pugmarks of two distinct tigers are similar, all the features need to be accurately extracted for effective discrimination. For this purpose it was also thought necessary to plan an automated feature extraction process by means of photographing and/or scanning the plaster-casts or the traces made out of these. This approach can be useful to try out additional features easily, without reexamining the casts or traces.

For a larger data set, a single combination of features would not be adequate. However, the above analysis shows that the conventional features extracted from pugmarks have reasonable potential for identifying tigers. A definite assessment of the discriminating power of pug marks can only be made on the basis of a mere complete analysis of a more substantial amount of data.

Three decades of tiger monitoring following this unrealistic method essentially failed in India- despite being backed by massive investments and the best of intentions. This failure has, inevitably, led to poor conservation practices. For example, field managers initially reported an increase in tiger population during the late 20th century, despite mounting evidence of deteriorating reserve protection and increasing poaching pressure. For example, there were so many discrepancies in the past. Seven months after the December 1992 census was conducted, the FD reported 251 tigers in the Indian Sundarbans. In the 613 paw-print tracings collected, 92 individual tigers, 132 tigresses, and 27 cubs were counted. Compared with the figures released before, the FD's figures were actually 265 in 1988 and 269 in 1990. In 1979, the authorities estimated the presence of 205 tigers. In 2004, the population of tigers rose to 274, which was, in the view of present findings, too optimistic. This number was unpalatable to many critics. During a personal trip or two, some wildlife experts evaluated that there were not more than 50 tigers in the mangroves.

The block-wise census figures of STR in 2004, reveal that all the compartments hold a number of tigers, but if it is categorised as high(20+), medium (10+) and low (below 10) grades, Pirkhali, Bagmara, Chamta, Panchamukhani and Arbesi were high zone, whereas Matla, Chhotohardi, Chandkhali, Gosaba, Harinbhanga, Khatuajhuri, Gona and Mayadwip formed the medium zone and Jhilla and Netidhopani low zone. All the previous censuses (1976-2004) in STR recorded a population of ±250 tigers on the basis of pugmark analysis.

Karanth *et al.* (2003), by evaluating the 'pugmark census method' employed by wildlife managers for three decades, identified three critical goals for monitoring tiger populations, in order of increasing sophistication: (1) distribution mapping, (2) tracking relative abundance, (3) estimation of absolute abundance and demonstrated that the present census-based paradigm does not work because it ignores the first two simpler goals, and targets, but fails to achieve, the most difficult third goal. They pointed out the utility and ready availability of alternative monitoring paradigms that deal with the central problems of spatial sampling and observability and proposed an alternative sampling-based approach that can be tailored to meet practical needs of tiger monitoring at different levels of refinement.

Bangladesh Sundarbans

Vicomte Edmond de Poncins reported from a trip to the Sundarbans in the year 1892 that there were many Bengal tigers in the Sundarbans mangrove forests (de Poncins, 1935). It is known from the Working Plans that 452 Bengal tigers were killed from 1912 to 1921 (Curtis, 1933). Past populations of the Bengal tiger and their prime prey species during the late 20th century were highlighted by different authors as shown below, but these figures seemed to be guess estimates.

species	Hendrichs (1975)*	Salter (1982)#	Gittins and Akand (1984)*	FD (1992)□	Tamang and Dey (1993)◇
Bengal tiger	350	425	430-450	359	362
Spotted deer	80,000	-	-	-	94,000
Wild boar	20,000	-	-	-	45,000
Rhesus macaque	40,000	88,000	1,26,000	-	40,000

Methods of estimates: *Field survey in compartments 3, 4, 5, 6, 29, 30, 31, 46, 47, 48, 49, 50

*Field survey in 110 km². area of Sundarbans South WLS

#Sample field survey

□Report of the people working in Sundarbans

◇Pugmark census in 350 km² of different compartment

Status of recorded tiger kills (number) in different months during January, 1999-March, 2000 (IUCN, 2004)

Month	Kills	Spotted deer			Wild boar			Monitor lizards		
	Number	A	B	C	A	B	C	A	B	C
Jan	4	1	1*	1	-	-	-	1	-	-
Feb	3	2	-	-	-	1	-	-	-	-
March	5	1*	2	-	-	-	1	-	1	-
April	2	-	1	-	-	-	1	-	-	-
June	4	-	1	2	-	1	-	-	-	-
July	2	-	1*	-	-	-	1	-	-	-
Sep	1	-	-	-	-	-	1	-	-	-
Oct	4	1*	2	-	-	-	-	-	1	-
March	1	-	-	1	-	-	-	-	-	-
Total	26	5	8	4	0	2	4	1	2	-

A: Very fresh kills and recorded within 12 hours of killing.

B. Moderately fresh kills and recorded within 48 hours of killing.

C. Old kills and recorded only the bones and other hard body parts.

*Deer doe.

Proportion of different prey species in tiger diets as derived from the scats and kill data (IUCN, 2004)

Prey species	Average body weight (kg)	Scat (%)	Kill (%)	Average (%)
Spotted deer	55	69.23	65.38	67.30
Wild Boar	38	15.66	23.07	19.36
Rhesus macaque	8	5.00	-	2.50
Monitor lizard	6	-	11.53	5.76
Unidentified	-	10.18	-	5.09
Total	-	100	100	100

Tigers make deep, distinctive tracks in the soft mud bank of *khals*. According to the tiger census conducted by the government in 2004, the Bangladesh part of the Sundarbans was home to 440 tigers, including 21 cubs but counting tigers by using pugmarks is not actually a reliable and scientific method. So, it didn't give the exact figure of Bangladesh Sundarbans tigers.

According to Khan (2007), a camera trap survey was first conducted in Bangladesh Sundarbans to know the absolute density of tigers in a high density area of the Sundarbans. A total of 105 km² area in the southern part of the Sundarbans East WLS was effectively covered in the survey. The entire survey area was crisscrossed extensively in order to find good trap-points, i.e., the spots where there is a good chance of getting the tiger in the camera-trap photo. The trap-points were about two km apart from each other, so there was no 'gap' area large enough to contain a tiger's movements during the sampling period and within which any tiger had a zero capture probability. The effectively sampled area was calculated by adding up to two km boundary strip width (since the two recaptured tigers crossed trap-points of about two km distance) with the outer boundary of the trap-polygon (straight lines connecting the outermost trap-points). Since the survey area was covered by estuary and large rivers on three sides, the boundary strip was added up to the land, but on the fourth (northern) side the boundary strip of two km was added.

The camera-traps were set on the basis of tiger cues, such as the presence of earlier tiger signs (tracks, scats, scrapes, scent deposits, etc.), intersections of trails (Karanth and Nichols 1998), and near the kills. During the sampling period the camera-traps were systematically shifted in three trapping sub-plots in order to cover all the potential trap-points by a limited number of camera-trap units. Out of the 90-day (here 'day' means day-night of 24-hours) sampling period (06 September to 04 December 2006), the first 45 days were treated as Occasion 1 (when the photographed individual tigers were 'marked') and the second 45 days were Occasion 2 (when the photographed individual tigers had two individuals that was 'marked' in Occasion 1). In each Occasion the camera-traps were deployed in three sub-plots (Kochikhali, Katka and Chita Katka sub-plots), one after another, for 15 days in each subplot. In order to get photos of both sides of a tiger, two cameras were placed in each trap point (since the stripe pattern is different in two sides of the same tiger), one almost facing another (not face-to-face so that the flash light of one camera does not distract the photo of the opposite camera). Therefore, five trap-points were covered simultaneously to cover each sub-plot.

The camera-traps were checked once every day in order to record the date and location of each photographic 'capture'. Individual tigers were identified from photographs by comparing the stripe patterns. Along with the small camera-trap survey, all the fresh (maximum five days old) tiger tracks were also surveyed for 16 months (October 2005 to January 2007) in riverbanks in order to estimate the relative density of tigers. In the Sundarbans, tigers frequently cross the rivers, especially those that are not too wide. The survey took place from a dinghy, which was driven slowly at a relatively constant speed while the observers searched for fresh tracks on both banks of the river. However, the same track, i.e. the same crossing, on two sides of the river was treated as one observation. In order to calculate the correlation between the relative density and absolute density of tigers, track survey in riverbanks in the Sundarbans East WLS was conducted in the same plot covered by camera-trapping. Based on this correlation the absolute densities have been estimated for the other five plots. The average of the absolute densities of the six different plots in six parts of the Sundarbans represents the average density of tiger population in the Bangladesh Sundarbans.

A total of seven photographs of tigers were obtained from the Katka-Kochikhali plot during the survey period (three in Occasion 1 and four in Occasion 2, with two 'recaptures' in Occasion 2), of which there were five different tigers. Using the 'capture' history data in CAPTURE2 software program it was estimated that the absolute density of tiger population (adult and subadult) in the 105 km² survey area in the southeastern end of the Bangladesh Sundarbans is 5 (standard error = 0.96, capture probability or p-hat = 0.70).

This means that the tiger density in the area covered by camera-trap survey is 4.8 tigers/100 km². Due to the complexity and lack of correctness of estimating the variance of estimated area sampled by camera-trapping, the standard error for this density estimate was not calculated. However, due to the fact that the sampled area (105 km²) was very close to 100 km², it is assumed that the standard error for the density estimate is very close to 0.96. This is the first scientific estimate of the tiger population density in any part of the Bangladesh Sundarbans.

Other than the tiger there were many photos of two common prey species of the tiger, i.e., spotted deer *Axis axis* and wild boar *Sus scrofa*, and a few photos of humans, red junglefowl *Gallus gallus*, ring lizard *Varanus salvator*, rhesus macaque *Macaca mulatta*, lesser adjutant *Leptoptilos javanicus*, greater coucal *Centropus sinensis* and jungle myna *Acridotheres fuscus*.

Based on tiger track counts, the relative density of tigers in six different plots was estimated in the table below. These six different plots were selected from six different areas and the average result of these six plots represents the average for the entire Bangladesh Sundarbans, which is 0.44 tracks/km of riverbank surveyed. The three plots in three sanctuaries clearly had higher densities of tiger tracks than the three blocks outside the sanctuaries.

In Katka-Kochikhali, the relative density of tigers was estimated from the same plot that was covered by a camera-trap survey to estimate the absolute density. Based on the correlation of absolute and relative densities of tigers in this plot, the absolute densities in other five sites were calculated.

The relative densities of tiger prey (excluding humans) in six plots were estimated by counting them along the two banks of rivers while the observers were on dinghy, counting tiger tracks along riverbanks. The estimated relative densities of tiger prey and the absolute densities of tiger in six plots show strong correlation [Pearson correlation = 0.951, significance (2-tailed) = 0.004, N = 6]. This strong correlation supports the correctness of the calculations of absolute densities of tigers in six different plots. The average of the absolute density of tigers in six plots, which represents the average density in the entire Bangladesh Sundarbans, is 3.7 tigers/100 km².

Species captured in camera-trap photos in the Sundarbans East WLS

Camera-trapped species	Number of photographs	Number of individuals	Average number of individuals/photo
Tiger	7	7	1
Human	39	134	3.44
Spotted deer	606	1,063	1.75
Wild boar	128	175	1.37
Rhesus macaque	10	10	1
Lesser adjutant	8	8	1
Red junglefowl	6	7	1.17
Greater coucal	5	5	1
House crow	7	9	1.29
Jungle myna	1	1	1
Ring lizard	12	12	1
Total	829	1,431	1.73

Relative density of tiger population in the Bangladesh Sundarbans

Survey area	Location	Status	Relative density*	Absolute density of tiger#
1. Katka-Kochikhali	South-east	WLS	0.58 (± 0.12)	4.8
2. Hironpoint	South	WLS	0.51 (± 0.14)	4.2
3. Mandarbaria	South-west	WLS	0.54 (± 0.15)	4.5
4. Harintana	East-Central	RF	0.42 (± 0.17)	3.5
5. Chandpai	North-east	RF	0.29 (± 0.19)	2.4
6. Burigoalini	North-west	RF	0.32 (± 0.16)	2.6
Average Bangladesh Sundarbans			0.44	3.7

*Average no. of tracks/km riverbank (± SE)

#no. of individuals/ 100 km²

Notes: In Katka-Kochikhali the tiger density was estimated by camera-trap survey in the same plot where the relative density of tiger was estimated on the basis of track counts. Based on the correlation between the relative density and absolute density in this plot, the absolute density has been calculated for the other five plots.

Relative density of potential tiger prey in the Bangladesh Sundarbans

Survey area	Relative density of potential tiger prey (average no./km observation from river)					
	Spotted deer	Wild boar	Rhesus macaque	Lesser adjutant	Human	Total minus human
1. Katka-Kochikhali	3.97	0.17	2.42	0.09	0.85	6.7
2. Hironpoint	3.66	0.15	2.36	0.09	0.87	6.3
3. Mandarbaria	3.62	0.15	2.40	0.07	0.00	6.2
4. Harintana	1.18	0.13	1.18	0.05	1.09	2.5
5. Chandpai	0.76	0.11	0.83	0.03	4.83	1.7
6. Burigoalini	0.96	0.12	0.85	0.05	4.98	2.0
Average Bangladesh Sundarbans	2.36	0.14	1.67	0.06	2.10	4.2

During the period 2007 and 2012 (Tiger Survey Report, 2012), the enumeration approach in the Bangladesh Sundarbans relies on counting tiger track sets along the banks of *khals* (creeks) to give an index of relative tiger abundance for each of the surveyed sample units covering 80-95% of SRF across its four ranges Satkhira and Khulna (under the West Division), and Chandpai and Sarankhola (under the East Division) with total number of compartments 55.

The survey does not record the number of individual tigers because there is no direct evidence from the SRF to quantify the relationship between the number of tiger track sets/km of *khal* surveyed and the actual size of the tiger population. To allow for detection of changes in abundance, sample units need to be large enough so that they can be occupied by several tigers. A total of 65 sample units was created with a mean area of 63.3 ± 14.1 km^2 (range = 40-100 km^2), which means they are large enough to encompass at least more than one female tiger territory (Barlow *et al.*, 2008). Female tiger territories, however, may be smaller (range 12.2-16.2 km²) larger in different parts of the Sundarbans. Also, male tigers are likely to have larger territories, possibly three to seven times the size of females and transients can move over relatively large distances. However, more than one tiger may be operational in the same area at any given time.

There is strong evidence from a large-scale survey in India, that the frequency of tiger tracks is a very good indication of the actual tiger population (Jhala *et al.* 2011). Therefore, if track numbers increase, it can be assumed there has been an increase in the absolute tiger abundance and likewise, if track numbers decrease it can be assumed there has been a decrease in the absolute tiger abundance.

The first survey was conducted in 2007, followed by surveys in 2009, 2010 and 2011 and the results have been compared with those of the most recent survey conducted over three weeks in April, 2012. The SRF was divided into 65 sample units with an average size of 63 km² ± 1.75 SE based on preliminary data on tiger home ranges collected in the SRF (Barlow *et al.*, 2011). In 2007 (January-February 2007) a sample of 441 *khals* (average 8 km ± 3 SD khal/sample unit) was selected randomly within the 65 sample units, with each surveyed once over a two-month period. In 2011 (February to April), the survey was conducted over a sample of 2 or 3 randomly selected *khals* for each of the 65 sample units. A total of 156 *khals* were selected. All environmental conditions were roughly similar between survey years and surveys were conducted in the same season.

In 2012 a total of 224 tiger track sets were recorded from 853 km of surveyed creeks with an average of 0.24 tiger track sets/km of creeks surveyed/sample unit. Between 2007 and 2012 the index of relative tiger abundance (tiger track sets/km of creeks surveyed/sample unit) for the compared 52 sample units has declined by 69%. Furthermore, no tiger track could be detected in 11 surveyed sample units indicating a loss in habitat occupancy of up to 15% between 2007 and 2012. It was assumed that the current poaching level of deer may have a serious impact on the prey population, which in turn would have a negative impact on the tiger population.

In 2012 the highest relative tiger abundance was found in Satkhira range with a median 0.45 tiger track sets/km of *khal* surveyed/sample unit, followed by Khulna range with a median 0.27 tiger track sets/km of *khal* surveyed/sample unit. The lowest relative tiger abundance was found in Sarankhola range, with a median 0.21 tiger track sets/km of *khal* surveyed/sample unit, and Chandpai range with a median 0.10 tiger track sets/km of *khal* surveyed/sample unit.

The comparison of results for 52 sample units (80% of the SRF) between 2007 and 2011 revealed a 76% decrease in recorded tiger tracks sets/km of *khal* surveyed (from 924 to 224), where the lengths of surveyed *khal* decreased just slightly from 973 to 853 km (12%). Further analyses showed positive correlations between the lengths of *khals* surveyed and recorded tiger track sets/km of *khals* surveyed, indicating that in sample units with reduced *khal* survey lengths in 2012 lower relative tiger abundances were found, and that in increased *khal* survey lengths in 2009 relative higher tiger abundances were found. No correlations could be found for the 2007, 2010 and 2011 data sets.

The comparison of tiger track sets/km of *khal* surveyed/sample unit, revealed a highly significant 69% decline in the relative tiger abundance from 2007 to 2012. The 2007 median 0.77 tiger track sets/km of *khal* surveyed/sample unit initially increased in 2009 to a median 0.99 tiger track sets/km of *khal* surveyed/sample unit, but then decreased in 2010 to a median 0.45 tiger track sets/km of *khal* surveyed/sample unit, and respectively in 2011 to a median 0.50 tiger track sets/km of *khal* surveyed/sample unit. In 2012 it showed another decline to a median of 0.24 tiger track sets/km of *khal* surveyed/sample unit. Comparing the relative tiger abundances for the different ranges between years, a significant decline in relative tiger abundances in three of four ranges was found.

The highest decline was found in Chandpai range, where the relative tiger abundance decreased from a median 0.67 tiger track sets/km of *khal* surveyed/sample unit in 2007 to a median 0.10 tiger track sets/km of *khal* surveyed/sample unit in 2012. This is equivalent to an overall 85% decline in relative tiger abundance between 2007 and 2012 in the Chandpai range.

In the Sarankhola range, the relative tiger abundance decreased from a median 0.75 tiger track sets/km of *khal* surveyed/sample unit in 2007 to a mean 0.21 tiger track sets/km of *khal* surveyed/sample unit in 2012. This is equivalent to an overall 72% decline in relative tiger abundance between 2007 and 2012 in the Sarankhola range.

In the Satkhira range, the relative tiger abundance decreased from a median 1.47 tiger track sets/km of *khal* surveyed/sample unit in 2007, to a median of 0.45 tiger track sets/km of *khal* surveyed/sample unit in 2012. This is equivalent to an overall 69% decline in relative tiger abundance between 2007 and 2012 in Satkhira range.

Only in Khulna range the relative tiger abundance remained stable over the last 5 years, but with consistently low relative tiger abundance compared to the other three ranges. The relative tiger abundances differ significantly between areas, with highest relative tiger abundance in Satkhira range and lower relative tiger abundance in Chandpai and Sarankhola ranges.

Khan (2007) opined that different 'estimates' of the total number of tigers in the Sundarbans of Bangladesh mainly based on 'pugmark census' or interviewing, which are not scientific and does not fit to any of the conceptual framework of population sampling methods. The pugmark census assumes that tigers are individually identifiable from their pugmarks, which is not the case, so these 'estimates' cannot even be considered as indices of relative abundance.

Hossain *et al.* (2018) also reported that the tiger's relative abundance in Bangladesh Sundarbans was higher in 2007 and declined by 2011 with changes in landscape factors. In the 2007 survey, tiger tracks were recorded in all sample units, while in 2011 tiger tracks were not detected in nine sample units. Overall the median tiger tracks/km surveyed was 0.88 tracks/km $\pm$ 0.86 SD in 2007 and 0.47 tracks/km $\pm$ 0.64 SD in 2011. Overall 2011 had a significantly lower tiger track rate median compared to 2007 (Wilcoxon rank sum test with continuity correction, $W = 2900.5$, $p < 0.001$). In 2007, the median number of tiger tracks per survey between PAs and non-PAs was 1.76 tracks/km2 $\pm$ 0.62 SD and 0.76 $\pm$ 0.82 SD respectively. The difference was significant (Wilcoxon rank sum test with continuity correction, $W = 546$, $p < 0.001$). In 2011, the median number of tiger tracks per survey between PAs and non-PAs was 0.57 tracks/km^2± 0.50 SD and 0.48 ± 0.68 SD respectively.

Double-sampling method

This is the new method adopted by WII in the tiger census. The primary stage involved a ground survey by the FD. Under this FD staff collect evidence of the tiger's presence like pugmarks, scat, scratches on trees, or other such unmistakable signs of tiger presence. The next stage involves the camera trapping.

Camera trapping

The mangrove habitat of Sundarbans is unique. The normal approaches to tiger density estimation from camera trap population estimates are not applicable here. It is not possible to derive the effectively trapped area calculations from the usual half mean maximum distance moved by recaptured tigers. Therefore, home ranges are estimated from tagged tigers. The radius of home range is used to determine the effectively sampled area from the camera trap polygon to calculate density estimates from camera traps, which is applied to all tiger-occupied areas of Sundarbans. The extent and relative abundance of tigers throughout the TR is found through sign surveys in channels.

Due to the difficulty of walking in the Sundarbans mangrove forests and locating game trails for setting camera traps, camera traps could be deployed in a systematic grid based approach used across India. Instead, camera traps were set up at strategic locations, near fresh and brackish water ponds, using attractants to lure tigers to our camera stations. Initially, fishing nets were also used to orient the approaching tigers to get proper flank photographs for uniquely identifying each tiger from its stripe patterns.

The camera trapping technique using the mark-recapture framework is statistically more reliable than the traditional method of counting pugmarks. But, in practice, population estimates of tigers based on the above technique suffer from problems such as high cost of equipment (Tk 3.27 crores for 200 special cameras in Bangladesh), risk of camera theft and low precision of density estimates especially in areas of low tiger density because the technique relies on sampling tigers at only a few predetermined locations where camera traps are set.

Installation of cameras

Cameras may well be left in dense forests for several days to capture images of individual tigers. But impractical to put in cameras at every place that's likely to possess tigers, and even in places where they're installed, there's no certainty that the tiger would walk into a camera's range.

Supported the bottom survey locations were chosen for installing cameras. These cameras are heat and motion-sensitive. They lie idle till they detect any motion or a sudden change in temperature which implies they capture nearly anything that moves- other animals, even birds. All these get captured by the camera. Each tiger is understood to possess a really unique stripe pattern. This can be accustomed to differentiate one tiger from the other.

Indian Sundarbans

No tiger estimation was done in the Indian Sundarbans in 2006 as at that time, the new protocol for sampling this hostile and unique tiger habitat had not been developed. The second and third assessments were carried out in 2010 and 2014 which included the Sundarban tigers. The information generated by three earlier cycles of tiger status evaluation exercises resulted in major changes in policy and management of tiger populations and provided scientific data to fully implement provisions of the Wildlife (Protection) Act 1972, as amended in 2006, in letter and spirit. The major outcomes that were direct or indirect consequence of information generated by the monitoring exercises were-

(1) tiger landscape conservation plans,

(2) designation and notification of inviolate critical core and buffer areas of TRs,

(3) identification and declaration of new TRs,

(4) recognition of tiger landscapes and the importance of the corridors and their physical delineation at the highest levels of governance,

(5) integrating tiger conservation with developmental activities using the power of reliable information in a Geographic Information System database,

(6) planning reintroduction and supplementation strategies for tigers and (7) to prioritise conservation investments to target unique vulnerable gene pools (Kolipakam *et al*. 2019).

All these provide an opportunity to incorporate conservation objectives supported with sound science based data, on equal footing with economic, sociological, and other values in policy and decision making for the benefit of the society.

In the Sundarbans, the earlier enumerators relied on footprints alone. But the Sundarbans is a different tiger habitat. Here the tides wash away the footprints. The soft clay soil distorts the shape and size of the pugmarks. Human errors magnify the inbuilt mistakes. And, above all, in the sloppy, slushy silt, most pugmarks look like formless holes punched in the mud.

Even under logistic constraints and difficult terrain in the tigerlands like the Sundarbans, densities of tigers is derived annually using encounter rates of tiger-sign (e.g. number of tiger track sets or scats encountered per 10 km, the proportion of 1 km trail segments in which tiger tracks were detected, number of fresh tiger tracks crossing the path of a boat or vehicle). It appeared that during April even with strong spring tides and partly rainfall fresh track sets lasted to an average of 7.74 days.

This is, however, not translated into accurate predictions of tiger populations. It is argued that when the field staffs sailing the water, they cannot see beyond a certain point on the land and many areas thus remain uninvestigated. Thus, this is not a total count but only a tidal channel search and the inner mangrove forests were excluded due to lack of proper animal trails and fear of tiger attacks. In 1995, while conducting the census survey in the Sundarbans based on pugmark method one of the FD staff fell prey to a tiger. Fear of the tiger is so overwhelming that the pronunciation of 'Bagh' (tiger) is taboo.

The first study was carried out in STR by Karanth and Nichols (2000) from October 1998 to February 1999 and the results are shown in the table below.

Efforts	Results
Total number of trapping points	62
Sampling efforts	1,086 trap-nights
Number of sampling occasions	18
Camera trap polygon area	539.9 km^2
Estimated buffer width W$^\wedge$	2.00 km
Estimated sampled area A$^\wedge$ (W$^\wedge$)	832.0 km^2
Total of effective photographic captures of tigers	8
Number of individually identified tigers (Mt+1)	6
Catch per unit effort	1.10 tiger captures/100 trap-nights
Capture-recapture model used to estimate population size	Mb
Estimated number of tigers in the sampled area N$^\wedge$ (SE [N$^\wedge$])	7 (3.82)
Estimated animal density for tigers in the sampled area D$^\wedge$(SE [D$^\wedge$])	0.84 (0.46) tigers/100 km^2

Table: *Results of camera trap survey of tigers in STR during October 1998 to February 1999*

In 2010 camera trapping was done in only a few points on an experimental basis when the tiger population in the Indian Sundarbans was estimated in a mark re-capture framework with closed population estimators in an area of about 200 km^2. This set-up allowed us to estimate population size reliably. But due to the small number of camera stations (12) and uneven geographical spread of camera traps, it was not possible to obtain a reliable estimate of mean maximum distance (MMDM) moved by recaptured tigers nor use the spatially explicit models effectively. Models estimating effective trapping area attempt to estimate home range radius either by estimating MMDM or through centres of activity, in the case of the Sundarbans direct estimates of home ranges based on telemetry data were attempted. Home range radius was used from 95% fixed kernel area estimates of tiger home ranges as a buffer to the camera trap polygon for estimating effectively trapped area. The telemetry data suggested that though tigers do cross wide channels, crossing of channels >1 km in width was rare. A habitat mask was, therefore, used wherein channels >1 km in width were considered barriers to movement over the short term duration of the camera trapping exercise. 10 adult tigers and two cubs were photo-captured. The best model selected by CAPTURE was model Mh (incorporating individual heterogeneity) and the population estimate was 11 (se 3) tigers. The home range radius of four satellite-radio tagged and camera trapped polygons, giving an area of 438 km². After applying a habitat mask bounded by channels >1 km the effective camera trapped area was 257 km². Tiger density was computed to be 4.3 (se 0.3) tiger per 100 km². Since the tiger occupied area of the STR was 1,645 km² and the tiger signs were found throughout this area with a similar variation across STR as found within the camera trapped area, it would be possible to extrapolate this tiger density across the reserve without much loss of accuracy.

Ideally, 2-4 additional camera trap replicate areas need to be sampled and additional data from radio collared tigers are needed to provide more accurate and precise estimates of tiger density. But till these are obtained, this first quantitative assessment estimates the number of tigers to be around 70 (64 to 90) tigers for STR (in 1,645 km²). Further refinement in methodology, involvement of other institutions is needed and mention must be made that the 2010 estimate is subject to further study and by better methodology.

Mallick (2013) has recorded some interesting results. The most important outcome is that maximum tigers were sighted in Pirkhali block, but the frequency was highest here during the four months from January to April and then the sightings started reducing from May onwards recovering only in December. This may indicate a periodical fluctuation of population in this block. Another remarkable feature is that Netidhopani is the second important sighting area in the region, but here also the sighting records were not uniform throughout the year, but fluctuates during the rainy and winter months. Arbesi block is the third important sighting area. Here most sightings were recorded in December. On the contrary, the sighting record in the adjacent Panchamukhani block is low throughout the year. Matla block may be termed as very low in terms of sighting, where from August to December no sighting record was available. Experience is almost similar in Chamta, Chhotohardi, Chandkhali and Gosaba blocks. But in the southern blocks of Gona, and Bagmara, sighting was almost negligible and in Mayadwip it was nil. Jhilla, Khatuajhuri and Harinbhanga blocks were not very important in terms of sighting. So, in the northern belt of forest blocks the presence of tigers was mostly felt, whereas only one block in the central zone, i.e. Netidhopani, holds most of its residents throughout the year.

Unfortunately, the tigers are notoriously elusive, particularly so in the Sundarbans, for which direct sighting is rare. In most of the cases, only the loners were sighted and occasionally a pair of two (male-female during mating time or mother-cub during rearing). However, rarely, up to five tigers have been sighted in the Sundarbans.

Roy (2019) has given the sampling details of camera trapping exercises in STR during 2010-2014.

Year	Location	Effective trapping area in km²	Number of camera traps	Trap nights
2010	West Range	270 (SE range 251–291)	11	407
2012	West Range	518 (SE range 500–535)	29	1,073
2014	West Range	518 (SE range 498- 537)	76	2,763
	Sajnekhali Range	347 (SE range 328- 366)	40	960

The sampling efforts comprised 37 occasions (in both 2010 and 2012) and 57 in 2014. Number of unique tigers captured (Mt+1) was 10 in 2010, 22 in 2012 and 14 in 2014 in the West Range and 14 in the Sajnekhali Range in 2014. Tiger densities estimated by these models were 4.08 (SE 1.51) in 2010 and 5.81 (SE 1.24) tigers/100 km² in 2012. For 2014, tiger density was 3.15 (SE 0.88) in the West range while in Sajnekhali the density was 4.79 (SE 1.31).

24-Parganas (South) FD

Whereas the total forest area in South 24-Parganas FD is 1,679 km², during January-March, 2012 first camera trap survey outside STR was conducted over 982.56 km² forests in two ranges (Ramganga and Raidighi) under this division to know the habitats used by the tigers, estimate the minimum population and density of tiger (Das *et al.* 2012). Based on the presence of these pugmarks and scats, 41 heat motion sensitive cameras were laid in the Ramganga and Raidighi Ranges. The exercise for Ramganga was done between 23 January and 21 February, while trap cameras were laid and taken out in Raidighi between 11 February and 12 March.

This exercise has found the presence of 17 tigers. The exercise has photographed eight individual tigers in the Ramganga range and nine in the Raidighi range. The presence of tiger was also found in Dhulibhashani and Chulkathi compartments in Ramganga, while Ajmalmari and Herobhanga compartments in Raidighi range also reported good tiger density. A decent tiger density beyond STR shows that the Sundarban still holds the potential to sustain a healthy tiger population.

Density estimation of tigers in 24-Parganas (South) FD in 2012

Variables	Ramganga range		Raidighi Range	
	Estimation	Standard Error	Estimation	Standard Error
Total land mass	204 km²	-	192 km²	-
Effective Trapping Area	184.5 km²	-	141.3 km²	-
Camera points	20	-	21	-
No. of trap nights	600	-	714	-
Density Estimate with ETA#	4.3 individuals /100 km²	-	7.08 individuals /100 km²	-

Variables	Ramganga range		Raidighi Range	
	Estimation	Standard Error	Estimation	Standard Error
Density Estimate with total land mass	3.9 individuals /100 km²	-	5.24 individuals /100 km²	-
Density estimates in MLSECR*	3.8 individuals /100 km²	1.5	5.2 individuals /100 km²	1.7

#Effective Trapping Area

*Spatially Explicit Maximum Likelihood Methods

It is expected that the captured individuals within the minimum convex polygon were also present outside. The average capture probability per occasion was found to be 0.06 (p-hat). At Herobhanga RF, two individuals were captured but the data was not sufficient to estimate the range and density. Density was calculated conventionally considering the total landmass of the study area. To reduce the possibility of overestimation, density was also calculated through MLSECR.

Based on the camera trap study in South 24-Parganas FD conducted in 2012 it appears that these habitats have immense potential to harbour a healthy tiger population as the area has comparatively higher density of tigers in spite of being a RF. A wide array of factors is likely to contribute to this. For example, change in weather, lunar-tidal cycle, coverage of vegetation, plant type, water availability, human presence and prey-base activity impact the activity of the tiger in this unique landscape. Hence, this potential tiger habitat over 556 km² has been declared as a new PA, namely the West Sundarbans WLS.

SBR: Comparative analysis

The tiger population dynamics of SBR between 2012 and 2016 are discussed below in five different segments (Roy Chowdhury *et al.*, 2018).

24-Parganas (South) FD

In a total sampling effort during 2012-2013 [Ramganga 562 km² (20 locations x 30 days), Raidighi 229 km² (21 locations x 34 days) and Herobhanga 58 km² (two locations x 34 days)], 21 adult tigers were photo-captured. The next survey in 2014-2015 (73 locations x 41 days), carried out in the entire area, photo-captured 17 adult tigers and 4 cubs. A total of 10 individuals were common between these two years and 7 new adult individuals were identified. A total sampling effort (50 locations x 40 days) in 2015-2016, photo-captured 24 adult tigers, 2 sub adults and one cub. Among these 24 adult tigers, 18 individuals were common when compared with 2014-2015 and 6 new adult individuals were identified. A total of 3 tiger individuals found in 2014-2015 were not photo-captured in 2015-2016.

NP (East) Range

A total sampling effort (56 locations X 49 days) in 2012-2013 photo-captured 20 adult tigers. The next survey (60 locations x 39 days) that was carried out during 2013-2014, photo-captured 18 adult tigers. A total of 13 individuals were common between these two years and 5 new adult individuals were identified. In a sampling effort (60 locations x 19 days) during 2015-2016, a total of 14 adult tigers were photo-captured. Among these 14 adult tigers, 7 individuals were common with those photo-captured in 2014-2015 and 2 individuals were common when compared with the findings of 2012-2013. A total of 5 new adult individuals were identified. A total of 11 tiger individuals found in 2014-2015 were not photo-captured in 2015-2016.

NP (West) Range

A total of 21 adult individuals along with two cubs were reported by the FD during 2012-2013. The next survey (60 locations x 38 days) carried out during 2015-2016 photo-captured 19 adult tigers and 3 cubs.

Sajnekhali WLS

In a sampling effort (56 locations x 46 days) during 2012-2013, a total of 14 adult tigers were photo-captured. Out of these 14 adult individuals, one male was rescued and is now in Jharkhali rescue centre. In the next survey (60 locations x 30 days) carried out during 2015-2016, 9 adult tigers and 2 sub-adults were photo-captured. Among these 9 adult tigers, 5 individuals were common with those found in 2014-2015 and 4 new adult individuals were identified. A total of 8 tiger individuals, which were found in 2012-2013, were not photo-captured in 2015-2016.

Basirhat Range

A total sampling effort (56 locations X 31 days) during 2012-2013 photo-captured 13 adult tigers. The next survey (63 locations x 53 days) was carried out during 2014-2015 and photo-captured 14 adult tigers and 3 cubs. Out of these three cubs, two were found dead just after the exercise. A total of 7 individuals were common between these two years and 7 new adult individuals were identified. In a sampling effort (44 locations x 40 days) during 2015-2016, a total of 11 adult tigers were photo-captured. Among these 11 adult tigers, 7 individuals were common with those found in 2014-2015 and 4 new adult individuals were identified. A total of 7 tiger individuals, which were found in 2014-2015, were not photo-captured in 2015-2016.

The All India Coordinated Tiger Census was held in the Indian Sundarbans in two phases from December 5, 2021 to January 6, 2022, under STR area and from January 8 to February 10, 2022, under the jurisdiction of South 24-Parganas FD. The exercise was carried out by analysing data from 1,307 camera traps installed at 726 locations in SBR. In total, the ground survey covered 315 trails over a length of 1,339 km. The number of camera trapped tigers is 100 including four cubs. Cubs are not considered under the survey. Near about 108 tigers are found through Camera Trapping in the year 2022 – 2023, in ocular estimation of Phase – IV Tiger Estimation.

Month wise direct sightings and indirect evidences of adult tigers and cubs in STR during 1916-1917

Month	Direct sighting		Vocalisations heard	Sighting of fresh pugmarks	
	Adult	Cub		Adult	Cub
April	8	-	3	66	-
May	4	-	2	37	-
June	14	-	5	35	-
July	11	-	5	38	-
August	10	-	4	58	-
September	24	-	2	47	-
October	23	-	5	72	-
November	29	-	1	122	-
December	18	-	3	64	-
January	18	2	2	94	-
February	14	-	2	35	-
March	5	-	1	36	-
Total	178	2	35	704	-

Initiation in Indian Sundarbans

To demystify the mangrove tigers, for the first time in 2000, remote cameras were deployed in the Indian Sundarbans (Karanth and Nichols, 2000). But the scenario of camera trapping in Sundarbans is not like it is in other parts of the globe, where it has already been completed with success. The pioneering study based on camera trapping by conservationists was not able to give any convincing result about tigers- the study had limitations as camera traps were set up around freshwater ponds due to lack of forest trail in thick mangrove vegetation. On the basis of field surveys, carried out from October, 1998 to February, 1999, Karanth and Nichols (2000) reported the tiger density of 0.84 tigers/100 km². The photographs of six different tigers obtained by camera traps showed differences in stripe patterns that permit unambiguous identification of the individuals. But no estimates of prey abundance were provided. The data set of captures was small (SBT 101–106= 6 inclusive of a cub less than one year of age [SBT 105]). In the Sundarbans, periodic tidal phases are the biggest threat for camera trapping as high tides always create the risk of inundation of camera-trap equipment. Rough weather conditions are also an issue to address before starting similar investigations.

During this first camera-trap survey, 3-4 cameras were set around each sweet water pond (n= 15 [Jhingakhali, Jhillamukh, Burirdabri, Balkhali, Duttachera, Sajnekhali, Sudhanyakhali, Choragaji, Deulbharani, Dhopni, Chamta, Maraboni, Keorasuti, Haldibari and Begukhali), but they could not set traps at 14 other ponds within or surrounding the same polygon either for logistical reasons or disturbance by intruders with attendant risk of theft to the equipment. They made two estimates of the sampled area for Sundarbans: the first one using a 'small' value of W^ = 2.00 km and a second one using a 'large' value of W^= 5.00 km. The first value corresponds to a situation where the home range size of tigers is expected to be approximately 12 km², with mutually exclusive ranges; the latter value corresponds to a situation where the home range size is expected to be approximately 78 km², with possibly overlapping home ranges. The estimates of the area effectively sampled by our camera-trapping system, derived using these two alternative approaches are 832 km² and 1240 km², respectively. They termed these two values as the 'lower' and 'higher' estimates of sampled area size. The estimated mean tiger density N^ derived using the smaller sampled area estimate works out to 0.84 tigers/100 km², with a 95% confidence interval CI(N) of 0.84-3.60 tigers/100 km² (this higher density value is the one reported in the summary provided earlier). The corresponding values of lower tiger density derived using the larger estimate of the sampled area are N^= 0.56 tigers/100 km² and CI(N)= 0.56-2.42 tigers/100 km².

Karanth and Nichols (2000) further noted that the statistical inferences, on which the tiger population density estimates for the Sundarbans were derived, are relatively less robust than those they had made in other study areas, for the following reasons:

1. The capture probability per sampling occasion was relatively lower in the Sundarbans when compared to other study areas: Sundarbans $p= 0.077$ <$p= 0.11$-0.22 at other study areas except at Bandipur ($p= 0.055$).

2. The inability to select from among competing capture-recapture models due to lack of recaptures and relatively few individual animals captured. They could only use the simple removal model Mb with their data.

3. The lower number of tiger captures per unit effort (CPU) of 1.10 tiger captures/100 trap nights at the Sundarbans as opposed to CPU's of 14.5 in Kaziranga, 10.83 in Kanha, 8.69 in Ranthambore, 8.12 in Nagarahole, 5.32 in Pench, 4.94 in Bhadra and 3.37 in Bandipur. Only Namdapha with 0 captures/100 trap-nights had a lower catch per unit effort than the Sundarbans.

4. Problems were associated with the estimation of the sampled area due to the sampling design and the lack of data on distances between recaptures of individual tigers. If the sizes of home ranges were small and the ranges were exclusive in the Sundarbans (buffer distance <2.00 km), there might be a possibility that some individuals in the sampled area had zero probability of being photo-captured. In that case, the true population size in the sampled area might be higher than the estimated size, by a few individuals.

According to them, the data for Sundarbans were suggestive of a relatively low-density population with a mean tiger density (D^) value in the range of 0.6-3.6 tigers/100 km². The approximate prey densities required to support these tiger densities may vary in the range of 2.4-14.8 ungulate prey/km. According to them, the data for Sundarbans were suggestive of a relatively low-density population with a mean tiger density (D^) value in the range of 0.6-3.6 tigers/100 km². The approximate prey densities required to support these tiger densities may vary in the range of 2.4-14.8 ungulate prey/km². They were unable to obtain data on the prey densities for physical constraints. However, they suggested the estimated densities of major tiger prey species (cheetal, wild pig and rhesus macaque)possibly using thermal imaging or any other alternative technique. They were unable to obtain data on the prey densities for physical constraints.

To avoid controversies, the forest authorities handed over the 2004 survey data for statistical analysis to a government department dealing with statistical analysis. However, instead of adopting scientific procedures, the department published the report without consulting the forest officers. In 2006, Indian Statistical Institute claimed that the number of tigers is not more than 75, the report, however, was not accepted and the results were not brought out officially. The findings were full of errors. For example, a tiger cannot be simultaneously present at two locations at a distance of a hundred kilometres.

Accordingly, such monitoring began in STR from 5th January 2006 to 10th January 2006 (Phase I). Thereafter, WII deputed one research scholar to carry out "camera trapping". This exercise was done by him from July to September 2006. Later, thirty-five teams, comprising 250 forest personnel and representatives of NGOs, took part in the next phases of ecological monitoring. According to the census in 2008, occupancy of tigers in the Sundarbans was about 1,586 km², though the number of occupants could not be assessed due to lack of reliability. The exercise in 2008 was unsuccessful due to difficult terrain of the mangrove forest. Hence, since 2004, the tiger population in western Sundarban was debatable and could not be known for the two year's time series (2006 or 2008). However in 2010, WII used home-range information obtained from radio-collared tigers and combined this information with camera-trapping data.

In STR, during the year 2010-2011, an attempt was made to estimate the tiger population and density by using camera (Moultrie Digital) traps in a mark recapture framework with closed population estimators at Netidhopani and the area covered was 220 km². Since the Sundarban is a unique habitat with six hourly tidal effects, it is extremely difficult to locate regularly used game trails for setting the camera traps. Therefore, it was decided to use lures and baits (meat or egg) to attract the tigers to camera traps. Due to this limitation the cameras could not be set systematically across the study area but were sparsely spaced near attractants of fresh water ponds. In all, 102 photos were taken using camera traps recording the presence of 12 different tigers (10 adults and two cubs). But, due to limited sampling, a reliable estimate of mean maximum distance moved could not be made for density estimation. Nor the camera configuration provided a robust design for using the modern approaches of spatially explicit likelihood and Bayesian approaches to density estimation. After the study, density was estimated at 4.3 individuals/100 km^2 and the number of tigers estimated to be 70 within a range of 64-90 tigers in STR after extrapolation (Jhala *et al.* 2011).

In 2011, the first initiative was taken when WWF-India, Sundarbans programme, entered the forest of Lothian WLS, located at the southern corner of SBR. The entire division was divided into 4 km^2 grid, each grid having a pair of cameras facing each other for better and additional frames that will help identify the animals being captured. The cameras run on battery and are fitted with heat and motion sensors. They switch on automatically when animals tread near them. The cameras need to be installed in places having a reasonable possibility of being crossed by the animals. Relatively high ground is required to install them so that these are not damaged by the high tides. Moreover, the camera traps have to be installed at a height of less than four feet to shoot the animals because they generate infra-red rays that turn the camera on whenever they hit an obstacle. Recapturing is the essence of the exercise. It helps know how frequently a tiger has been passing a particular area. Even though it usually takes three weeks to recapture an animal, it could take up to a month in the Sundarbans.

As a part of the study, first camera traps were deployed on an experimental basis. However, the exercise was washed out with loss of 20 units of camera traps due to inundation by high tides as an effect of the 'supermoon'. This huge loss is a reminder of the supremacy of nature and the might of the tides. It is necessary to understand the tidal cycles before preparing the final design of a camera-trapping exercise.

From the learning and experience in Lothian, the concept of 'highest point of high tide' was established, which states that camera-trapping should be started during the high tides because it facilitates finding places with low risk of submergence. With the experiences gained in Lothian, in January 2012, camera traps were first deployed in the tiger forest by the WWF staff in 24-Parganas FD of SBR. Sometimes, a strange smell in the air and fresh pugmarks on ground during camera deployment confirms the presence of the felid species. In each trap station, olfactory attractant is applied as it is safe to attract a tiger with the use of olfactory lure. Otherwise, trying to track a tiger in Sundarbans is very dangerous, probably impossible.

Eventually, the camera-trapping exercise in the RFs became the biggest breakthrough in tiger conservation history in Sundarbans. Safe recovery of 98% camera traps after the entire exercise and the estimate of a minimum of 20 tigers from 24-Parganas FD explains the success of the study (Das *et al*. 2012), as the area was earlier believed to be a non-tiger habitat. This unique study provides necessary information to NTCA to establish standard protocol for camera trapping in Sundarbans and plays a vital role in the declaration of the 556 km^2 area of 24-Parganas South FD as 'West Sundarbans WLS'.

Ultimately, SBR has adopted the method of camera trapping, and the assessed figure under new methodology is available for 2014, i.e. 76 (62-96) tigers. In 2016-17, 87 tigers were photo-captured after an exercise conducted in the entire mangrove habitat. This camera-trapping study not only gives the numbers of tigers, but also investigates the health of the dense mangrove forest.

Results of camera trap survey of tigers in the Sundarban landscape since 2012

Year	Minimum number of adult	Minimum number of cub	Total	Remarks
2012-2013	89	2	91	Survey in entire Sundarban by WWF
2013-2014	62	-	62	All-India Tiger Estimation, 2014
2014-2015	31	7	38	Survey in 24-Parganas (South) FD and Basirhat Range of STR
2015-2016	81	4	85	Survey by WWF in entire landscape
2016-2017	87	4	91	Survey in entire landscape

Demography of tiger individuals captured in SBR (2016-17 to 2018-2019 and July 2021) is tabulated below.

Unit	2016-2017	2017-2018	2018-2019	July, 2021
STR				74
Sajnekhali WLS	14	15	10	
Basirhat Range	14	13	19	
NP (West) Range	19	18	20	
NP (East) Range	16	20	24	
24-Parganas (S) FD	24	22	23	22
Total	87*	88	96	96

*Two individuals captured in Basirhat Range were recaptured in Sajnekhali WLS (89–2= 87) and cubs were excluded from the total count.

Sampling efforts for ground surveys in the Indian Sundarbans during 2022 all-India tiger estimation, Phase I, conducted from December 5, 2021 to January 6, 2022, in STR and from January 8 to February 10, 2022 in the 24-Parganas (South) FD

Number of trails	Total length (in km)	Number of plots	Images of individual tigers captured
315	1,339	595	100

Initiation in Bangladesh Sundarbans

Khan (2012) conducted a camera-trap survey in the Sundarbans East WLS (total area of 312 km², covering only the southern part of the sanctuary, for more than 200 days from October, 2005 to January, 2007. The results were extrapolated from the core study area in Katka-Kochikhali, southeastern Sundarbans, to five additional sites- Hironpoint, Mandarbaria, Harintana, Chandpai, and Burigoalini (each survey site approximately 170 km²)- using indices of abundance. With the use of 10 camera-traps at 15 trap-points, field data provided a total of 829 photos, including seven photos of five individual tigers. A total of 5.0 (SE = 0.98) tigers (adults and sub-adults) are thus estimated in the core area with an estimated density of 4.8 tigers/100km². Distance sampling surveys conducted on large mammalian prey species obtained an overall density estimate of 27.9 individuals per km² and a biomass density of 1,037kg per km². By combining the estimates of absolute density with indices of abundance, an average of 3.7 tigers per 100 km² across the region was estimated, which given an area of 5,770 km², would predict a minimum of approximately 200 tigers in the Bangladesh Sundarbans.

Later, Bangladesh-India Joint Tiger Census Project conducted the tiger census 2015 examining some 1,500 images and footprints of tigers taken from the Sundarbans through camera trapping and found the horribly low figure of tigers. Experts observe that the loss of habitat, unchecked wildlife poaching, animal-human conflict in the forest and lack of management of forest are the main reasons behind the rapid fall in the tiger population.

According to the FD's data, at least 49 tigers were killed in the last 14 years (2001-2014) since the illegal poaching of wildlife and tiger-human conflict is on the rise in the Sundarbans.

In the first phase of the Bangladesh-India joint tiger census project, completed in April this year beginning November 1, 2013, some 89 infrared cameras were used to capture tigers' movements within a 3,000 km² area in the Bangladesh part of the Sundarbans. The second phase of the tiger census project using camera trapping methods began on November 12, 2014.

Aziz *et al.* (2017a) recorded 121 tigers (95% CI: 84-158 tigers) for the total area of the Bangladesh Sundarbans and provided a density estimate of 2.85 ± SE 0.44 tigers/100 km² (95% CI: 1.99-3.71 tigers/100 km²) for the areas sampled (Satkhira RF Block 342 km², West WLS 715 km², Chandpai RF Block 554 km², East WLS 383 km² = 1,994 km²) between November 2014 and February 2015.

Current Indo-Bangladesh status

As per the camera-trapping distance sampling (CTDS), the current tiger population in the mangrove landscape is 210 (114 in Bangladesh in 2018 and 96 in India in 2021). Mannan *et al.* (2023) also recorded the current tiger population as 114. However, Aziz *et al.* (2019) estimated a total population up to 254 in the contiguous mangrove forests of both India and Bangladesh.

Tiger cubs are categorised as per age- (a) 2-3 months, (b) 4-5 months, (c) 5-8 months and (d) juvenile (1-1.5 yr) in relation to their mothers. Cubs at and above mid-shoulder height of mothers (d) were considered for tiger population estimation since 2006.

Factors affecting the wild tiger status are:

- targeted killing on account of Human-Tiger Conflicts (HTC)
- poaching owing to vulnerable geographical location, with proximity to several international borders
- depletion of prey base of on account of subsistence poaching
- forced forest resource dependency of locals for want of livelihood options, viz., wood and non-timber forest produce
- proximity to human-dominated land parcels.

The ecological integrity of tiger habitat is also threatened by rise in sea-level on account of climate change, apart from reduced availability of fresh water owing to diversion.

STR

A comparative estimate of tiger population in STR during 2016-2017 and 2019-2020 is shown below.

Forest Range	Year wise number of individuals and variation		
	2016-2017	2019-2020	Number increased (+) or decreased (-)
Sajnekhali WLS	14	10	-4
Basirhat	14	19	+5
NP (West)	19	20	+1
NP (East)	16	24	+8
24-Parganas (South) FD	24	23	-1
Total	87*	96	Growth rate: 0.10

*Two individuals captured in Basirhat Range were recaptured in Sajnekhali WLS (89 - 2 = 87) and cubs were excluded from the total count.

During 2016-2017 the following direct and indirect evidences of tiger in STR are recorded:

Month/Year	Direct sighting		Vocalisations heard	Adult pugmarks
	Adult	Cub		
April 2016	8	-	3	66
May 2016	4	-	2	37
June 2016	14	-	5	35
July 2016	11	-	5	38
August 2016	10	-	4	58
September 2016	24	-	2	47
October 2016	23	-	5	72
November 2016	29	-	1	122
December 2016	18	-	3	64
January 2017	18	2	4	94
February 2017	14	-	2	35
March 2017	5	-	1	36
Total	178	2	35	704

Out of 96 tigers, 23 were identified as male and 43 as female, while the sex of 30 adults could not be determined. The survey also revealed the presence of 3 tiger cubs.

SBR

23 tigers were found in 24 Parganas (South) Division on the western side of STR, 73 adult tigers were recorded inside the four divisions of the Sundarban STR. There are individual tigers that use both sides of the border in case of India and Bangladesh.

Tigers do not follow the same route very often. Tigers prefer core habitats and generally avoid areas with high edge density. On the Indian side, tiger sign and sighting encounter rates were higher, especially in the core zone of the NP as compared to other areas.

Bangladesh

The Bangladesh Sundarbans, by virtue of its larger area (4,832 km²), has few more tigers compared with the Indian side (1,841 km²), but the scientists, after collating findings at the mangrove forests of both countries, have said that the Indian Sundarbans has four tigers per 100 km², almost twice that of the Bangladesh side (2.17 tigers per 100 km²). The reason is poaching of the tiger and the prey animals. Between 2001 and July 2020, altogether 38 tigers died in the Bangladesh Sundarbans, 22 in East Division and 16 in West Division. Aziz *et al.* (2017b) recorded six poisoned baits set to kill tigers and 1,427 snare loops in 56 snare sets to kill tiger prey in four sites (SB = Satkhira Block, CB = Chandpai Block, WS = West WLS, ES = East WLS). The only tiger poaching method detected was poisoned baits. Spotted deer, the principal prey of Sundarbans tiger, was used as bait in all cases. Poisoned bait was typically attached to a tree trimmed to the approximate height of a tiger, and placed next to tiger trails (indicated by tiger tracks). A single spotted deer was used to create two bait stations, with body parts being prepared by removing the intestines, dismembering, skinning, and coating in poison identified as a carbamate pesticide (Carbofuran). Of the six tiger poison baits, four were recorded in the WS and two in the ES. No bait stations were found in the CB and SB. Overall higher tiger and prey poaching activities were recorded in WS (37%) following ES (25%), SB and CB (19%) areas. The occurrence of poaching activity was significantly positively correlated with the distance from forest guard posts and significantly positively correlated with the distance from the nearest river. In 2013 in the SRF a tiger was seen with the loop of a nylon snare tightly constricting its forearm. Another tigress was rescued from a village adjacent to the SRF in 2012 that had probably escaped from a prey snare and had lost its right hind leg.

The previous census using camera traps, however, has identified the tiger population to be between 83 and 130 individuals using 63 tiger's presence points in the Bangladesh Sundarbans (Dey *et al.*, 2015).

The encounter rate of human sign and sighting was higher in Bangladesh Sundarbans which is further exacerbated by the usage of river channels for transportation of commercial vehicles. The tiger population in the Bangladesh Sundarbans is much below the actual carrying capacity, while the Indian side has reached carrying capacity. The findings point to better management in the Indian Sundarbans.

The area weighted tiger density from eight camera trapped blocks [Bangladesh: Block I Sarankhola (309.18 km²), Block II Satkhira (366.44 km²), Block III Khulna (588.08 km²), India: Block IV Basirhat (325.4 km²), Block V Ramganga (228.76 km²), Block VI NP East (485.45 km²), Block VII NP West (420.33 km²) and Block VIII Sajnekhali (188.51 km²)] was 2.85 (2.10-3.61) tigers/100 km². The highest tiger density was recorded in Block VIII (Sajnekhali) while the lowest density was recorded from Block III (Khulna).

Tiger density was found to have a positive relation with tiger sign intensity ($r = 0.464$; $P = 0.008$); and encounter rate of prey ($r = 0.447$, $P = 0.012$), while it had a negative relationship with encounter rate of human disturbance ($r = -0.554$, $P = 0.0012$).

Parameters	Satkhira	Khulna	Sarankhola	Overall
Total number of camera trap stations	253	96	187	536
Number of days sampled	102	69	78	249
Number of such stations captured tiger images	146	21	98	265

Parameters	Satkhira	Khulna	Sarankhola	Overall
Total number of tiger images captured	1,675	78	713	2,466
Total identified adult tiger individuals	36	4	23	63
Male tigers	6	1	2	9
Female tigers	22	3	19	44
Sex-unidentified tigers	8	0	2	10
Number of cubs	2	0	3	5
Number of juveniles	3	0	1	4
Maximum occasion of captures in a station	14	9	9 (n=2)	14
Maximum individual captured in a station	3	3	4	4
Maximum capture of an individual	30 (n=2)	18	33	33
Maximum relocations of an individual	21	12	22	22
Total camera-trap nights	10,965	4,762	8,681	24,408
Minimum camera-trap area (km²)	1,208	165	283	1,656
Effective camera-trap area (km²)	1,421	516	1,405	3,342

Table: Summary of the camera trapping efforts and outputs in three forest blocks of Bangladesh Sundarbans in 2018 (Bangladesh FD, 2019)

Sample block	Year	Area (km²)	Individuals	Density/100 km²
Satkhira	2018	1,208	36	2.74±0.46
	2015	366	13	2.77±0.78
Khulna	2018	165	4	1.20±0.64
	2015	588	7	1.08±0.04
Sarankhola	2018	283	23	3.33±0.71
	2015	309	18	3.70±0.91
Overall	2018	1,656	63	2.55±0.32
	2015	1,265	38	2.17±NA

Table: *Comparison of sample-area wise and overall tiger density of 2018 with those of 2015 (Dey et al., 2015)*

NA: Not available

Parameters	Basirhat	Sajnekhali	NP East	NP West	Ramganga
Minimum bounding polygon area in km²	325.4	188.51	485.45	420.33	228.76
Total number of camera trap stations	56	40	60	76	30
Number of days sampled	31	23	31	32	36
Total number of tiger individuals captured	14	14	20	14	5
Density of tigers per 100 km²	1.08	4.79	3.77	3.15	1.57

Table: *Summary of the camera trapping efforts and outputs in five forest blocks of Indian Sundarbans (Jhala et al., 2016)*

Comparative estimates of tiger densities in Indian Sundarbans during 2012-2015 (Roy Chowdhury *et al.*, 2018)

During the period January 2012 - April 2013, to estimate tiger numbers, WWF-India and SBR collaboratively carried out camera trapping sampling in SBR to establish a population baseline for the Sundarbans. In 24-Parganas (South) FD, the exercise was carried out from January 2012 to March 2012. Twenty unique individuals were identified from this division and range wise density was calculated. Tiger density was estimated to be 3.8 (±SE1.5)/100 sq. km. and 5.2 (±SE1.7)/100 km² for the Raidighi Range. In the NP (East) range, the exercise was carried out from November 2012-January 2013. 21 unique individuals were identified from this range. Tiger density was estimated to be 3.69(±SE0.82)/100 km². In Sajnekhali WLS, the exercise was carried out from January – March 2013. 13 unique individuals were identified from this range. Tiger density was estimated to be 2.36 (±SE 0.64)/100 km². In Basirhat Range, the exercise was carried out from March 2013 to April 2013. Thirteen unique individuals were identified from this range. Tiger density was estimated to be 2.57 (±SE 0.76)/100 km². In session 2013-2014, camera traps were installed in Ramganga Range of 24-Parganas (South) FD and in NP (East) Range of STR. The same dataset was used in the All India Tiger Census. Five unique individuals were identified from Ramganga Range. The estimated tiger density was 1.57 (±SE 0.74) individuals/100 km² for Ramganga Range. Twenty unique individuals were identified from the NP (East) Range. The estimated tiger density was 3.77 (±SE 1.03) individuals/100 km² for the NP (East) Range. Camera traps could only be deployed in 20 locations in Basirhat Range for a short period of time due to rough weather conditions. The data from this exercise was not taken into account. In 2014-2015, camera traps were again placed in Basirhat Range of STR and in the entire tiger habitat of 24-Parganas (South) FD comprising three ranges (Ramganga, Raidighi and Herobhanga Range). Seventeen unique individuals along with four cubs were identified from 24-Parganas (South) FD. The estimated tiger density was 3.42 (±SE 0.74) individuals/100 km² for 24-Parganas (South) FD. Sixteen unique individuals along with three cubs were identified from Basirhat Range. The estimated tiger density was 3.33 (±SE 0.09) individuals/100 km² for the Basirhat Range.

A detailed data metrics showing total efforts and capture summaries during the period December, 2015 to April, 2016 in SBR is shown below following Roy Choudhury *et al.* (2018).

Parametres	24-Parganas (South) FD	NP (East) Range	NP (West) Range	Sajnekh ali WLS	Basirh at Range
Total area (km²)	454	850	890	430	466
Total grids of 4 km² each	116	155	138	103	94
Camera trap grids	50	60	60	60	44
Trap Night Efforts	2,000	1,140	2,280	1,800	1,760
Trapping area [km²] (Extent of trapping grid)	200	240	240	240	176
Total camera trap stations with tiger captures	37	23	38	27	16
Tiger Individuals Identified					
*Previously encountered	18	9	-	5	7
New	8	5	19	6	4
Total	26	14	19	11	11
Grand total	81*				

*Out of these 81 individuals four individuals were sub adults (less than 2 years old) and were omitted from analysis. No individual was found common between the ranges. A new individual means it has never been encountered in previous camera trapping sessions.

Unit (Range)	Adult Male	Adult Female	Unidentified	Total Adult	*Cub and Sub-adult
24-Parganas (South) FD	9	15	0	24	3
NP (East)	2	12	0	14	0
NP (West)	4	13	2	19	3
Sajnekhali WLS	1	6	2	9	2
Basirhat	3	6	2	11	0
Total	19	52	6	77	8

Table: *Demography of tiger individuals in SBR*

*A tiger individual is considered as a cub or sub-adult if less than two years old (pre-dispersal juveniles).

Most of the sampling blocks (ranges) are separated by large and wide river channels. No individual was found common between the ranges although some crossed from one range to another over the study period (December, 2015 to April, 2016). Capture of four cubs and four sub-adults (less than two years old) were not incorporated into density estimation, following the recommendation of Karanth and Nichols (2002). Sub-adults and cubs were classified after comparing the dataset across sessions. Because left and right flank pictures for all tigers are not available, unique individuals were identified using all available pictures of both flanks and the right flank, while omitting left flank only pictures.

In 24-Parganas (South) FD there were no tiger captures on 22.5% occasions. A total of 27 unique tigers were identified. Out of these 27 individuals, 3 individuals- SB110 (Sub-adult male), SB111 (Sub-adult male) and SB118 (a cub) were not used for analysis as all of them were less than two years old. The capture history file used for data analysis had 86 records of tiger captures of 24 tigers. This included fifteen captures of SB90, thirteen captures of SB83, seven captures of SB92, six captures of SB10, four captures of SB7, SB13, SB18 and SB114, three captures of SB14, SB91 and SB94, two captures of SB20, SB93, SB112, SB113, SB115, SB116 and SB117 and single captures of SB3, SB9, SB15, SB96, SB97 and SB98.

In the NP (East) Range there were no tiger captures on 46.7% occasions. A total of 14 tigers were individually identified. The capture history file used in data analysis had 35 records. This included six captures of SB21, five captures of SB44, four captures of SB123, three captures of SB38 and SB120, two captures of SB35, SB40, SB86, SB87 and SB119 and single captures of S24, SB84, SB121 and SB122.

In the NP (West) Range there were no tiger captures on 10.5% occasions. A total of 22 tigers were individually identified. Out of these 22 individuals, 3 were cubs (SB153, SB154 and SB155) and not used for analysis as all of them were less than one year old. The capture history file used for data analysis had 93 captures and recaptures. This included thirteen captures of SB134 and SB140, eleven captures of SB139, eight captures of SB141, seven captures of SB138, six captures of SB135 and SB142, five captures of SB144 and SB152, four captures of SB137 and SB 143, three captures of SB 148, two captures of SB 136 and single captures of SB145, SB146, SB147, SB149, SB 150 and SB151.

In Sajnekhali WLS there were no tiger captures on 36.7% occasions. A total of 11 tigers were individually identified. Out of these 11 individuals, 2 individuals — SB132 (Sub-adult female), and SB133 (Sub-adult female) were not used for analysis as all of them were less than two years old. The capture history file used for data analysis had 37 captures and recaptures. This included ten captures of SB48, seven captures of SB57, six captures of SB124, five captures of SB64, three captures of SB 129, two captures of SB49 and SB 131, and single captures of SB50 and SB130.

In Basirhat Range there were no tiger captures on 50% occasions. A total of 11 tigers were individually identified. The capture history file used for data analysis had 31 captures and recaptures. This included six captures of SB61, four captures of SB124, SB125 and SB126, two captures of SB69, SB72, SB104, SB105, SB106 and SB 127 and a single capture of SB101.

Tiger mortalities

Tiger mortality often goes undetected and unreported even in terrestrial habitats that are more accessible than the mangroves. Extensive use of snares to trap tiger-prey and a few snares for tigers was reported in the Bangladesh Sundarbans. Reliable information on tiger and prey poaching is lacking.

Undivided Sundarbans

Between 1881 and 1912 more than 2,400 full grown tigers were killed by the hunters in the area under official patronage, i.e. rewards (Chakrabarti, 2010). The author has not given any break up, but put up a note that he derived this figure from the Annual Reports of the FD, which do not take tally of those killings that took place outside the forest area or were not reported to the Department.

There is another record of the period 1881-1916, giving a figure of 150 tigers killed or trapped in the undivided Sundarbans, the year wise break of which is shown in the table below (Gupta, 1966).

Year	Number of tigers killed
1881-1882	3
1883-1884	8
1907-1908	9
1911-1912	61
1914-1915	36
1915-1916	33

Table: *Tiger killing records in the undivided Sundarbans during late 19th and early 20th century*

Kailash Sankhala in his book 'Tiger' writes that 86 tigers were shot in the Sundarbans alone in 1914. During his visits to the Sundarbans, Sankhala observed fresh pug marks of tigers in every part of the forest. According to another source a total of 1,259 tiger deaths were recorded from 1881 to 2006: 233 tigers in Bangladesh over 42 years; 6 tiger deaths in India over 6 years;

and 1,020 tiger deaths over 33 years for the whole Sundarbans, where country was not specified (Barlow, 2009). Curtis (1933) recorded that FD killed 452 Bengal tigers from 1912 to 1921 (Curtis, 1933) and 269 tigers were killed between 1947 and 1971, or an average of 11.2 tigers per year (Salter, 1984).

Indian Sundarbans

Table: Straying tigers killed by the fringe villagers during the period 1990-2001

Date of Occurrence	Village	Civil block	Remarks
08.12.1990	Dayapur	Gosaba	One male killed
23.01.1993	Hemnagar	Hingalganj	One male electrocuted
05.01.1994	Sudhanyakhali	Gosaba	One male killed; floating carcass detected by a private launch.
12.05.1994	Jamespur	Gosaba	One male killed in self-defence
26.09.1994	Hemnagar	Hingalganj	One male poisoned; carcass found in a paddy field
05.11.1994	Jamespur	Gosaba	One male carcass found in a paddy field
08.03.1995	Luxbagan	Gosaba	One male killed
29.07.1998	Rajat Jubilee	Gosaba	One male poisoned; carcass found in a paddy field
29.08.1998	Kalitala	Hingalganj	One male killed.
06.03.1999	-	Gosaba	One male detected by fishermen in river Raimangal.
30.07.2001	Pakhiralaya	Gosaba	One male killed

Date of Occurrence	Village	Civil block	Remarks
02.10.2001	Kishorimo-hanpur	Pathar-pratima	Killed
03.12.2001	Tush khal, Khatuajhuri-1	Basirhat range	One male killed
15.12.2001	Kumirmari	Gosaba	Killed

Some of the straying tigers in the reclaimed areas died due to poisoning or electrocution. This negative attitude is intensified for their resentment against the strict enforcement of the laws regarding entry into the forests. In the late 20th century, at least 12 tigers were killed by the villagers. Sometimes thousands of people from the neighbouring villages gather to drive the straying tiger back to the forest by using drums, crackers, fire, etc. Sometimes the tiger goes back to the forest on its own. The forest staff and local Panchayat members always try to protect the animal and persuade the villagers not to harm the animal. But, when the mob becomes arrogant, the situation often goes beyond control and they kill the strayed tiger. Sometimes all the evidence of killing was quickly removed or destroyed by the villagers to avoid punishment.

On the night of 26th August, 1998 one tiger strayed into the village of Kalitala opposite to Jhingekhali station of STR and killed two cows in a cowshed. Patrolling was intensified to prevent the re-entry of the tiger into the village. Subsequently, on 29th the body of a dead male tiger was found floating in a pond at Kalitala village and the carcass was subsequently recovered, post-mortem was done and the visceral parts were sent for forensic examination. The case was reportedly of poisoning using Endosulfan. The poison was mixed in the half-eaten body part of the cows which the tiger had eaten upon re-entering the cow shed on the night of 28th.

Three incidents of 29th July, 2001 at Pakhiralaya, 2nd October, 2001 at Kishorimohanpur and 15th December, 2001 at Kumirmari in the transition zone of STR were the last cases of such retaliatory killings of tigers by the villagers. Previously, in isolated cases, the straying tigers were killed by the villagers in self-defence, although all the straying animals were not man-eaters.

As recorded above, on 30th July, 2001 a male tiger was killed at Pakhiralay village. He crossed the river Gomor and entered the tree patch of Dayapur village at night on 29th July and again crossed the other part of the river and finally took shelter in a thick plantation of *Avicennia* on the village side, which is about one km away from Sajnekhali WLS. On 30th staff of Pakhiralaya Range went to the spot near the ferry *ghat* but were blocked by the mob. With the help of the Panchayat members they entered the village but could not detect the tiger in the plantation area. On 31st a thorough search was undertaken and then a gunny bag containing few bricks and hind portions of the tiger was recovered from the river side, which is about 200 m from the place of occurrence and near the forested part of the village. An FIR was lodged at Gosaba PS. The post mortem of the tiger was done by a team of veterinary officers. They found that the anterior part of the tiger had three smooth-edged spherical puncture wounds on the head, ear and neck region, which was doubted to be because of gun-shot wound, and lesions that might have developed as a result of traumatic injury by some blunt object.

It appeared that a few poachers had taken away the tiger's tail, tongue, canine teeth, paws, claws, lungs, liver, kidney, testis, etc. Thereafter they removed all the signs of evidence from the spot. They cut the body into two pieces, put the anterior part (head with part of front legs and body) and the hind part with legs in two separate gunny bags along with bricks, tied with ropes and threw those bags in the river. Three persons were, however, injured by the tiger and hospitalised. The police arrested them.

On 3rd December, 2001 a male tiger was detected by the patrolling staff near the Tush khal, about 100 m wide. The highly rotten carcass was found about 20 feet away from the bank. The body was brought to Khatuajhuri camp and measured 1.40 m from skull to tail base. The hind and soft portions containing testis were eaten up by the scavengers like water monitor, etc. The skin on the body was intact except the skull portion. All the canines were removed from the socket. No abnormal perforation of the bullet mark was noticed on the entire body. The body was under putrefaction and full of maggots. The tiger appeared to have died 10-12 days ago. The forest site was composed of mixed vegetation of *Excoecaria* and *Aegiceros* trees. There were dragging signs on the field and blood-stains were found within a distance of about five-six feet from the place where the carcass was found. Following samples were also found at the place: 1. Some detached lower mandible teeth; 2. Four-six pieces of scats of tiger; 3. Deep tiger claw marks were found on one *Excoecaria* tree. Incidentally no human foot-print were found in the area. The post mortem of the body was done on 4th December and since the abdominal region was found to be almost decomposed or eaten up only parts of liver, skin and hind limb bone were collected for forensic examination.

A tiger carcass was found by the forest staff at Narayantala pond of Gosaba-3 block under NP (West) Range on 9th May, 2000. The body was found inside the bushes of *Ceriops* and *Tamarix*. The body was almost decomposed without having any visceral or muscular parts. The skin was all dried up and some traces of skin were found on the skull; however, hair was found scattered all over the site. Skin was also found on patches on the limbs, tail, rib bones etc. Several bones were found detached from the skeleton due to tidal action of the water. The length of the tiger from skull to the tip of the tail was 1.95 m and the tail length of the available portion was 0.80 m. No sign of gunshot was found on the body. The molar teeth were found to be old and worn out and were yellow in colour which is indication of old age. Bone marrow in vertebral columns and fluid in limb bones, major joints were found to be completely dried up and this indicated that the carcass appeared to be a male tiger. The post-mortem of the body was undertaken on 11th May. This case appears to be a natural death due to old age.

On 4th November, 2001 a report of a roaring tiger in a trap laid by a poacher in Arbesi-3 was received by the station officer at Jhingakhali. The secret information was received from the adjacent Kalitala village. The station officer reached the spot but could not get down at night due to high tide. Early morning the next day the staff could locate an *Excoecaria* tree of about 10 feet height of which all branches were broken by the tiger as there were pugmarks of a tiger and one small piece of nylon trap rope was also traced about half a km inside the forest. In all 71 numbers of traps made of nylon rope could be located.

In fact, one tigress was trapped and could escape by releasing herself from the rope. Incidentally there was also evidence of the presence of a male tiger that probably visited the tigress when she was trapped. The pugmarks of both the male and female tigers were taken. It was found that the poachers of Kalitala had fixed nylon rope traps for deer poaching in which a tiger was accidentally trapped. Poachers on seeing the tiger ran away and the information was leaked. The nylon rope trap was not strong enough to hold a tiger.

On receipt of information from the Hentalbari FPC in the buffer area of STR under Basirhat range that traps were laid for prey species, a forester went to Banabibi *bharani* of Jhilla-3 on 20th February, 2002 along with the FPC members. The area was very close to rivers Duttar and Melmel as well as the villages like Imlibari and Hentalbari. After searching for about an hour isolated traps made up of nylon with circumference of 45 cm x 60 cm were found hanging from the trees with height varying from 15 cm to 75 cm from the ground floor because the ground floor was not of uniform height and non-availability of strong trees. The trees were mainly of Gewa (*Excoecaria agallocha*) and Math Garan (*Ceriops tagal*). Seizure of 30 traps was done on a single day. It appeared that one deer was trapped and a tiger dragged the deer along with the trap for about 30 m as pugmarks and other evidence like blood etc. were revealed. About 80% of the deer was eaten by the tiger and only part of the skin, skull and other bones were left. In another trap a wild boar was trapped, which was not taken away from the spot by the poacher but the tiger also had eaten 80% of the boar's body. This was a small boar as appeared from the bones etc. Another wild boar was also found eaten by the tiger. A deer was also found trapped, which was later released.

Information was received from a member of the Hentalbari FPC that some crab collectors saw a dead tiger in the Chamta forest block. The collectors said that they could only see the leg and part of the body of a tiger lying on the forest floor about 10 days ago. Staff of Basirhat range along with the staff of Chamta camp jointly searched the site on 17th February, 2002. On thorough search at Chamta-3 they found a fishing cat instead on 18th February, which was comparatively of a bigger size.

Year	Male	Female	Unidentified	Age class
2008	1	-	-	3-10: 1
2011	2	0	1	3-10: 2; Unknown: 1
2012	1	0	2	<1: 1; >10: 1; Unknown: 1
2013	1	0	0	>10: 1
2015	2	1	0	1-3: 1; 3-10: 1; Unknown: 1
2019	1	-	-	3-10: 1
2021	1 (Length – 8.4 ft, Weight 101 kg)	1	-	11-12: 1 The male was found on 30th May at Harikhali camp in Harinbhanga-2; cause of death: Cardio-respiratory failure. The female was found on 26.2.2021 at Pirkhali 4-5; cause of death: Cardio-respiratory failure.

Table: *Distribution of tiger mortalities, based on age classes and gender in the Indian Sundarbans since 2008*

Comparatively, maximum cases were reported from the Central Indian and Eastern Ghats followed by Western Ghats, Shivalik hills and Gangetic plains, Northeast hills and Brahmaputra flood plains and lastly Sundarbans (Nigam *et al.*, 2016). In the year 2008-2009 last tiger poaching (1 no.) took place (Mallick, 2013). But there may be a possibility of unrecorded cases of poaching because the chronicles of seizure from Sundarbans between March 2009 and November 2010 revealed that poaching seems to be a silent killer. At least six persons were arrested from east Goraberia and Baruipur with tiger skins and skulls, with the seizures showing from full grown big cats to cubs all fell prey to the poachers.

On April 8, 2019, a tiger was found dead with a snare made of metal wire around its waist in Ajmalmari in the South 24-Parganas FD. It was the first reported unnatural tiger death in the Sundarbans after 2008. The tiger could have become a victim because of a snare set for Chital and wild boar. "Both animals are prized for their flesh in the Sundarbans.

The straying of the big cats has often been ascribed to the "high population density". Hence arguably the surge in human-animal conflict. The increase in the number of tigers could arguably be one major reason for the rise in straying incidents. When there are too many tigers in a particular forest, the "young adults" might be compelled to move out in search of new territory. At least eight tigers have strayed near villages in the Sundarbans in the two months since December 2021. They were captured and released into the wild.

WII has directed West Bengal's FD to release the tigers that are captured after they stray near villages to areas that have a relatively lower density of the tiger population. Three tigers captured over the past few days in a "forest compartment" called Chamta in the NP of STR, have been released. WII has identified Chamta as a forest block with a notably low density of the tiger population. However, the problem of relocation is rooted in the fact that the "optimum carrying capacity" of a forest is not a static concept. Therefore, relocating tigers to a forest that is different from where they came is a fraught proposition. The "peak density" of a forest is dependent on several parameters, such as prey base, human interference, and the male-female ratio of tigers. The mangrove delta has been ravaged by a welter of developments. The pandemic is said to have resulted in a manifold increase in human interference. The parameters have changed over time. There is no reason to believe that the parameters in 2022-2023 after the onslaughts of repeated super cyclones should be the same as in 2018. Hundreds of thousands of villagers, who used to work in other states, have come over to the Sundarbans. To many or the most, the loss of livelihood has forced them to switch over to forestry in search of fish and crabs. Since 2020, more than 30 people have been killed in attacks by tigers in the forests. In many cases, villagers enter forests after breaching the nylon nets that mark their boundaries.

Tiger mortalities in the Bangladesh Sundarbans

Following independence in 1971 the Bangladesh Wildlife (Preservation) (Amendment) Act, 1974 banned tiger killing in Bangladesh for the first time. But poaching continued surreptitiously in the village as well as the forests. In 2012 the new Wildlife (Preservation & Protection) Act of Bangladesh introduced tougher punishments for tiger killing (7 years imprisonment); however, this law is not sufficient to combat poaching without effective enforcement measures.

Reza *et al.* (2002), based on the records registered by the Bangladesh FD and field visits between January 1999 and June 2001, reported that 41 tigers (about 3 tigers per year) were killed by humans during 1984 to 2000. The FD declared four tigers as man-eaters between 1981 and 1992 in the Bangladesh Sundarbans and subsequently killed by them. From January 1999 to March 2000, a total of 6 tigers, which were not declared as man-eaters, were killed by humans. Again, this is an official figure; the actual figure of tiger killing may be several times higher than these.

Place	Hunter	Month/year	Punishment	Source
Burigoalini, Satkhira Range	A Major in the Bangladesh Rifles	August, 1998	No punishment	The Daily Star (July 2, 1999)
Bogi, Sarankhola Range	Two Bangladesh FD officials	November 6, 1999	Arrested by Police and suspended by FD	TIGERLINK, News, 5(1), January 1999, 31 p.
Khasiabad, Khulna Range	Trapped and found dead	May 19,1999	-	Station Officer, Khasiabad Forest Station
Datinakhali, Satkhira Range	Local poachers	First week of June,1999	-	The Daily Star. July 2,1999

Place	Hunter	Month/year	Punishment	Source
Katakhali, Chandpai Range	Villagers	July 22,1999	-	The Daily Observer, Ittefaq, Sangbad, Mukto kantho, Janakantha, etc July 24, 1999
Ashirsina, Chandpai Range	Local poachers	August 2, 1999	Lawsuit by FD	The Daily Mukto kantho, Prothom Aalo, August 1, 1999

Table: *Tigers killed (only those were reported) by humans between January 1999 and March 2000*

Gani (2002) recorded 23 tiger mortalities between 1996 and 2000 from the Sundarbans. 14 tigers entered villages from the adjacent forests by crossing rivers or canals. Villagers killed 12 of them directly and an official of the Bangladesh Rifles (BDR, a paramilitary force responsible for border security) killed one. One tiger came out of the forest and took shelter in an isolated place close to the village on the berm of a flood protection embankment. It was very sick and unable to move due to paralysis. The villagers did not kill this tiger as it could not enter the village. It was later rescued and brought to a Military Zoo, where it died under treatment. One tiger was killed in the farmhouse of a feudal lord on the grounds of self-defence, which was considered as poaching. It was observed that 65% of the tiger deaths occurred in the villages and 35% occurred in the forest. Moreover, the villagers directly killed 12 tigers, which is 52% of the total killings, followed by poachers who killed 10 tigers (44%) and one tiger died from old age sickness, which may be considered as a natural cause of death. The data on the prey that the tigers went after in villages was also analysed by Gani (2002). The tigers killed four cows, six domestic pigs, 16 goats, three dogs and two ducks in the villages between 1996 and 2000. In addition, the tigers that entered the villages killed two people and injured another 12 during the same period, mostly during daytime. The people were killed and injured while the tiger was being attacked by the villagers, who assembled from all directions with sticks and spears, giving the animal no space to escape. So the tiger jumped on the mob, killing and injuring some people as for example at Datinakhali.

The skin length (without tail) of the tigers killed in the villages was analysed by Gani (2002) and on that basis the animals were divided into sub-adult and adult categories. Skins less than 1.37 m long were considered as sub-adult and those longer as adult. It was observed from the data that six (43%) sub-adult and eight (57%) adult tigers entered villages in a period of five years. It is assumed that the sub-adults entered the village after they had dispersed from their natural area and were searching for a new territory. The sub-adults were in good health except for one animal that was sick from nematode infestation. Of seven tigers autopsied, four (one sub-adult and three adults) had huge nematode colonies in their stomachs. Veterinary surgeons also reported that 2 tigers were infested with ticks and had liver problems.

The time of day that the tigers entered the villages was recorded from the reports prepared by the field forest officials on human/tiger conflicts. In thirteen cases, it was observed that the tigers entered the villages either at night or very early in the morning. The majority of tiger crossings (62%) of rivers and canals were at near high tide or within two hours of high tide, when the velocity of the current was low.

FD's records indicate that professional poachers killed nine tigers during the study period. This is evidenced by the recovery of four carcasses from the forest and five skins from adjacent villages and urban areas. Only one skin showed signs of a gunshot wound, suggesting that poison was used on the other four tigers.

According to Khan (2009) an average of 2.6 tigers were killed in Bangladesh Sundarbans per year during 2002-2006.

Year	East FD	West FD	Total of the year
2002	2	2	4
2003	2	1	3
2004	2	2	4
2005	1	0	1

Year	East FD	West FD	Total of the year
2006	1	0	1
Grand total	8	5	13

Table: *Number of tigers killed by people in the Bangladesh Sundarbans over 2002-2006*

An average of three tigers are killed per year by the villagers in retribution for previous human and livestock killings or for the fear of future incidents (Barlow *et al.*, 2009; Gani, 2002). Though the Divisional Forest Officers concerned exerted themselves to ensure the destruction of as many tigers as possible, their efforts were only moderately successful. During 2008-2012, a total of eight tigers were killed in the villages (WildTeam, mimeo.).

Year	Number of tigers killed	Number of tigers injured	Number of tiger straying
2008	1	0	0
2009	2	0	4
2010	2	0	14
2011	1	0	2
2012	2	0	73
2013	0	0	16
2014	0	1	5
2015	0	0	6

Table: *Summary of tiger moralities and injuries in and around the Sundarbans against tiger straying into the villages*

Source: Bangladesh Tiger Action Plan 2018-2027

Ferdiousi and Khan (2021) compiled the tiger killing data in Satkhira, which are reproduced below.

Age group	Proportion of sex %		Seasonal killing %	
	Male (n= 8)	Female (n= 2)	Summer (n= 4)	Winter (n= 6)
Young	0	10	0	10
Middle aged	80	0	30	50
Old	0	10	10	0
Total (%)	80	10	40	60
Grand total	100		100	

Table: *Proportion of tigers killed by local peoples during 1990 to 2018 in and around the edge of Sundarbans mangrove forest*

In the Bangladesh Sundarbans and the adjoining region, ten tigers were killed in the course of 1990 to 2018. Our study result shows that in winter, tigers were killed by domestic people much higher (60%, n=6). As no remarkable data about tiger kill were found in different months, no statistical analyses were done to look into their significance. Male tigers were killed more (80%, n=8) than females (20%, n=2) (see table above). Out of ten tigers, eight were killed in the villages close to the Sundarbans and two were on the inner side of the Sundarbans.

People killed tigers for several reasons. Among them, the prime reason was the attack on people along with domestic animals. During the period, villagers killed 80% (n=8) tigers due to the charge on humans and livestock. Illegal hunting (poaching) is another major danger to the tigers and a major cause of killing tigers. During this study, 20% (n=2) tigers were killed inside of Sundarbans by the poachers.

The most common way of killing tigers by poachers is poisoning the partially consumed food of the tiger. Along with this, firearms like guns also are used directly or set in the way of the tiger trail which is activated by the tiger itself. Most of the extirpated tigers were young and in good condition, which notifies that, at least in the Sundarbans, a tiger can become a man-eater for a reason that falls in none of the properly-believed three main reasons: wounds and infirmity, old age, and loss of home range to other tigers.

In this perspective, Saif *et al.* (2018) identified five categories (village residents, poachers, shikaris, trappers and pirates) of people involved in killing tigers, each with different motives, methods and networks. Village residents kill tigers predominantly for safety, whereas others kill in the forest professionally or opportunistically. Poachers kill tigers for money, but for others the motives are more complex. The motives of local hunters are multifaceted, encompassing excitement, profit, esteem and status arising from providing tiger parts for local medicine. Pirates kill tigers for profit and safety but also as a protection service to the community. The emerging international trade in tiger bones, introduced to the area by non-local Bangladeshi traders, provides opportunities to sell tiger parts in the commercial trade and is a motive for tiger killing across all groups.

A total of 13 cases have been filed against 59 people for killing tigers in the eastern side of the Sundarbans in the 21st century, among whom only nine have been apprehended. Trial of six cases has ended till now, and in three of the cases, all the accused have been acquitted due to the FD's inability to produce evidence. In the three other cases, a total of seven accused have received jail terms ranging from six months to four and a half years.

According to data from the FD, a total of 28 tigers have either died or have been killed between 2001 and 2022. Among these dead tigers, 14 were killed by hunters, five died due to mob beating, one died in cyclone Sidr and eight others died naturally. These numbers are based on joint or separate drives against hunters by the FD, RAB and police. During these drives, hides, bones and other body parts of the tigers had been recovered and in rare cases, bodies of dead tigers were found inside the forest.

Lethal control of problem tiger

Tigers that kill humans accidentally need to be treated differently from those that have become dangerous to human life. Dangerous to human life needs to be removed immediately and should not be reintroduced into free ranging conditions. While tigers that have come into conflict situations accidentally can be captured and reintroduced into vacant/low-density habitats. Preferably, tigers that are released after relocation should be equipped with satellite telemetry collars so that they can be tracked remotely. In case such tigers continue to be problem animals they should be recaptured and put in captivity or eliminated. Release into a new area should preferably be done by soft release protocol. Hard release should only be considered when logistic constraints do not permit soft release. The release site should have good habitat, prey density and protection measures. Human density at the release site should be low and composed of communities that have experience of living with large carnivores.

Officially a problem tiger may be hunted as per provisions of the Wildlife (Protection) Act, 1972.

Only government officials or agents authorised by the Chief Wildlife Warden can authorise such killings or captures. In situations involving tigers cornered by uncontrollable mobs (with the imminent prospect of human deaths), or with injured tigers, lethal control is the only practical option. However, urban advocacy groups often oppose such tiger killings. Furthermore, in a free-ranging population, it is very difficult to specifically target the individual problem tiger- several tigers may have to be killed before the problem animal gets eliminated. Sometimes the park managers attempt to capture a 'problem tiger' and move it away from the spot of the conflict because this approach has wider social acceptability. However, according to Dr. Ullas K Karanth, such translocations are rarely practical, and may not have satisfactory conservation outcomes:

- In a free-ranging tiger population it is rarely possible to identify the individual problem animal for removal, unless it enters human settlements or is injured.
- Furthermore, safe chemical capture (or driving away) of tigers is usually rendered difficult because of crowd-control problems, injuries to the animal, lack of technical skills, scarcity of resources and other logistical problems.
- Even after safe capture, the problem tiger has to be permanently housed in captivity or relocated into the source population from which it came or into a new habitat.
- Wild tigers do not adapt well to life in captivity, and the capacity of Indian zoos to hold tigers is already saturated. Most zoos simply cannot afford to house an ever-increasing number of problem tigers.

- Most problem tigers that undergo capture and handling are injured in the process, particularly by losing their canine teeth in steel transport cages that are commonly used. Many are either old or weak animals evicted from their ranges by more vigorous rivals. Such tigers are unfit for relocation into the wild.

There are several ecological arguments against translocation of even healthy problem tigers into new habitats.

First, most such relocations simply result in transfer of the problem to a new location leading to a new situation of conflict, because high-quality tiger habitats devoid of conflict potential are scarce.

Second, even after translocation into a large reserve with an adequate prey base, the introduced animal will compete for space and prey with other individuals in the local tiger population. Because tigers are territorial animals, competition is likely to lead to elimination of either the introduced tiger or of another individual from the local population.

Translocation of tigers into established, high-density tiger social land-tenure systems as a way to maintain gene flow can lead to tiger deaths as was reported earlier in STR about 40 years ago (Seidensticker *et al.* 1976).

Estimation of Prey base

For estimation of selected mammalian prey base of tiger i.e. deer, wild boar and monkey which are distributed over the islands having both open and dense mangrove vegetation and for the census studies it was presumed that they normally do not cross to other islands. The islands were grouped into four area classes as shown below.

Number of islands/compartments	Area of the islands (km²)
32	15-30
26	31-50
11	51-70
2	71 and above

An island of Pirkhali-3 with a total land area of 17.72 km² was selected for total enumeration for two days using line transects and point counts. Eleven enumeration teams were formed of which five teams were to enumerate the entire periphery of 18.48 km. in country boats (as these boats do not cause much sound) and six teams were set in *machans* inside the forest, strategically located in the island. The entire area of the island was thus covered.

The major findings of field enumeration are:

(1) The hoof marks were available in the entire area or island, no concentration differences were located along the riverbanks and islands.

(2) Animal dung was not found significantly as the areas were inundated by high tide.

(3) Enumeration of STR for all islands need not be taken at one time, because intermigration in islands of major prey bases can be assumed to be nil for enumeration exercises. This will help the field managers use the limited resource and thus can undertake the work without causing added financial impact.

(4) The total count of all the islands will not be possible.

So, two sample counts in 3 major mangrove dwellers i.e. upper, middle and coastal mangrove habitats were undertaken.

Month	Wild Boar	Otter	Crocodile	Chital	Dolphin	Water Monitor	Water birds	Fishing	RhesusMonkey
April 2021-2022	225	35	55	871	93	178	995	6	1,627
April 2022-2023	221	15	15	879	70	175	182	6	1,627
May 2021-2022	190	20	61	905	137	556	562	3	1,476
May 2022-2023	179	18	13	911	81	131	125	3	1,476
June 2021-2022	195	47	49	1,003	151	207	131	9	1,398
June 2022-2023	189	20	11	1,011	83	147	53	9	1,398
July 2021-2022	215	55	57	885	229	107	95	7	1,103
July 2022-2023	213	9	18	931	87	141	25	7	1,103

Month	Wild Boar	Otter	Crocodile	Chital	Dolphin	Water Monitor	Water birds	Fishing	RhesusMonkey
August 2021-2022	230	49	34	1,011	182	71	83	11	927
August 2022-2023	295	19	14	1,019	91	149	65	11	927
September 2021-2022	235	29	91	703	146	142	359	5	801
September 2022-2023	306	17	25	873	109	145	96	5	801
October 2021-2022	185	30	83	916	93	137	857	13	1,521
October 2022-2023	341	19	59	951	121	131	111	13	1,521
November 2021-2022	239	59	65	1,205	111	260	904	4	1,207
November 2022-2023	347	20	69	1,231	135	153	357	4	1,207

Month	Wild Boar	Otter	Crocodile	Chital	Dolphin	Water Monitor	Water birds	Fishing	RhesusMonkey
December 2021-2022	195	36	51	1,178	109	233	1,364	8	1,009
December 2022-2023	358	14	73	1,191	141	131	398	8	1,009
January 2021-2022	431	71	89	991	114	87	1,201	6	1,679
January 2022-2023	431	16	93	1,301	138	121	478	6	1,679
February 2021-2022	415	81	76	729	121	13	931	11	1,509
February 2022-2023	419	11	98	1,106	145	111	598	11	1,509
March 2021-2022	395	59	81	809	127	127	1,042	14	1,349
March 2022-2023	193	20	91	759	125	127	298	14	1,349

Table: *Records of sightings of prey animals in STR during 2021-2022, and 2022-2023 (Annual Reports)*

Hunting and poaching

The crucial distinction to be made between poaching and hunting is where each sits in the eyes of the law. Illegal wildlife hunting is more commonly referred to as poaching. There is no legal definition of "Poaching" as per the Indian Wildlife (Protection) Act, 1972. There is no distinction between illegal hunting and poaching in India as per the law. Put simply, poaching is hunting without legal permission from whoever controls the land. Hunting is regulated by the government, and hunters must obtain permits authorising them to kill certain animals.

'Hunting', according to the Wildlife Protection Act, 1972, includes:

(i) Capturing, killing, poisoning, snaring and trapping any wild animal and every attempt to do so.
(ii) Driving any wild animal for any of the purposes specified i.e. Trading, commercialising.
(iii) Destroying or taking any parts of the animal or any species.

According to Section 9 of the Wildlife (Protection) Act, 1972, Prohibition of hunting has been explained. "No person shall hunt any wild animal as specified in schedules, I, II, III and IV except as provided under section 11 and section 12." Post-Independence hunting was banned by the government under the Wildlife (Protection) Act, 1972, except for some purposes which are defined in section 11 and section 12 of Wildlife (Protection) Act, 1972. In the case of the State of Bihar vs. Murad Ali Khan, 1989, it was decided that hunting of wild animals is to be permitted in certain cases and gave an example of self-defence that in order to protect ourselves from wild animals in any circumstances and killing or giving any harm to that animal will not fall under the provision of Wildlife Protection Act.

Following independence in 1971 the Bangladesh Wildlife (Preservation) (Amendment) Act, banned tiger killing in Bangladesh for the first time. But overall, an estimated minimum of 41 and a maximum of 55 tiger skins have been seized in Bangladesh, accounting for 2.2% of the global total. Bangladesh reported the highest number of seizures in 2015.

Pirate gangs were the dominant poachers from 1980 to 2017, but following an extensive campaign, the Sundarbans was declared pirate free in 2018.

Uddin *et al.* (2022) analysed the spatial distribution of tiger poachers among the unions and used crime script analysis of the dominant poacher type to identify intervention. Because pirates opportunistically poached tigers, the government's successful counter-pirate campaign inadvertently removed the dominant tiger poaching type. However, a temporary reduction in poaching was rapidly cancelled out by the emergence of at least 32 specialist tiger-poaching teams. With the risk of extortion and robbery from pirates gone, other groups increased the frequency of opportunistic and targeted tiger poaching. Based on expert interviews, it was estimated that 341 tiger poachers of all types are active throughout the unions, with 79% of specialists concentrated in 27% of unions. The highly focused counter-pirate campaign reduced motivations and opportunities for piracy but left intact the opportunity structure and trade connections for tiger poaching, and with insufficient enforcement officers trading has flourished.

Uddin *et al.* (2023) described eight distinct tiger product types, four of which were described as solely consumed by elite local consumers for medicinal, spiritual and/or ornamental purposes. The details are given below.

Commodity	Value in USD	Purpose	Consumers/centres
Ornamental tiger skin (n = 163)	$15,120–$17,450 per skin	Symbol of prestige	Bangladeshi elites in Dhaka, Sylhet and Chattogram
Ornamental bones (n = 163)	$581–$1160 per kg	Showpieces	Bangladeshi elites in Khulna, Dhaka, Sylhet and Chattogram

Commodity	Value in USD	Purpose	Consumers/centres
Tiger bone powder (n = 11)	$1163–$1745 per kg	Wine and tonic for physical power	Local Bangladeshi elites and expatriates, especially in the United Kingdom. local communities in Myanmar and India
Tiger canine amulet (n = 40)	$230–$350 per canine	Source of energy and courage and to keep away evil spirits	Bangladeshi local and international elite and tribal community members in Khulna, Dhaka, Chattogram and Sylhet
Tiger meat (n = 12)	$60–$120 per 10 grams (dried)	For energy and increasing body weight	Local tribal community members in Sundarbans fringes
Tiger hair (n = 19)	$35–$58 per gram of hair	As a symbol of power to keep away evil spirits and reduce waist pain	Local tribal communities living around the Sundarbans, Chattogram and Sylhet

Commodity	Value in USD	Purpose	Consumers/centres
Tiger claws (n = 40)	$117–$175 per claw	As a symbol of power used in pendants and amulets	Bangladeshi local communities in Khulna, Dhaka, Chattogram, Sylheti diaspora
Tiger milk (n = 2)	$233–$350 per 100 grams	To increase milk production in pregnant women	Local communities of Khulna, Sylhet, Chattogram, Cox's Bazar diaspora

The number of deaths/killing/poaching of deer in Bangladesh Sundarbans is 387 during the period from 2004-2005 to 2019-2020 with an average of 24 numbers/year, while mortality of about 51 tigers are recorded during the same period with an average of 3 numbers/year (Mahmood *et al.*, 2021).

Illegal trade

Poaching is the number one threat to tigers globally, and China is the biggest overall driver of demand, largely for use of their body parts in traditional medicine. Research from big cat conservation group Panthera and the Chinese Academy of Sciences said tiger parts harvested in the Sundarbans have been exported to 15 countries, with India and China being the most common destinations, whereas Bangladesh plays a much more significant role in the illicit tiger trade (Uddin *et al.*, 2023). Elite Bangladeshis were the most important consumer group, and tigers were sourced from populations in NE India, Myanmar and Bangladesh Sundarbans to supply domestic demand.

Pirate groups operating in the Sundarbans found a lucrative trade in tiger poaching before a government crackdown starting in 2016. At least 117 pirates were shot dead and hundreds more were detained, according to official figures, while many others surrendered as part of a government amnesty. The vacuum created by the crackdown had been filled by more than 30 specialist tiger poaching syndicates and opportunistic poachers. Traders operated through their own logistics companies and in some cases concealed their activities through licences for legal wildlife trade. Wealthy local buyers were purchasing medicines using tiger parts "as well as large ornamental items for display such as skulls and skins.

Twenty-one of the 32 interviewees familiar with the tiger trade in southwest Bangladesh described tiger trafficking flows in the region as bidirectional between India and Bangladesh. Wild tigers were poached from the Bangladesh and Indian Sundarbans before their parts were processed at Kaikhali, Satkhira, Mongla, Bagerhat, Patharghata and Khulna. The tiger parts were described as being sold directly to Indian buyers or Bangladeshi distributors in Dhaka with international connections. In addition, the tiger parts were described as being frequently smuggled by road through the Benapole land port (n = 39) into Kolkata and other parts of West Bengal in India. Interviewees also described tiger parts being smuggled by air via Dhaka airport to Kolkata and New Delhi in India (n = 33). In some cases, tiger skins were described as being dried and processed inside the Sundarbans before being smuggled into India on fishing boats through poorly patrolled waterways along the international border. Interviewees also described the supply of tiger parts from Kolkata and West Bengal into Bangladesh.

Chapter 8
Morphological Adaptations

"There is nothing like the thrill of walking through the jungle looking for
a tiger and knowing they could be watching you already."
– Ashlan Gorse Cousteau

The very beauty of the tiger, the beaute du diable, in truth, and the vivid
combination of black, yellow, and white on its glossy skin, is terrible to look
upon, let alone the malignant cunning shot from the eyes, and the cannibal
hunger expressed in their curling lips and flashing white teeth. The tiger
shows considerable variation in its size, colouration and markings,
reflecting the variety of habitats it occupies throughout its very wide
geographical distribution.

The living tiger subspecies' morphology (physical structure and
appearance) exhibit a cline. A cline occurs when a single species gradually
begins to look different over its geographic distribution as it adapts to
varying climates and habitats. Therefore, the species at the northern end of
their geographic distribution may look very different in size, colour, hair-
density, etc. than their southern counterparts. The tiger cline depicts
subspecies that decrease in size and have darker stripe colouration the
further south their range extends.

Evolutionary biology of Sundarbans tiger

From an evolutionary biology perspective and in the context of biodiversity
conservation, the endangered Sundarbans mangrove tiger population is
becoming ever smaller and more isolated due to human-induced habitat
loss. This process often leads to decreased genetic diversity and reduced
evolutionary potential, ultimately implying increased extinction risk.
Moreover, there are problems of low densities and elusive behaviour of the
Sundarbans tiger, which hamper broad-scale sample collection and long-
term monitoring.

Out of three haplotypes recently detected within the Sundarbans tigers, one is unique to this population and the remaining two are shared with tiger populations inhabiting central Indian landscapes (Aziz *et al.* 2022). Considering the rapid decline in the Bengal tiger population throughout its range, the critical mangrove forests along the Indo-Bangladesh border, would potentially be the last refuge of the big cats because no other place like the Sundarbans is left on earth.

The mangrove tigers are different from the mainland tigers in many ways, which are significant for their conservation in terms of ecological role played by them. Following Singh et al. (2014), a comparison between Sundarbans and mainland Bengal tiger landscape in India may be made hereinbelow.

Ecological parameter	Sundarbans tiger landscape	Mainland tiger landscape
Prey species	Small size prey (Chital and Wild Boar)	Large size prey (Sambar and Nilgai)
Habitat	Mangrove forest	Tropical forest
Competitor	None	Leopard
Tiger density	4.3/100 km²	16/100 km²

Table: *Comparison between Sundarbans and mainland Bengal tiger landscape in India*

Male

Location	Body mass (kg)				Total length (cm)				Reference
Sundarbans	4	97	150	115	3	252	261	255	Chattopadhyay and Banerjee, 2008; Mukherjee, 2011
Central India (Panna and Kanha)	3	197	250	229	-	-	-	-	Chundawat (in litt. 2005)
Nilgiris BR	3	209	227	217	3	289	311	298	Karanth, 1993;

Location	Body mass (kg)				Total length (cm)				Reference
									Sunquist and Sunquist, 2002
Sariska TR	1	-	-	220	-	-	-	-	Sinha, 2008
Chitwan NP, Nepal	7	184	260	221	2	292	310	301	Smith *et al.*, 1993; Sunquist, 1981

Female

Location	Body mass (kg)				Total length (cm)				Reference
Sundarban	5	72	80	76	3	228	239	234	Barlow *et al.*, 2006; Barlow, 2009; Mukherjee, 2011
Central India (Panna, Kanha and Pench)	4	115	145	127.5	-	-	-	-	Chundawat (in litt. 2005); Mazumder *et al.*, 2012
Nilgiris BR	2	102	177	139.5	2	248	260	254	Karanth, 1993
Nagarjunsagar- Srisailam TR	1	-	-	150	1	-	-	236	Messias, 2001
Sariska TR	3	135	170	149	-	-	-	-	Sinha, 2008, 2009; Sebastian, 2010
Chitwan NP, Nepal	19	102	150	130	5	251	282	266	Smith *et al.*, 1993; Sunquist, 1981

Table: *Sex and location specific morphological distinction of Bengal tigers in the Indian subcontinent*

Based on the above mentioned populations it is calculated that the average weight for an adult male tiger in mainland is 222.3 kg (n= 13) and 136.5 kg (n= 28) for the females. All the specimens captured from the Sundarbans were frail underweight and some of them incapable to hunt but those specimens (males and females) were not baited.

Generally, the male tiger has more weight compared to a female tiger in all the stages of its life. The male tiger attains its maximum weight at around 5-6 years and then it stabilises. The average weight of a male tiger of more than 5 years at Panna TR is more than 200 kg. The female tiger attains its maximum weight a little earlier than the male tiger at around 4.5 to 5.5 years and then it stabilises. The average weight of a female tiger of more than 5 years at Panna TR is around 140 kg.

Naturally, the male tiger is also slightly bigger in body length compared to the female tiger in all the stages of its life. The male tiger attains its maximum body length at around 2 years of age and then it stabilises. The average body length of a male tiger of more than 2 years at Panna TR is around 200 cm. The female tiger attains its maximum body length at around 2.5 years of age and then it stabilises. The average body length of a female tiger of more than 2 years at Panna TR is around 175 cm.

Tail

A tiger's tail is less than one metre in length. Tail length of the captured tigers in STR was measured from 80 cm to 84 cm against the mainland tigers measuring about 1 m. The tail length of the male tiger is just slightly bigger compared to the female tiger in all the stages of its life. The male tiger attains its maximum tail length at around 20 months of age and then it stabilises. The average tail length of a male tiger of more than 2 years at Panna TR is around 100 cm. The female tiger also attains its maximum tail length at around 20 months of age and then it stabilises. The average tail length of a female tiger of more than 2 years at Panna TR is around 90 cm.

The tail may play a part in their visual communication. The tail of the Sundarbans tiger is comparatively more muscular and consistently thicker at the base than that of other tigers, which is used as a limb for swimming, in the sinuous, sweeping, side-to-side manner of the crocodile (Montgomery, 2008). At the time of swimming they put the tail in the water to physically determine the intensity of current as well as direction. The tail gathers a lot of momentum when swung from side to side. They use their tail for balance when making sharp turns in pursuit of prey. It also plays a part in tiger's communication. A tiger's tail swishing madly from side to side or held low with just an occasional twitch indicates aggression. A relaxed tiger has a droopy tail.

Total length

The total length of the male tiger is slightly bigger compared to the female tiger in all the stages of its life as total length is addition of body length and tail length. The male tiger attains its maximum total length at around 2 years of age and then it stabilises. The average total length of a male tiger of more than 2 years at Panna TR is around 300 cm. The female tiger attains its maximum total length at around 2.5 years of age and then it stabilises. The average total length of a female tiger of more than 2 years at Panna TR is around 275 cm.

Many sportsmen, European and native, believe that the Sundarbans tiger is in some respects unlike his kindred of the hills and plains; that they are longer and thinner in build, and looser in their skin. The tigers of those woods and salt or brackish waters are longer from tip of nose to tip of tail, but not so high, as those of other localities; their footprints assuredly seem to be smaller in general than those of others, and such individuals as have been seen appeared certainly rather lower in stature (Baker, 1887). Living, as the Sundarbans tigers do, an amphibious sort of life in thick and to man impenetrable forests and thickets, constantly crossing and recrossing creeks and rivers of salt or brackish water, and often obtaining none other to drink, they might reasonably be expected to differ more from the rest of their kind than they actually do, as a matter of fact.

Ahmad (1981) has differentiated the two isolated regional big cats in the then Bengal:

North Bengal	Sundarbans (Khulna)
Grow bigger in length, lighter in colour- almost yellow with black stripes and white from the throat down the body.	Thick-set, dark almost brown-red in colour and in the forest, one can hardly distinguish the black stripes; much smaller in size but stocky.

Naturally, the Sundarbans tigers living in isolation in the closed and open mangrove forests, located at the southern tip of West Bengal (India) and Bangladesh (Khulna Division), bordered by the northern Bay of Bengal, are quite different from any other regional tiger because of its adaptability to the unique mangrove habitat. They are somewhat shorter in length than their counterparts in other areas, but they have bigger heads and splayed paws. Fayrer (1875) wrote: "...amongst the finest specimens are those of Saugur (Sagar) Island and the Sunderbunds (Sundarbans) at the mouth of the Hooghly". For example, the Bengal tiger in Assam of North-East India is featured by broadhead, conspicuous maine, a large and heavy body, stout limbs and minimally tapering tail, whereas the Bengal tigers in the Sundarbans mangroves are distinguished by light body, small to longish head, scanty fur and mane, a conspicuously tapering tail, the largest canine teeth of any living cat, and perfect retractable claws for climbing trees.

The behaviour of the Sundarban tiger is highly individual specific and can't be generated or replicable (Mukherjee, 2004). It is combined with a highly developed sense of hearing and vision. The tigers use infrasound to communicate over long distances or dense forest vegetation because the sound is capable of passing through a variety of mediums.

In the opinion of Baker (1887), it does not appear that the colour of tigers varies much according to their habitat; some may be a little darker and brighter than others, but the general tone is rarely departed from, and thus tigers from youth to maturity display the brightest tints; from maturity to old age the colours fade, the brilliant yellow becomes paler, the stripes grow brownish and fainter, till at extreme old age the ground is a pale tawny, and the stripes are faintly marked.

Jim Corbett and several other hunter-naturalists suggested an obtuse olfactory sense in the tiger; but there were also many remarks contradicting this opinion. Close observation on tame or semi-tame tigers proved that they have a fairly keen sense of smell. The sense of olfaction in tiger cubs seems to develop after the age of four months. However, its power of smell has been found to be not as powerful as the hearing ability.

Skeleton

The tiger has a flexible spine that allows it to pounce on its prey. The flexibility of the tiger's skeleton and its long hind legs, allows them to jump to the height of 10 to 12 feet. A short rounded skull and the jaw muscle attached to a bony ridge on the top of the skull called the sagittal crest, gives power to the tiger's lower jaw. In total tigers have thirty teeth that measure up to 2.5 to 3 inches long, which makes for the largest canines of all big cats.

In May, 2009 the skeleton of an adult tiger was recovered from the Ajmalmari-3 forest in between Jharkhali and Sandeshkhali in the Indian Sundarbans, which might be a case of poaching. Officers probing the seizure of a tiger skin on 25th March recovered the skeleton. The seizure had been followed by a spate of arrests, including that of poaching kingpin Panchan Giri on April 7.

Size

Previously, the Siberian tiger was considered the largest of the subspecies, however recent studies indicate that the Bengal tiger is currently the largest. An outsized male Bengal tiger shot in northern Uttar Pradesh, India, in November 1967, measured 3.22 m between pegs - or 3.37 m over the curves- and weighed approximately 389 kg. However, it must be taken into account that this particular tiger had killed a buffalo the night before, so had a very full stomach! This specimen is now on display at the US Museum of Natural History at the Smithsonian Institution in Washington DC, USA. Baker (1887) quoted Captain Williamson about a large tiger recorded from Murshidabad:

"...the tiger proved to be the largest ever killed on the Cossim-bazaar island. The circumference of the joint of his wrist was twenty-six inches, he was thirteen feet and a few inches from the tip of his nose to the end of his tail, and in a right line, taken as he lay, from the sole of his fore paw to the tip of his withers between the shoulders gave very nearly four feet for his height." Another record of a tiger shot in the same vicinity many years afterwards is available, but on the right bank of the Bhagirathi, and not on "the island of Cossim-bazar," which then boasted of no jungles capable of harbouring tigers. That skin, measured thirteen feet in length, was of proportionately vast breadth, and therefore had not been unduly stretched longitudinally; the tail was rather long but not in an extraordinary degree; the breadth across the back was immense; the colour was a pale tawny with faint brown stripes widely marked, as if Nature allowing only a certain number of stripes to each individual, the unusual growth of the hide had caused them to be separated at greater intervals than was common. This was the skin of a very old and extraordinarily huge tiger; on the shoulders it was as thick and tough as that of a wild bull, and it exhibited several old scars of cuts and bullet wounds".

The tiger population in the mangrove habitat is isolated from other mainland populations for ±1,000 years and exhibits strong distinctive morphological adaptation to mangrove forests. The Sundarbans tiger differs from the mainland tigers and is also genetically one of the most divergent groups amongst Bengal tigers, suggesting insular dwarfism. It is the smallest among all Bengal tigers found elsewhere on the Indian mainland. Its size has probably evolved to suit the difficult mangrove habitat of the islands. The stress factor associated with changes in their natural habitat and the availability of the smaller prey species is often related to such a phenomenon. They are unique and, therefore, of global importance.

Dr. Kalyan Chakraborty (late 1970s) had given the average measurements of a male tiger as follows:

Length with tail... 261 cm (205-275 cm);

Length of only tail... 77 cm;

Nose-tip to tail-base... 184 cm;

Tail-girth: Base... 23 cm; Tip... 4 cm;

Head-girth... 76 cm; Length 8-9 cm, Width 7-8 cm;

Neck-girth... 63 cm;

Chest-girth... 110 cm;

Teeth: Front... Length 5 cm; Width 10 cm; Back Length 4 cm; Width 4 cm;

Foot-print... 14 cm;

Dr. Kalyan Chakraborty took measurements of 400 pugmarks collected from different forest blocks and summarised the difference of length between the front and hind legs into nine sets in table below.

Front length and width in cm	Hind length and width in cm	Difference between two in cm
7.5 x 8.5	6.5 x 7.5	75
11 x 9	9 x 8	112
9.5 x 12.5	10.5 x 11	115
13 x 11	13 x 11	122
13.5 x 12.5	12.5 x 11.5	125
11.5 x 14	12.5 x 14	130
14 x 14.5	13 x 14.5	135
14 x 16	14 x 16	140
17 x 16	17 x 16	145

Table: *Hind pugmark measurements for individual tigers in Bangladesh Sundarbans (IUCN, 2004)*

Identification	Maximum length (cm)	Maximum pad width (cm)	
		Male	Female
Cub	6.6-6.9	5.5	5.1
	7.0-7.4	5.6	5.2
	7.5-7.9	5.7	5.3
	8.0-8.5	5.8	5.4
	8.6-8.8	6.8	5.9
	8.9-9.1	6.9	6.0

Identification	Maximum length (cm)	Maximum pad width (cm)	
		Male	Female
	9.2-9.4	7.0	6.1
Juvenile	9.5-9.7	7.1	6.2
	9.9-10.1	7.2	6.3
	10.2-10.4	7.3	6.4
	10.5-10.7	7.4	6.5
	10.8-11.0	7.5	6.7
	11.1-11.3	8.4	7.6
	11.4-11.6	8.5	7.7
	11.7-11.9	8.6	7.8
Adult	12.0-12.2	8.7	7.9
	12.3-12.5	8.8	8.0
	12.6-12.7	8.9	8.1
	12.8-13.0	9.0	8.2
	>13.1	>9.1	8.3

The Sundarbans tigers are sometimes found relatively smaller in body size, but very light in weight, compared with the mainland tigers. Although there was such a trend in the past, during the past decade the difference has been more acute (Mallick, 2013). There are possibilities that the specimens weighed were undernourished when captured because the animals here have to constantly struggle for food in a tough hunting ground, daily flooded by two high tides at an interval of six hours. These tigers are living, literally, on the edge. A table showing the body size (in cm) range of the mainland and the Sundarban tigers are given in table below.

Criteria in 2010	Sundarban		Mainland	
	Male	Female	Male	Female
Nose tip to tail-base	168-188	157-161	165-221	143-179
Total length including tail-base to tip	252-283	228-250	250-313	222-282
Neck girth	53-58	48	65-75	60
Chest girth	89-110.5	87	119-160	104-107
Shoulder height	88-94	67	88-114	86-91

Comparatively, the largest male Siberian tiger (*Panthera tigris altaica*) found in the Russian Far East can weigh up to 300 kg, and measure about 3.3 m in length followed by the Bengal tiger in the Indian mainland weighing up to 260 kg, whereas the tiger in Peninsular Malaysia, Sumatra and Indonesia is smallest, rarely exceeding 142 kg but tigers in the Sundarbans vary in weight from 97 to 115 kg in case of male and 72-80 kg for female.

A smaller size means lesser requirement of food, and easier movement through muddy terrain, rivers, and creeks. A minimal mass and body size is beneficial to overcome the prey above a certain size, additional body mass will be a disadvantage as the net energy gain diminishes.

Shoulder height

A male tiger has slightly more shoulder height compared to a female tiger in all the stages of its life. The male tiger's shoulder height keeps on increasing though slightly till the age of 10 years with a maximum shoulder height of around 125 cm. The female tiger attains its maximum shoulder height at around 2.5 years of age and then it stabilises. The average shoulder height of a female tiger of more than 2.5 years at Panna TR is around 100 cm as against 88-94 cm and 67 cm in case of the male and female Sundarbans tigers respectively.

Neck girth

The neck girth of the male tiger is more compared to the female tiger in all the stages of its life. The male tiger's neck girth keeps on increasing at a gentle pace from around 70 cm at 2 years of age to 80 cm at the age of 10 years. Female tiger neck girth remains stable at around 60 cm from the age of 2 years till the older age. Comparatively the corresponding figures for the Sundarbans tigers are 53-58 cm and 48 cm respectively.

Chest girth

Chest girth of the male tiger is more compared to the female tiger in all the stages of its life. The male tiger's chest girth remains stable at around 130 cm once it attains this at the age of 2 years. The female tiger's chest girth remains at around 110 cm from the age of 2 years till 7 years of age. It drops to around 100 cm beyond 7 years of age. The comparative figures for the Sundarbans tigers are 89-110.5 cm and 87 cm respectively.

Abdomen girth

Abdomen girth of the male tiger is more compared to the female tiger in all the stages of its life. The male tiger's abdomen girth remains stable at around 140 cm once it is attained at the age of 2 years. The female tiger's abdomen girth remains stable at around 110-115 cm once it is attained at the age of 2 years. In the Sundarbans a large, dominant male has a full belly hanging heavier than that of the female, which could not be measured.

Skull

Skull of tigers differs from other carnivorous species in shape. The strong bones are designed to absorb the stress of its powerful lolling bite. Skull index (Breadth/Length x 100) of fully grown Sundarbans tigers varies from 67 to 71.64 which indicates it to be brachycephalic. The skull of a tiger is articulated by 29 different bones connected with fine fissures. It is stout and rounded in shape which provides more support for their powerful jaws. Surprisingly, the skulls of the Sundarbans tigers were found to be smaller than those of the mainland tigers. The Sundarbans male tigers differ significantly (p<0.01) from the mainland subspecies. The Sundarbans male tiger skulls were smaller in overall size (skull length, width, and jaw size), and had proportionally narrower muzzles and mastoid regions and the Sundarbans male tiger skulls also differed significantly in shape from two Sunda Island subspecies (*P. t. sondaica*, *P. t. balica*), primarily by having a relatively broad occiput (Barlow *et al.*, 2010). The Sundarbans female skulls differed significantly (p<0.05) in shape from *P.t. altaica*, *P.t. jacksoni*, *P.t. tigris*, *P.t. balica* and most markedly from *P.t. sondaica*. Barlow *et al.* (2010) suspected that the small skull and body size of the Sundarbans tigers could be a consequence of having no sambar (*Cervus unicolor*) sized or larger prey available.

Moreover, the ratio of their skull circumference to neck girth is almost equal or could be less than one, as compared to the mainland tigers which is greater than one. This may be the reason that in some cases, the standard radio collars brought to track the Sundarbans tigers did not actually fit them.

Tigers' powerful jaw muscles are attached to a bony ridge that lay on top of the skull called the sagittal crest. These muscles function to rapidly clamp down on prey with crushing force.

Mukherjee (2004) conducted osteological studies of four skull specimens in STR. Sample - 1 was of the tiger which entered into the Samsernagar village adjoining Jhingekhali Station of Basirhat range on 28th August, 1998. The tiger's body was later found to be floating in the village pond. The analysis of the visceral parts confirmed the presence of endosulfan which is a commonly used insecticide for farm crops. The age of this male tiger was approximately 14 years, the length of tiger up to the tail tip was 2.5 m and without the tail was 1.7 m, the width of the sample was 9 m and the breadth was 1.2 m.

Sample - 2 was of the tiger which was found on the bank of Pargumti village close to Jhingekhali Station of Basirhat range. The dilapidated body was found on 5th March, 1999 and it appeared to have come with the high tide. The length of the body was 2.2 m and breadth 1.1 m and body weight about 8 kg. The age of the tiger as per the size of the skeleton was estimated to be about 3 years.

Sample - 3 was of a dead body of a tiger which was found in the Narayantala pond in Gosaba 3. The body was highly decomposed and was found on 9th May, 2000. Only the skull with hair patch and part of limbs, tail and bones were found to contain dry and wrinkled skin flaps which were without hair. Several bone parts were found detached from the skeleton by the tidal water action. No. of bones found were-

1. Skull of which mouth edges were decayed and with few teeth in upper jaw, the lower jaw was partly broken;

2. Vertebral columns were with only 24 no. of vertebrae, where the vertebral formula of *Panthera tigris tigris* is C_7T_{13}, $L_7S_3Cy_{21}$; (C for Cervical, T for Thoracic, L for Lumbar, S for Sacral, and Cy for Coccygeal);

3. Limb bones 9 in number;

4. Scapula and ilium 2 in nos.;

5. Left and right rib bones 7 each where 13 ribs on each side present in tiger;

6. Tailbone 19 in nos.

There are 21 Coccygeal vertebrae in the tiger which form the skeleton of the tail. As only a skeleton was found, the measurements of the skeleton was found to be as follows- length from nose tip to tail tip 1.95 m, tail length of the available portion was 0.8 m. As per the ilium and pelvic bones the carcass appeared to be that of a male tiger.

Sample - 4 of a tiger was recovered from the Naubanki area of Sajnekhali WLS. The sample appeared to be killed by poaching.

The observation during the study pointed out some indications regarding general idea about the parameters of major bones which are shown in the following table:

Osteological Measurements of Sundarbans tigers

Sl.no.	Bone	In cm	Sample 1	Sample 2	Sample 3	Sample 4
1	Skull	Length	28.3		26.1	
		Width	19.1	16.5	18.7	18.1
2	Scapula	Length	22	19.4	22.3	22.4

Sl.no.	Bone	In cm	Sample 1	Sample 2	Sample 3	Sample 4
		Width	11.9	11.3	12.9	12.6
		Maximum height of spine at acromion process	4.5	3.7	4	4.2
3	Hume-rus	Length	29.2	26.7	29	29.1
		Girth at deltoid tuberosity	10.5	9.1	10	10
4.	Ulna	Length	28.6	16.2	30.9	
		Girth below coronoid process	9.8	8.4	9.3	
5	Radius	Length		20.3	24.6	23.4
		Girth at the base of radial tuberosity		5.5	6.1	6
6	Pelvic girdle	Length		24.2	26.8	26.3
		Distance below two wings			14	12.8
		Transverse diameter			10.8	9.5

Sl.no.	Bone	In cm	Sample 1	Sample 2	Sample 3	Sample 4
		Conjugate diameter		10.5	11	11.2
		Acetabular depth		2.5	2.8	2.7
		Acetabular diameter		3.9	4	4
7	Femur	Length	33	-	32.4	32.5
		Girth below lesser	9.6		9.5	9.6
8	Tibia	Length	28.6		27.9	28.4
		Girth	12.2		13	11.9

Axial skeleton

Axial skeleton of a tiger, like other vertebrates, consists of head, neck and tail which remain parallel to the ground held up in position by the appendicular skeleton of limbs consisting of two forelimbs and two hind limbs.

Hyoid Bone

Tigers have a small piece of cartilage at the base of their tongue which enables them to roar when they breathe out but are not able to purr. Hyoid apparatus comprises a single basihyoid bone and four pairs of short pieces of bones.

Appendicular skeleton

Forelimb

Forelimbs are more important in tigers for all its movement. Its size, thickness and strength is highly characterised.

Scapula

It is one of the main bones of the shoulder and is wide, flat and almost oval in shape. Its straight caudal border is thicker than the semicircular cranial border. The lateral surface of the scapula is vertically divided into two almost canal areas by a scapular spine. In the lower third, it is almost 4 - 4.5 cm high. The length and width of scapula of adult Sundarbans tiger are 22 - 22.6 cm and 12 - 12.9 cm respectively.

Humerus

It is a long bone that constitutes the skeleton of the arm. It is an irregularly cylindrical bone slightly curved and roughly 's' shaped. The length of humerus in adults varies from 29-29.2 cm and shaft girth at the point below deltoid tubercle carries from 10-10.5 cm. The shaft or body of the humerus is more or less triangular in proximal l/3rd & cylindrical in distal 2/3rd.

Radius and Ulna

They are two well developed separate bones and are in contact with each other by their ends. The radius is the main long bone of the forearm and is shorter than its companion bone i.e. ulna. Radius has two extremities and a shaft. The length of radius in adults is about 24.6 cm and girth at the base of radial tuberosity is about 6.1 cm.

Ulna is longer than radius. Unlike in other members of Felidae ulna is not the longest bone of the body of tigers. It has a big proximal end, somewhat flat or triangular shaft and a small distal extremity. The length of ulna in adult Sundarbans tigers varies from 28.6-30.9 cm and girth of the shaft below coronoid process varies from 9.3 to 9.8 cm.

Carpus

The carpus of the tiger consists of seven bones as in a domestic cat. These are arranged into two irregular rows namely proximal row and distal row.

In the proximal row are 3 bones named from medial to lateral side as radial carpal, ulnar carpal and accessory carpal.

Metacarpus

In tigers, like the domestic cats there are five metacarpal bones. All are long bones with very thick cortex. Each bone presents proximal extremity, shaft and distal extremity.

Digits

In the forelimbs, the Indian tiger has five digits out of which the first one is very short and does not come in contact with the ground. Digit number II, III, IV & V are well developed and contact with the ground producing the characteristic pug marks.

Pelvic girdle

The Pelvic girdle constitutes the hip regions of hind limb. This consists of 3 fused flat bones namely Ilium, Ischium and Pubic on each side of the body forming a fused mass of bone. The length of Pelvic girdle in adult tigers is 26.3-26.8 cm. The conjugate diameter (i.e. distance between body of sacral and cranial end of pubic symphysis) of Pelvic girdle of adult tiger is 11-11.5 cm. The transverse diameter is 7-7.3 cm. Distance between two tuber ischii varies from 7.5 to 10.8 cm.

Generally in case of female tiger (tigress) both the diameters arc more and the pelvic girdle is more inclined forward. Pelvic outlet and ischial arch is larger. The ischia of both sides join in a wider angle and make the cavity roomier. The depth and diameters of acetabulum are 2.7 - 2.8 cm. and 4-4.1 cm in adult Sundarbans tigers.

Femur

It is the longest bone in the body of Indian tigers and is a strong, heavy bone with a cortex of compact tissue. It has two extremities and a shaft. Length of femur in adult tigers of Sundarbans area is 32.4-33 cm. and girth below lesser trochanter is 9.5 - 9.6 cm. No nutrient foramen is found on the cranial aspect of the shaft as in other carnivorous animals. In tigers, both lateral and medial condyles in distal end of femur are almost equal in size.

Patella

The Patella is the largest sesamoid bone of the body and is placed on the trochlea of the femur. In tiger, it is more or less oval in shape with the proximal part slightly wider than the distal part.

Fabellae

These are small sesamoid bones on the caudal aspect of the stifle joint and a small half globular bone comes in contact with the facet on the epicondyle of femur.

Tibia and Fibula

Tibia is larger, stronger and heavier than fibula. It has two extremities and a shaft and is made up of thick compact tissues. The length of tibia of adult Sundarbans tigers is 27.9 - 28.6 cm and girth at tibial crest is 11.9 to 13 cm.

The hind paw of a tiger corresponds to the complete foot of a human and the skeleton consists of tarsus, metatarsus and digits, which resemble the domestic cat.

It can be said from the table that among the four skeletons (three adults and one sub-adult) first, third and fourth skeletons are almost equal in size and are comparable to each other. From the external appearance of decomposed carcass and measurements of long bones, it is revealed that skeletons are of adult tigers whereas the skeleton of the second carcass is of a sub-adult tiger in which all measurements are lesser than each of the other three.

Among the first, third and fourth skeletons pelvic girdle is bigger and is more roomy and hind part is more massive in first skeleton which indicates that it may be a skeleton of adult female tiger whereas third and fourth skeletons are of adult male tigers.

The right radius and ulna of third sample's skeleton showed a cauliflower like growth (length 7.8 cm, breadth 1.7 cm, thickness 1.4 cm) in its caudal part which may be metastasis form of carcinoma and some bones were somewhat palpable as fragile which indicates the probable cause of death of that tiger to be cancer. In the third skeleton the claws (anguishes) were withdrawn within the footpad, which revealed at the time of death that the animal was calm and quiet. But in other three cases, the anguishes were projected out extensively which indicates the irritated state of the animal at the time of death.

From laboratory examination of visceral organs (stomach content) of the carcass, the cause of death of the animal was reported to be endosulfan poisoning.

General osteological study of selected long bones of Sundarbans tiger reveals that different parameters of all bones are somewhat lesser than corresponding long bones of Indian tigers of other regions such as tiger of Project Tiger Melghat Amaravati. Skull size of Sundarbans tiger is lesser than that of Melghat tigers (Pandit, 1994).

Collarbone

Tigers have a reduced-sized clavicle (collarbone). This characteristic enables them to attain greater stride lengths because the smaller clavicle allows for a wider, unrestricted range of movement of the scapula (shoulder blade) when running.

Hair and Colouration

The tiger (*Panthera tigris*) is most recognised for its dark stripes against an orange background. At first glance, the subspecies are very similar animals, but they have very prominent differences. In general, Gloger's rule states that animals living in warm and humid areas are darker and possess brighter colours than those living in cold or dry regions. For example, the Amur tiger *Panthera tigris altaica*, living in the cool region of Far East, has a light orange pelage with dull black stripes; however, the Sumatran tiger *Panthera tigris sumatrensis* is red-brownish with bright black stripes. The volume of fur in the case of the Siberian tiger is greater than that of others, in order to protect them from the cold.

Morphological examination categorises tiger hairs into more durable long, straight guard hairs or less durable short, curled zigzag underhairs. Different tiger coat colour variants exhibit a consistent pattern in hair composition whereby 90% of hairs are zigzag hairs, which predominantly contribute to pelage colouration. Zigzag hairs from the orange background of a wild-type tiger are agouti-patterned with sub-apical pheomelanin bands and dark tips and bases, and the stripes on their face, sides, legs and stomach vary in width and length, whether they are single or double-looped, colouration from a light brown to dark black and are not symmetrical from one side of the tiger to the other. The stripe patterning on top of the tiger's head resembles the Chinese character of "wang" which means "king." They are the only large cat species to have distinctive striping located on both the hair and skin. They strut around forests and grasslands with these vertical-striped markings.

The hair of the tiger provides camouflage, warmth and protection for them. The guard hairs mainly function for protection purposes. The primary function of the tiger's hair is for warmth. The underfur traps air which insulates the tiger's body thereby keeping it warm. These stripes help them camouflage with filtered sunlight and the shadow of trees so that they are not easily spotted by its prey. While humans have trichromatic eyes, meaning our vision can process red, blue and green-based colours, giving us the ability to see tigers appearing as orange, the tiger's prime prey species like deer are only able to process blue and green, making them obliviously colour blind to orange. So for an unsuspecting deer, a hunting tiger would blend into the environment fairly seamlessly.

The Sundarbans tigers are variable, having a unique striped colour of deep yellow, orange and black while the mainland tigers possess the light yellow-orange to deep reddish-orange background colouration. Black or melanistic coloured tigers have been reported but further research is required before assessing whether these sightings were of true melanistic tigers or darker versions of the orange tiger (with few large broad dark stripes). Tigers with white background coloration are not considered albinos. An albino would be pure white in colour (no striping) and would have pink or red eyes. White tigers are leucocystic meaning that they have a recessive gene that causes them to lack dark colours. Therefore they usually have a white colour with light to medium brown striping and blue eyes. For unknown reasons, white tigers seem to grow bigger and at a faster rate than their orange counterparts.

Pennant (1798:153) stated that the colours of the Sundarbans tigers differ to their age or state of health; the ground colour of a young or vigorous beast is almost of a brilliant orange; the black intense, and the little white it has is most pure. In old or fickly beast the black is dull, and the yellow fades to a sandy hue.

Colour aberrations

Individual coat or pelage differences in the case of genetic mutations such as melanism, hypomelanism, mutant and albino are recorded in case of the Bengal tiger.

Colonial period

Black tigers are a rare colour variant of the Bengal tiger and are not a distinct species or geographic subspecies. Captain Guy Dollman of the British Natural History Museum wrote of two cases of melanism in the tiger in Central Provinces and Dibrugarh, Assam, in The Times, 14 October 1936. In response to Dollman, W.H. Carter wrote in the Times of 16 October 1936:

"In one of the official district Gazetteers of Bengal (Khulna or Backerganj Sundarbans 22.222^{0}N, 88.839^{0}E) there is mentioned a local variety of tiger which had lost its stripes as camouflage in the open sandy tracts of Sundarbans. The uniform colour scheme adopted was, however, brown and not black, but perhaps his cousin in the hinterland found black more suited to his background. The author of the Gazetteer in question is, I believe, dead."

Post colonial period

Black tigers were first officially recorded in the forests of Similipal in 1975-76, when forest officials along with two foreign tourists saw two full-grown black tigers on the road leading to Matughar meadow on a bright winter day, according to Lala Aswini Kumar Singh, a tiger expert. Similipal is home to the rare 'black' tiger, a pseudo-melanistic (pseudo = false; melanistic = dark/black coloured) tiger variant that was once thought to be a myth. These animals develop broader than average black stripes that look like they have spread or leaked onto the surrounding tawny background. All the black tigers of Similipal carry a single mutation in the gene Transmembrane Aminopeptidase Q (Taqpep) which seems to be the cause of this odd patterning. The Similipal mutation is either very rare or absent in any other captive or wild Indian tiger population. The only other black tigers outside of Similipal in India (from Nandankanan zoo at Bhubaneswar and Arignar Anna Zoological Park at Chennai), which are captive-born, share a common ancestor with Similipal tigers. Only three black tigers of the 12 tigers identified in Similipal have been seen through photo trapping. In July 2022, a black tiger was observed in Similipal. Due to its geographic isolation (the nearest tiger population is roughly 800 km away), the Similipal tiger population likely arose from a small founding group of tigers. This isolation combined with inbreeding and genetic drift were probably the driving forces behind the unique black tiger population of Similipal.

In March, 2020 a few fishermen went to catch crabs in Gosaba's Kumirmari forest. Then a tiger jumped on one of them. The tiger dragged the fisherman into the forest. The fishermen who went with him were shocked to see the colour and size of the tiger even though they could not save their mate. They claimed that the tiger was much bigger than the normal Royal Bengal tiger. Some other fishermen who reached the spot after receiving the news said that the footprint of the tiger is bigger than that of a normal tiger. The FD officials thought that the stories of black tigers being heard in the villages seemed to be rumours. Trap cameras were installed in various places of Sundarbans to monitor the movement of tigers. But no black tiger would have been captured so far. However, it is not really surprising if a black tiger is confirmed in the Sundarbans due to geographic isolation, inbreeding and genetic drift.

Jaws

Tigers have short, squat jaws designed to bite hard. Their jaw muscles attach to a bony ridge on the back of their skull called a sagittal crest. This attachment gives them immense power, enough to deliver over 74 kg/cm^2 of bite force with every bite.

Dentition

Tigers have fewer teeth than other carnivores such as dogs (42 teeth) with only 30 teeth- 12 incisors, four canines, 12 premolars, and two molars. Teeth of tigers are exceptionally stout, with long and slightly curved canines. The number and diversity of cheek teeth is reduced by loss of post-carnassial molars and the most anterior premolars, leaving a dentition that is highly specialised for carnivory. The teeth are subjected to unpredictably high load from struggling prey, which over time leads to wear and tear. As the age advances, tooth wear is manifested by change in colour from white to pale yellow and later darker yellow. Gum line shows recession with considerable insult to teeth that includes either missing or at times broken teeth.

Milk teeth

Tiger cubs are born toothless. Their deciduous (temporary) teeth come in within a week or two after birth. These teeth are referred to as milk teeth similar to humans' baby teeth. The milk teeth are eventually replaced by the permanent ones. This process starts at about six months of age. The baby teeth are not pushed out in the same way as occurs with humans. The adult set forms beside the milk teeth and once they are well grown, the first teeth will fall out. This prevents an animal dependent upon its teeth for killing from ever having gaps which might prevent it feeding properly. Therefore, they are seldom without a set of teeth.

Canines

As tigers age their teeth, especially the canines, grow and develop. A young cat has under-developed canines, a somewhat older cat has larger canines and is no longer snow white, but a creamy colour. The adult tiger is well evidenced by the prominent yellowed canines. Out of four canines, two are on the top and two on the bottom. In each jaw one is on the right side and the other is on the left side. Tigers have the largest canines of all big cat species, nearly the size of a man's middle finger. All canines follow the same pattern. Length of all canine teeth increases till the age of 2 years and stabilises after that till the age of 8 years and falls after that. The size of male canine teeth is more compared to female canine teeth. The size of the male upper right canine is maximum at around 6 cm in length while rest of the canines are of an average length of about little more than 5 cm after the age of 2 years. For females, the average length of all canine teeth is about equal to or a little less than 5 cm after the age of 2 years. For both male and female tigers after 8-9 years of age, the canine size starts reducing due to wear and tear.

The canines have abundant pressure-sensing nerves that enable the tiger to identify the location needed to sever the neck of its prey. These teeth are used in holding the prey. Tigers have periodontal mechanoreceptors and proprioreceptors in their maxillary and mandibular teeth that allow exquisite bite sensation. These sensations are transmitted via the V2 and V3 divisions of the trigeminal nerve to the thalamus and thereafter to the postcentral gyrus of the parietal lobe. In tigers this system is particularly developed in the canine teeth, where a high receptor concentration exists, allowing them to finely adjust their grasp during biting. These sensory receptors provide real-time feedback to the tiger allowing the animal to align its canine teeth between the cervical vertebrae, especially needed in larger animals with more robust cervical vertebrae, to facilitate cord transection during powerful hyperextension.

Premolars

Tigers have six premolars on the top, and six on the bottom. Beginning with the bottom teeth, the first premolar is smaller than the other two, and has three high peaks (crowns). The middle crown is the highest. The second premolar looks much like the first premolar, with three sharp crowns, the tallest of which is the middle crown. The third premolar, however, has only two crowns, and each are about the same size. In the upper jaw, the first premolar is very small and only has one crown. It resembles an incisor more closely than a premolar. The second and third premolars resemble the bottom premolars, with the third premolar being one of the largest cheek teeth.

Molars

Tigers have only two molars, located in their upper jaws, behind the third premolar. Their molars are so small that they're almost invisible. Together with the premolars, the tiger's cheek teeth are called carnassials, which enable the tiger to shear meat from their prey like knife blades. They swallow large-sheared pieces of meat whole. Tigers are capable of penetrating deeply into their prey because of the large gap between the carnassials (back teeth) and the canines holding prey tightly.

The small incisors located in the front of the mouth (between the two top and bottom canines) enable the tiger to pick off meat and feathers from their prey.

Tongue

The tiger's tongue is covered with numerous small, sharp, rear-facing projections called papillae. These papillae give the tongue a rough, rasping texture and are designed to help strip feathers, fur and meat from prey. An important part of hygiene and cleanliness is the tiger's rasping tongue. In a typical grooming action the tiger will lick its paws and use them to wipe over the face, ears and forehead. Any injuries are carefully licked and coated with antiseptic saliva from the tongue. Tiger saliva contains lysozyme enzymes, which attack the cell walls of wounds providing defence against infection. This aids in preventing infection entering into a wound. Tigresses constantly lick the cubs to keep them clean as well as strengthening the bond between mother and cubs. The tongue is also used for lapping water.

Eye

In addition to the upper and lower eyelids that protect the eye, the big cats have a nictitating membrane on each eye that helps keep it moist and removes dust from the surface.

Binocular vision

Eyesight plays a significant role in tiger's hunting technique. Tigers have Binocular Vision, which allows the animal to judge its prey's distance accurately. The opening line of William Blake's iconic poem 'The Tyger' (1794) is 'Tyger Tyger, burning bright' (it's sometimes modernised as 'Tiger, Tiger, burning bright') highlighting 'the fire' of the creature's eyes. The retinal specialisations that cause glowing eyes. A tiger's eyes are the brightest of any animal and the way they bulge gives a wide angle view.

Evolution has seen the cat develop the most accurate binocular vision of all carnivores, but it is very difficult to see a stationary object, particularly if the cat is in moonlight and the prey animal is standing in the shadow. A prey that is only five metres away may well be invisible unless it moves, though only an ear twitch is required for detection. Fortunately for the tiger, but unfortunately for prey species, in the presence of a prowling cat other animals tend to become nervous and start to move, drawing attention to them.

Colour vision

Rods and cones are light-receptive cells in the eye. The more, the better is the colour vision. The more rods, the better the low-light vision, but these are of no use for seeing in colour. The eyes of the tigers contain primarily rods. Tigers have circular pupils and yellow irises (white tigers have blue irises). Due to the lack of cones in the eye these see depth rather than colour. There is some debate about how much colour tigers can actually see. Until recent times, it was considered that felids were colour blind, but it has now been established that green, blue and yellow may be recognised, along with various shades of grey.

Pupil's shape

Big cats have very obviously round pupils in comparison to the small cat's narrow and vertical slits.

Eye position

In the case of predators like the tiger, the eyes are directed forwards rather than one on each side of their head. This provides binocular vision because each eye's field of vision overlaps creating a three dimensional image. Binocular vision enables them to accurately assess distances and depth which is extremely useful for manoeuvring within their complex environment and stalking prey. So they can easily keep track of the prey animals. Prey animals have eyes on each side of the head; this widens their field of vision and enables them to watch better for the predators.

Eye colour at night

Generally, tigers have yellow or amber eyes with black irises. Contrary to popular belief, the eyes of the tiger do not always burn red at night. Under illumination, and depending upon the angle of reflection, the reflected light may appear anything from a reddish-yellow to a bluish-green.

Tapetum lucidum

Tigers have an adaptation that reflects any light back to the retina. This means the night vision of a tiger is quite remarkable and six times better than that of a human. Called the *Tapetum lucidum* (translation: "bright carpet") this reflective layer is located at the back of the eye behind the retina. Tigers have more rods (responsible for visual acuity for shapes) in their eyes than cones (responsible for colour vision) to assist with their night vision. The increased number of rods allows them to detect movement of prey in darkness where colour vision would not be useful. Any light not absorbed by the rods and cones is mirrored back to these eye parts a second time; this allows more light to be absorbed. This, along with a high number of rods, gives the tiger the ability to see in very low-light conditions and allows them to see better during the night as well. The big cats in general have a broad horizontal line of nerve cells near the central portion of their eye that enables them to have better peripheral vision. This characteristic is especially useful for hunting prey that is running across a plain.

Ear

The tiger's hearing is its most highly developed sense. It is far more dependent upon hearing than sight or smell and is mainly used for hunting. Their ears are capable of rotating, similar to a radar dish, to detect the origins of various sounds such as the high-frequency sounds produced by prey in the dense forest undergrowth. It may hear sounds up to 60 kHz whereas a human's upper auditory range is about 20 kHz. This sensitivity enables them to detect the high-pitched sounds emitted by prey and their movements. Such acute hearing allows the big cat to distinguish the rustle of leaves in the breeze from the sound of an animal brushing through the undergrowth. It is thought that other animals can be identified from the sounds they make during movement, allowing the tiger to concentrate on preferred prey. Tests have shown that tigers can seek out a grazing animal with less than a five degree error. It seems that a twig-breaking beneath a human foot, or the sound of a man breathing from within a hide, is also readily identified.

An experiment was conducted in late 1970s at Haldi camp of STR by fastening a pig with one of the watchtower's posts. The animal cried during the whole night; suddenly just before sunrise it stopped oinking. It was anticipated that a tiger might be present nearby. Even the birds had stopped chirping. A wild boar was also seen running away afar. It was observed that on the opposite side of the watchtower a tiger was sitting by the side of the sweetwater pond. After about fifteen minutes the tiger went deep into the forests. It did not appear for hours afterwards. The birds had started screeching and the chital stag began bellowing. Again in the evening there was pin-drop silence in the surrounding. Yes, the tiger had returned to the same place. Suddenly, the tied pig made a low sound. Instantly, the tiger was alerted and took position, moved through the forests towards the pig; after coming at a distance of about 10 m from the watchtower, it started crawling, with a lightning speed attacked the prey and broke its neck. It could capture only half of the pig's body and entered the forests to return to the place of hunting at night for the remaining part of the prey. This experiment has no doubt established the more powerful hearing ability of the tiger. It could not detect the prey earlier when it was present there because the pig did not make any noise but remained silent.

White ear spots

The small, rounded ears of the tigers have a most prominent white circular spot or 'flashes' on the backside, surrounded by black. This contrast between the white patch and the black rest of the ear could well betray an animal from a distance, asserting that light colours tend to contrast with forest vegetation. The reasons for these remain mostly unknown. These false "eyespots", called ocelli, apparently play an important role in intraspecies communication. There are, however, a few ideas as to the probable functioning of these eyespots. One of which is that they function as "false eyes" to fool prey; making the tiger seem bigger and watchful to a potential predator attacking from the rear. There is, however, no such natural predator in the Sundarbans mangroves.

The other idea is that they play a role in aggressive communication because when threatened tigers may twist their ears forward against a rival that serves as intimidation. These ear spots are prominent during the attack threat. They perform this technique in such a way that the spots can be seen from the front, providing a visual warning to the enemies that the tiger is alert and watching them. Even when a tiger used to sleep or take rest in the undergrowth of a mangrove forest, it reveals that the white spots on the back of the ears discourage other animals from mobbing. It is probably the hunter's tale that the tiger's ear markings resemble "a watchful and unwinking eye". A tiger sleeps deliberately in such a position that displays eye-spots: the tiger "puts his head on his fore paws and turns his ears rather forwards."

Another suggestion is that young cubs follow these markings ("Follow-me" signal) in the wild to find and follow their mothers in dense forests or tall grassland. When a tigress drops into the stalking position, with ears flattened against her head, these spots are obscured making it more difficult for the cubs to follow and ruin the kill. This suggestion is weakened by the fact that these spots appear on both males and females and males have nothing at all to do with the raising of cubs. In this context, it is worth mentioning that the white spots are remarkable even on the ears of tiger cubs.

Nose

The approximate age of the tigers may be known through the colour of their nose. While in young its pink in colour and keeps darkening to blackish-brown as they grow older. For details see Ageing.

Sense of smell

The sense of smell in the tiger is much lower than in the dog and a very small segment of the brain is devoted to interpreting scent. Some disagreement exists over how much the tiger uses its sense of smell in hunting with most experts agreeing that smell plays only a minor part. Experiments have shown that the tiger is unable to detect human scent, and even rotting meat is often overlooked until it is well and truly high.

However, the ability to discriminate between one odour and another is still very strong as the following examples show:

Tigers can communicate within their territories.

It is possible for one cat to tell the difference between its own markings and those of another cat - even if the second cat has overlaid those of the original.

A tiger can tell whether a scent belongs to a male or a female, and whether or not the animal is from the local area, or a stranger.

A male is able to determine whether or not a female is in estrus and so ready to mate by using flehmen. It may also be used in an attempt to identify any unusual smell.

After smelling a scent mark, a tiger will often grimace in an action called flehmen. The flehmen response is most often exhibited by both males and females where other cats have marked their territory by spraying scent. Through reading the odour it's possible to identify which cat left the scent and whether or not it is a stranger entering the territory.

Whiskers

Vibrissae or tactile hairs are commonly termed whiskers. Tigers have five different types of whiskers that detect sensory information and are differentiated by their location on the body. Whiskers differ from guard hairs in that they are thicker, more deeply rooted in the skin and surrounded by a small capsule of blood. The root of the whisker displaces the blood when the whisker comes into contact with something thereby amplifying the movement. Sensory nerves detect this movement and send signals to the brain for interpretation.

Mystacial whisker

The mystacial whiskers are located on the tiger's muzzle (snout) and are used when attacking prey and navigating in the dark. The tiger uses these whiskers to sense where they should inflict a bite. When navigating through darkness the tiger's pupils dilate to let more light enter the eye to increase their vision. The dilated pupils of their eyes assist their night vision but makes focusing on objects up-close difficult. The tiger's mystacial whiskers help it feel its way through the dark.

Superciliary whiskers

Superciliary whiskers are located above the eyes on the eyebrows.

Cheek whiskers

Cheek whiskers are located just behind the mystacial whiskers on the cheeks.

Carpal whiskers

Carpal whiskers are located on the back of the tiger's front legs.

Tylotrich whiskers

Tylotrich whiskers are located randomly throughout the body.

Facial whiskers

The facial whiskers of the tiger are about 15 cm in length with those of males being the longest and heaviest.

Whiskers grow and fall out just like nails and fur, but they are around three times thicker than fur and rooted very deeply in areas of the face which have a plentiful supply of blood vessels and nerves. Slightly longer tylotrich is single hair randomly spread across a felid's coat, which operates in the same manner as facial whiskers to deliver sensory information, while the whiskers on the rear of the front legs assist in tree climbing and during contact with prey.

Touch

Tigers have a well-developed sense of touch that they use to navigate in darkness, detect danger and attack prey.

Legs, Feet and Claws

The hind legs of the tiger are longer than their front legs. This characteristic enables them to leap forward distances up to 10 m. The bones of the tiger's front legs are strong and dense to support the large musculature needed to take down large prey. The bones in each of the tiger's feet are tightly connected by ligaments enabling them to buffer the impact of landing from running, pouncing and leaping. Tigers have large padded feet that enable them to silently stalk prey in the forests.

Tiger claws are retractable in that ligaments hold them in a protective skin sheath when they're not being used. The ligaments are in a relaxed position when the claws are retracted thereby expending no musculature effort. Tigers retract their claws to ensure that they remain sharp for times when they are needed and to tread silently up to unsuspecting prey. Other ligaments will extract the claws when attacking prey or defending themselves which does require musculature effort.

The claws of the tiger are up to 10 cm in length and are used to grasp its prey and bring it down, until the jaws can go for the prey's neck. Each paw has four of these claws and one specialised claw called a dewclaw. A dewclaw is located farther back on the foot and thereby does not touch the ground when walking. Dewclaws function similarly to thumbs in that they are used for grasping prey and aid in climbing.

The Sundarbans tigers have claws adapted to strike and the somewhat curved canines, the longest among living felids with a crown height of up to 90 mm, designed to hold prey and used for biting and killing; short strong jaws are controlled by powerful muscles, soft pads for steadily approach make the tiger capable of sudden speed and burst of power.

Tiger claws are curved which enables them to superiorly grasp and hold large prey and climb trees head-first. However, the claws' curvature, the tiger's size and weight is a great hindrance in climbing down from trees. Tigers must either crawl backwards or jump down from trees, making them the most inferior climbers of the big cat family.

For cleaning the tigers repeatedly scratch. They mark out territory through the scratching of trees. Once a tree is chosen it is used repeatedly leaving deep longitudinal marks on the trunk. Not all tigers scratch at trees, some do it frequently while others don't bother. Secretions from glands in the feet leave deposits which are easily smelled by other tigers. For experts the height of claw marks on trees is an indicator of the tiger's size.

Stripes

The number of stripes in different subspecies of the tiger is shown below (Source: Seidensticker *et al.*, 1999 'Riding the tiger: tiger conservation in human-dominated landscapes').

Subspecies	Average mid-flank no	Range	Standard deviation
Javan *P.t. sondaica*	30.5	23-37	4.34
Bali *P.t. balica*	30.5	23-37	4.34
Sumatran *P.t. sumatrae*	27	20-34	5.94
Caspian *P.t. virgata*	26.8	24-28	1.89
Indochinese *P.t. corbetti*	26.4	21-31	2.54
Siberian *P.t. altaica*	22	22-26	2.52
Bengal *P.t. tigris*	22	21-29	3.23

Across the species, the average number of stripes mid-flank is around 26. Multiplying this by 2 for both flanks to make 52; add in the hind legs at 10 for each leg making 20; add the forelegs at 3 per leg making 6; add in the tail (ring) at about 10; finally add the stripes on the tiger's head at about 10; the total is 98 stripes. Bearing in mind (1) the difficulty in counting and (2) the variation between species and individual tigers, it would be wise to round it up to about 100. In case of the Bengal tiger the total number of stripes would approximately be 90 (22 x 2= 44 + 20 + 6 + 10 + 10) or maximum 97. Of course it is not a very precise process as a tiger's stripes are irregular in shape and fragmented.

Alleged cryptic properties of the tiger's stripes are the exemplum for aggressive camouflage among felids. Generally, it is supposed that the stripes match the natural background of the habitat of the tiger. Wallace (1870) gave a classical description of the function of the tiger's stripes:

"The tiger is a jungle animal, and hides himself among tufts of grass or of bamboo, and in these positions the vertical stripes with which his body is adorned must so assimilate with the vertical stems of the bamboo, as to assist greatly in concealing him from his approaching prey."

Baker (1890) links the colouration of the tiger directly with grasslands:

"The striped skin of the tiger harmonises in a peculiar manner with dry sticks, yellowish tufts of grass, and the remains of burnt stumps, which are so frequently the family of colours that form the surroundings of the animal."

On the other hand, Mottram (1915) considered the tiger as an excellent example of "reed-painting" (i.e. reeds being seemingly imprinted on the coat) and brought detailed description of this analogy: "(Dark stripes) are irregularly distributed, they for the most part increase in width from above, down, and on the belly they terminate in a large dark mass which, curiously enough, often shows a small white centre."

Cott (1940) refers to the tiger's stripes as a combination of both Baker and Mottram stating that "(…) the vertical tawny-orange and black stripes of the tiger assimilate with the tall parallel grass stems and reeds of the swamps and grassy plains where it lives."

As the outline of the tiger appeared to be hardly distinguishable against the background, the conclusion arises that tiger's stripes resemble the stems of grass or reed in the eyes of its prey. However, Perry (1965) mentions that most hunters reported the tiger's coat harmonises with either reeds, grassland or forest, making him difficult to be spotted both by day and by night. A common prey, the deer, is likely to be "colour-blind" (respectively dichromatic) and its sight is challenged during crepuscular light conditions. Even if it works under certain light conditions, the problem is of a secondary importance since deer rely mainly on the sense of smell and hearing.

There are basically three main hunting methods used by big cats which can be briefly called as ambushing, running, and stalking. Before attacking a victim from ambush, a felid is perfectly still, waiting for its prey to come as close to the cover as possible. Tigers occasionally use this method in the vicinity of water sources because prey frequents these areas relatively often. Stalking is also characteristic for the tiger. The method of running simply means that a cat is given a chance to run undetected at the start of its pursuit, for example when its prey suddenly emerges some 100 metres away, so there is no need for a hidden approach accomplished by stalking.

However perfect a tiger's camouflage may be, its value is negative the instant he moves, because any moving object excites the suspicion of the animal he preys upon. Deer (and other animals) are sensitive to moving visual stimuli and have trouble to recognise still objects. The grasses in the foothills of the *terai* are red or ochre or brown for half the year, and blend with a tiger's colouring, but this does not prevent his slightest movement being immediately spotted by deer or monkeys.

Nevertheless, tigers hunt predominantly at night. The lack of light causes impossibility to distinguish colour hues, most animals probably possess monochromatic vision at night in effect. Camouflage could serve well under such conditions, but it is more likely tigers simply benefit from a lowered level of light on the whole. Under an overcast sky, it can be very difficult for some species to merely orient in forest. Tigers themselves have excellent vision during the night.

In twilight only the white chest and ruff of a tiger are visible, and his stripes break up the outline of his body, as they do in deep day-shadow also and the crazy face-markings of staring white eye-spaces with their frowning angular black markings, surmounted by black-barred forehead, produce a blurring effect.

Facial markings

Tigers have mysterious facial markings on their forehead, cheekbones and under their eyes. The black coloured facial markings are the most common ones and they are found within orange fur and the white furred tigers. The reddish-brown coloured facial markings are only present within Golden Tabby tigers whose stripes are also reddish-brown coloured. Tigers are possibly the only big cats within the big cat family which have two colours of facial markings on their face. The pattern of the facial markings on the tigers also helps them create the camouflage impact just like their stripes do for them during the hunting.

In fact, these conspicuous facial markings are the identification keys of the big cats in the wild. First of all, the facial markings of the tigers are different from other big cats. Other big cats usually have spotted facial markings. Such big cats include the leopards, jaguars and the clouded leopards. On the other hand; the facial markings on the tigers' face (especially on their forehead) are linearly patterned. Furthermore, these linear markings appear to be in horizontal lines on their forehead.

Secondly, the facial markings on each and every tiger are unique and this way they can be uniquely identified against each other. No two tigers have the same set of facial markings. Therefore, just like the stripes of the tigers, the facial markings may act as the fingerprints which means that one tiger can always be differentiated from another tiger, no matter how identical the tigers look in their appearance.

Lastly, the facial markings of the tigers also help them against the bites of the biting flies. It's been proven scientifically that the pattern of the facial markings on their face or the stripes on their fur disrupts the landing of the biting flies, as the flies get dazzled and seek danger within them. They consider them as some insect-eating reptiles, therefore they also feel scared of them at first hand.

The combination of short convoluted stripes on the forehead, white eye patches and long stripes accentuating the outline of the head make up an imposing appearance of the big cat. This general pattern involves a number of varieties so individuals can be recognised by differences in facial markings, as was demonstrated by Schaller (1967) in Kanha. That no two individuals display the same pattern on the face has been substantiated in case of both males and females in the mainland, for example, Machali of Ranthambore with a fish mark on her cheek; and Sundarban mangrove e.g. Nantu with a trident mark on his cheek. In Kanha, a male tiger, named Munna, was born with black, letter-shaped markings on his forehead.

In the wild, the researchers always classify one tiger from another on the basis of their facial markings or the pattern of the stripes on their fur. Do the tigers identify each other on the basis of facial markings? Some observers do believe that the tigers are intelligent enough to recognise each other through their facial markings as well.

Gait

The gait of individual tigers varies from animal to animal, but certain things are common to them all- walk on their toes; their heels are positioned halfway up their legs. While walking, the tiger lifts both limbs on the same side together. Some people claim that the tiger has almost mathematically precise movement with the hind paw stepping in exactly the spot previously occupied by the forefoot. Others have done careful measurements and claimed a slight lag by the rear paws. Yet another idea is that the longer length of the hindlegs creates an impression of the hindfoot falling ahead of the forefoot. It now seems most likely that the overlap only occurs during the stalking of prey, but in a normal walk the separation can be measured accurately to find a gap of some centimetres. This separation between each stride is used in the field to estimate the size of a tiger. When running flat out the tiger's gait changes so that only a single paw touches the ground at any one time. These big cats can move extremely fast over short distances, having a top speed of around 56 km per hour, but they cannot sustain this speed in the way a cheetah can.

Though the Sundarban tiger remains one of the most famous wild animals in the world, very less authentic data about their body growth data are available.

Art of chemical immobilisation

Wild tigers may need to be captured and immobilised for various purposes like rescue, relocation, monitoring, diagnostic testing, management of disease, and euthanasia. However, wild tiger capture and immobilisation requires specialist training to ensure the process is carried out as safely and humanely as possible.

Of late, there is a growing management trend of rescue, capture, sedation (tentative dose of an adult tiger 150-200 mg Xylazine with 400-600 mg Ketamine), revival (recovery dose of antagonist @ 10 mg Yohimbine against 100 mg of Xylazine) and rehabilitation of the straying tigers in the wild. Also tigers are being radio-collared, micro-chipped or ear-tagged to study their behaviour, home range etc. For all these management practices, tigers need to be tranquilised with the help of chemical immobilising drugs and revived with the help of Antidote or Reversal drugs.

Problems (Lewis *et al.*, 2012)

When a tiger strays into a village, people often surround the tiger. As time goes on, the crowd grows as people arrive in the area to see the tiger. If the authorities cannot control the crowd then the tiger is normally killed. A large crowd and noise can make the tiger nervous and scared and so may provoke the tiger to attack nearby people. The crowd may be so angry that it is impossible or unsafe for the immobilisation team to approach and deal with the tiger. Furthermore, the people's situation may influence options for monitoring, moving, or releasing the tiger after it has been anaesthetised. So it is wise to manage the crowd from the beginning of the incident and keep the crowd at a safe distance from the tiger.

Prior to anaesthesia, there are several factors to be assessed: weight, sex, age, physical condition, and psychological condition from a distance, preferably by using a binocular. The weight needs to be estimated to calculate the amount of drug to use. Sometimes it is not always easy to get a close look at the tiger, but for the Sundarbans, female tigers will weigh roughly 75-85 kg and males may weigh 110-140 kg. Young tigers may require more and older tigers may require less drug per unit body weight than prime-age adult tigers. Furthermore, a starved, sick, exhausted, or malnourished tiger will usually require less dosage than a healthy, well-fed tiger. There is a higher risk of mortality for tigers in bad physical condition, despite everything being done correctly. Excited tigers require higher anaesthetic doses, take longer to become recumbent, and are prone to more anaesthetic complications. A critical period of stress for a tiger is the period between when it first detects a person nearby and when it is anaesthetised; stress levels increase dramatically when a tiger detects a person nearby and the tiger may hurt itself trying to escape.

When working with tigers the following precautions may reduce the risk of injury or death to people or the tiger. A forked stick can be very useful to hold or beat off a tiger or make yourself look more threatening. A gun can be used to either scare away the tiger with a blank fire, or to shoot the tiger as a last resort if human life is being threatened. It is also important to be aware that tigers will naturally attack animals that are afraid of them and are trying to escape. If you face the tiger together as a team then the tiger may be less likely to attack. Also, the more the tiger feels threatened or angry (for example if people are throwing stones at the tiger) then the more likely the tiger is going to attack. Likewise a tiger may feel threatened if someone is too close. It is therefore a good idea to keep at least 100 m away unless you have to get closer to save someone's life or capture the tiger. Surrounding the tiger may also give the tiger no option but to attack. Leaving an escape route gives the tiger the option of running back to the forest.

Aiming to dart the tiger in the large muscle masses present at the front of the upper hind leg or around the shoulder area will be safer than the head, chest wall, abdominal wall, lower legs and genital area. Injections in those areas are very painful and may injure vital organs like the eye, ear, heart or reproductive organs. If possible, animals should be darted when standing still. The loose skin folds present when sitting or lying down will often absorb much of a dart's impact energy, resulting in subcutaneous delivery or even failure to penetrate the skin at all. If an animal is moving when darted, there is a higher risk of dart deflection and inaccurate placement.

After the tiger is fully immobilised it is necessary to check "ABC" (Air way, Breathing and Circulation) to assess the initial vital signs. Then one should check the general health condition of the tiger. An immobilised tiger should be transported to a safe place at the beginning of the anaesthesia using a stretcher or a wooden box. Wherever possible release the tiger at the site of capture or close by.

Details of tranquilisation of tigers during 2020-2021 in STR

Date	Description	Location	Immobili-sation drug used/dose	Post-tranquili-sation status	Remarks
21.12.2020	Adult female, 10-11 years old, weight 90 kg, length 85 cm.	Harikhali post side, Harinbhanga-2	Ketamine 10 ml by Pistol at about 10.54 am	Tigress was taken out from the cage and medically checked-up.	Unable to hunt due to loss of canines and senile age and sent to Jharkhali camp.
26.12.2020	Adult male, 7-8 years; weight 120 kg; length 90 cm.	Bolkhali Khal side; Harinbhanga-1	Ketamine 4.5 ml and Xylazine 2.5 ml by Jab stick at about 11.20 am	Tiger was taken out from the cage and medically checked-up.	After treatment it became fit for release in nature with radio-collar.

Further extensive research must be undertaken before making any conclusive remark on insular dwarfism. Thousands of camera trap images available during the tiger monitoring might focus on this controversy righteously.

The tigers from northern and northeastern India have deeper reddish coats and those in central, southern and western India have a lighter coat, which has a clear correlation with the vegetation. Similarly, the Sumatran tigers are darkest in densest vegetation, and Siberians are lightest in sparse vegetation. The Sundarbans tiger is far stockier and thick, and almost brownish-red in colour. Its glossy coat is often damaged due to salinity.

However, recent sightings and photographic records of many wild tigers in the mangrove forests over the last few years have indicated that those tigers are apparently as big as those found in the mainland to be considered underweight.

Genetic studies continue to investigate whether the Sundarbans tiger is a different subspecies, but considering the mutation rates that led to a genetic change, usually an animal that was isolated for a period of one Myr was classified as different species and one that was genetically isolated in between 20,000 and 50,000 years was a different sub-species. In the case of the Sundarbans tiger, this criterion is still far from being fulfilled.

Haplotypes are a set of closely linked genetic markers present on one chromosome which tend to be inherited together. In 2022, generation of 1263 BP of mtDNA sequences across 4 mtDNA genes for 33 tiger samples from the Bangladesh Sundarbans after comparing these with 33 mtDNA haplotypes known from all subspecies of extant tigers, has detected three haplotypes within the Sundarbans tigers, of which one is unique to this population and the remaining two are shared with tiger populations inhabiting central Indian landscapes- that is to say, the mainland tigers in the Central Indian landscape i.e. Madhya Pradesh, Maharashtra, Chhattisgarh and parts of Andhra Pradesh (e.g. Bandhavgarh, Pench, Kanha, Tadoba and Nagzira) and the Sundarbans tigers have the same lineage and the latter have been found to be different genetically from North Indian tigers (Corbett NP). In that case, despite their outer differences, the Sundarbans tigers are genetically more close to the isolated peninsular big cat population. The sighting of tigers in Midnapore along the border of Singhbhum district of Jharkhand, Mayurbhanj district of Odisha near Nayagram and Haldi and other rivers visiting from the Sundarbans and from the western jungles has been reported (Bahuguna and Mallick 2010).

Chromosomes

Chromosomes are the carriers of genes. The chromosomes of the tiger are arranged in 19 pairs giving a total of 38 chromosomes altogether.

Genetics

Each tiger has two sets of genes inherited from their mother and father. So there are two variations (alleles) for each kind of gene. Individuals have a random mixture of homozygous and heterozygous genes. When a different allele for a particular gene comes from each parent, it is said to be heterozygous for that gene. When the same allele for a particular gene comes from each parent it is said to be homozygous.

Table: Genetic diversity indices and effective population size (NE) of Sundarbans, peninsular and northern India tiger populations based on nine microsatellite markers and mtDNA sequences. N = sample size, MNA = mean number of alleles, AR = allelic richness, HO = observed heterozygosity, HE = expected heterozygosity, FIS = inbreeding coefficient, S = number of segregating sites, h = haplotype diversity and π = nucleotide diversity (Singh, 2017)

Population	Microsatellites						
	N	MNA	AR	H_o	H_e	F_{Is}	N_e (95 % CIs)
Sundarbans	13	3.33	3.24	0.491	0.587	0.109	−241 (10-infinity)
Peninsular	73	7.33	4.80	0.492	0.707	0.300	73 (48–128)
Northern	62	4.88	4.18	0.401	0.674	0.394	48 (30–92)

Population	Mitochondrial DNA				
	N	S	h	π	N
Sundarbans	8	3	0.679	0.001	8
Peninsular	15	19	1.00	0.002	15
Northern	8	3	0.820	0.001	8

Population genetic structure and gene flow on Sundarbans tiger landscape

Alignment of the 2,600 bp sequences of Sundarbans tigers (n = 6) with the sequences from six tiger subspecies revealed three nucleotide substitutions that were specific to the Bengal tiger (*Panthera t. tigris*) and two that were specific to the Sundarbans tiger. Of these, Sundarbans specific (or unique) substitutions were located in the ND5 gene and were present in all the six Sundarbans tiger samples, while the other substitution in the cytb gene was only found in three samples. The three nucleotide substitutions, specific (diagnostic) to the Bengal tiger but found also in the Sundarbans tiger, indicate that Sundarbans tigers belong to the Bengal tiger (*P. t. tigris*) subspecies. The Sundarbans tigers were a monophyletic group with other Bengal tigers in the median-joining (M-J) network. Further evaluation of Sundarbans specific substitutions with the 1026 bp sequences (cytb, ND2, ND5 and ND6) of other Bengal tiger populations (northern and peninsular) reveals that these haplotypes are unique to the Sundarbans tigers. In addition, the M-J network with this 1026 bp data revealed that the haplotypes typical to Sundarbans tigers comprise a separate group that stems from peninsular India and had further split into two lineages. This indicates that the Sundarbans tiger population contains two new closely-related mitochondrial lineages, which had not been identified hitherto. The lower number of mitochondrial lineages in Sundarbans in comparison to mainland tiger populations suggests that genetic drift and founder effects may have had an extensive impact on the Sundarbans gene pool. The absence of the haplotype unique to Sundarbans, in other tiger populations indicates that there has been practically no emigration of females from

Sundarbans to peninsular India after its geographical isolation. This is probably because there has been no suitable habitat for tigers to disperse to from Sundarbans.

Singh (2017) found significant genetic differentiation between the Sundarbans and mainland tiger populations with both microsatellite and mitochondrial markers.

Table: Pairwise FST-values between Sundarbans, Peninsular and Northern India tiger populations using microsatellite (below the diagonal) and mtDNA markers (above the diagonal)

Bengal tiger populations	Northern	Peninsular	Sundarbans
Northern	0	0.058*	0.250*
Peninsular	0.030**	0	0.116*
Sundarbans	0.070**	0.069	0

$*p < 0.05$, $**p < 0.001$

Microsatellite-based Bayesian clustering analysis using STRUCTURE with a large geographical scale (Sundarbans, northern and peninsular) suggested five genetic clusters with the mean $LnP(K) = -2882.64$, whereas K, calculated from the output of STRUCTURE results, proposed that the most likely number of clusters is four. However, many studies have argued that K does not always produce a proper resolution of population structure. Clustering analyses performed with the program TESS gave results similar to STRUCTURE with five genetic clusters. Thus, it is found that $K = 5$ would be more appropriate and justifiable. At $K = 5$, Sundarbans individuals were assigned together to a distinct cluster with high probability ($q > 0.8$) and there was no evidence of admixture. These analyses also suggested that the Sundarbans tiger population is an isolated population.

The Sundarbans tigers are separated by birth and for survival adapt to their unique mangrove habitat and climatic conditions, where land and hydrological regimes (brackish or freshwater) constantly change the dynamics of the environment. Moreover, their linkage with the northeastern tigers is not established. The earlier study that pegged all

Indian tigers in one group concluded the northeastern ones shared similarities with ones from central India. A later report using many more genetic samples determined that the northeast was a unique cluster in its own right. Therefore, tigers from Northeast India are also distinctly different from the rest of India in their genetic composition. This is likely as the Northeast tiger population probably forms a zone where there has been historic gene flow from Myanmar. Therefore, the tigers of the Northeast have more chances to share their genes with the Southeast Asian tigers of Myanmar, Thailand and South China because of their habitat connectivity in the trans-boundary areas. The Northeast, Dibang and Namdapha form one population cluster, while Manas, Kaziranga, Nameri and Buxa formed a second cluster.

Most of the time in the Sundarbans you cannot see the tiger, but you can see impressions in the mud. Reading pugmarks is a traditional art of the professional trackers or hunters. The pugmarks on soft soil or sand help trackers determine the type of a tiger and also the sex. Pugmarks usually have two parts. One is a pad with three lobes on the rear end. The other part is composed of four fingerprints. The fifth finger of the tiger is called Dew Claw, found only on the front limbs, it is placed slightly upwards and does not touch the ground; so its mark is not usually seen on the ground. This dew claw is retractable and is used during hunting. Like our longest left and right middle finger, the tiger's paws are similar. The tiger's front footprint is slightly larger in size than the rear one. Whereas in the front paw, the forwardmost points of the two middle toes are almost at the same level, in the hind one, the tips of the two middle toes are distinctly at different levels. The fresh footprints are intact and look fresh. Old impressions are dry and the edges of the impressions are fragile. In the sloppy, slushy silt, most tiger pugmarks look like formless holes dipped in the mud.

The shape and size of a tiger's pugmark is the indicator of its sex. For this purpose, you have to measure the length [PML - Pugmark Length] by drawing an imaginary line from the tip of the tiger fingerprint to the base of the pad and the width [PMB - Pugmark Breadth] by drawing an imaginary line along the two farthest points of the first and fourth fingers (not the fifth fingerprint). When a tiger saunters along a creek in the Sundarbans and leaves its pugmarks and if you note that the pads are comparatively smaller, slightly rectangular, each measuring about two and three-quarters inches, with elongated toes and less space between the toes than that of the male, the individual may be taken for a female because the pad marks of a male would be larger, roughly squarer and broader with rounded toes and more space between the toes than that of the female. The pug profile of a female is oval against the circular outline of a male. In a male, the difference between PML and PMB is <1.5 cm, while in case of a female the same difference is >1.5 cm.

The tiger's toes and pad are spread unlike those of the leopard (in the absence of the leopard in the Sundarbans, there is no scope of confusion in the field). The tiger's stride, i.e. the gap between the left front and rear or right front or rear is about ten times that of the PML (length) in case of a tiger cub. If the measurement between any two pugmarks is more than 30 cm, the stride is about three times of it.

In order to distinguish between the frontal and hind foot prints the gap (lengthwise) of the middle pug and the ring pug is smaller for the frontal paws while it is relatively more in case of rear or hind paws. If you consider the left and right pug marks, the middle digit is second when measured from left to right, then that is the right paw; for the left paw, the middle toe is third when measured from right to left. If you see no cracks in the pads and rings, the individual is <three years old, but in the case of the old tigers, they will have lots of cracks in the pad and rings. Because tiger cubs stay in the mother's area and move with her until they grow up, their pugmarks could easily be identified. By noticing those impressions on the muddy soil, we could assess the changes caused after the individual left them behind, such as weathering. We can also estimate the approximate time when that individual crossed a particular bank.

Aging

Ageing of the tigers is done using body size, characteristics, measurements, teeth eruption, wear and colouration, gumline recession and nose pigmentation and, as such, are divided into six age groups (stages: cubs <12 months, juveniles 1-2 years, sub-adults >2-3 years, young-adults >3-5 years, prime-adults >5-10 years, old-adults >10 years)(Jhala and Sadhu, 2017). These groups are described below.

1. Cubs (<12 months)

Born toothless, the cubs develop milk dentition by 1-1.5 months. They are usually observed after they are about two months old. At 2-3 months, the cubs look like a large domestic cat, when they are usually restricted to their birth site. The colour of their iris is blue-grey and begins to turn amber by the age of 3-4 months. The cubs are born toothless and develop their milk dentition by one to one and half month age. They start to eat some meat by two-months of age but largely depend on their mother's milk, which changes to a more meat diet by 4-5 months. By 5-6 months, the cubs are usually weaned to start consuming meat. At that time they become as large as a jackal and reach the belly of their mothers at this stage and accompany the mother to nearby kills to feed. The scrotum of the male cub is visible from early on, and there is a distinct size difference between male and female cubs by the age of 6-8 months.

2. Juveniles (1 to 2 years): Juvenile tigers accompany their mothers to larger kills and are usually not photographed alone, making size comparison easy. Tigers at this stage are roughly half the size of their mothers (about the size of a leopard, 50-120 Kg). Male tigers show faster growth than females and are seen to be substantially larger. Face proportions are cub-like, with a shorter snout and smaller face and has developed or partly developed permanent dentition (which begins at the age of about 9-10 months and is completed by 12-14 months.

3. Sub-adults or transient males or females (>2 to 3 years)

At the sub-adult stage, the males are substantially larger than the females. Often sub-adult tigers move around with their siblings, but by 30 months they become more solitary. The body is almost as large as adults and can no longer be used for size comparisons as most of the camera trap images are of solitary tigers. In exceptional cases, sub-adult males can weigh as much as 200 Kg and females as much as 120 Kg by 30 months. However, most sub-adult males are usually between 130-170 Kg and females 80-100 Kg.

The body proportions begin to fill up like adult tigers after 20-22 months, the face is close to that of the adult, losing its juvenile proportions as the snout elongates. Close inspection with binoculars by an experienced observer can still distinguish facial features of sub-adults from that of adults. The belly of sub-adults is flat and taut compared to a more rounding of the belly in adults. Skin flap on the belly is missing in sub-adults. The definitive feature to identify sub-adult tigers from adult tigers is their teeth observed with binoculars or photographed with a telelens.

In sub-adults the permanent dentition is fully formed, sharp, intact and the canines are milk white often with a pinkish tinge, tips are pointed without any wear, and there are no signs to show a receding gum line. A prominent ridgeline on the inner side of the canines and a groove on the outer edge of the canines is clearly visible. The nose is usually pink with no black specks or pigment. Facial hair in the form of a short mane below the lower jaw/cheek is usually seen even in camera trapped photos of males.

4. Young-adults (>3 to 5 years)

By three years most tigers are close to full adult size, but continue to accumulate weight up to 4-4.5 years of age. Adult males range from 200 to 260 kg, while adult females range from 110 to 180 kg showing a pronounced sexual dimorphism in size. By this stage, the face is no longer cub-like with full snout and adult skull proportions. Belly gets rounded, often with a slight sag which increases with age. Often a skin fold on the belly begins to show. Teeth start to turn cream colour to yellowish by three to four years and are no longer milky white. By five years of age, the yellow canines begin to get brownish stains. The canine ridge and groves are visible, with little or no wear on teeth. The nose is usually pink, but sometimes a few black specks or pigmentation are seen.

Pregnant and lactating tigress: A heavily pregnant tigress can be distinguished from a fully fed tigress by the visibility of prominent teats and udders. After birth, the belly is normal with full udders where nipples show signs of intense suckling. For un-bred females and early days following first births the nipples are pink in colour and become pigmented, darkened grey and keratinised after cubs suckle intensely. Nipples subsequently retain this grey coloration throughout life.

5. Prime-adults (>5 to 10 years)

The belly is sagging and rounded, often belly fold is visible. Teeth are brownish-yellow and begin to show wear which is visible on canines (no sharp points but rounded) and incisors with a binocular or through telephoto photographs. The canine ridge is almost indiscernible, and the grove is highly worn out. Careful inspection shows a receding gum line on canines, making the canines appear larger. Black spots on the nose and slightly sagging lips on the lower jaw are often seen.

6. Old-adults (>10 years)

Belly and belly skin fold are sagging; hip bones tend to protrude; nose shows pigmentation. Canines and incisors are worn down, often broken or missing and with dark brown stains, jaws and lips are often sagging, the lips show a fold. Close observation with binoculars/telephoto-photographs show receded gum-line on worn canines (if any). In their last few months some become blind and may experience partial paralysis.

Body condition of tigers is usually not a good parameter to use for ageing. Often very old tigers that have lost their canines can be in poor condition but can regain condition with just a couple of good meals.

By using the criteria described above, age of an adult tiger may be determined with an error margin of about a year, and younger tigers (sub-adults and juvenile) to about 3-4 months. Cubs can be aged with the accuracy of about a month by field observations.

Longevity

Tigers in the wild face a great many dangers and the odds are against them living past even their early months. One cub often dies at birth, this is usually the runt of the litter. The other cubs are prime targets for predators, including their own father. Once they become adults, the defence of territories, particularly by males, places the tiger at risk of injury or death. Chronic shortage of prey can also result in starvation and some are killed by the hunters, poachers or villagers in retaliation.

Usually the Bengal tigers live up to 20 years or more in captivity and when in the wild, their lifespan is around 12-15 years. In reality, a Bengal tiger was observed to live up to 26 years. It was a Sundarbans tiger, later named Raja in captivity. He is reported to be one of the longest surviving (25 years 10 months) tigers in India. He died at the South Khairbari tiger rescue and rehabilitation centre at Alipurduar's Madarihat on 11th July, 2022. Raja was brought to South Khairbari in August 2008, after being injured by a crocodile in a territorial fight in the Indian Sundarbans with more than 10 injuries. He was almost 11 years old when brought to South Khairbari, and there he survived for another 15 years. Raja's age has been estimated using the gum-line recession of the upper canine teeth, a physiological aspect of ageing. The success story of Raja will always be remembered as one of the greatest and rarest examples of successful *ex-situ* conservation of the Bengal tiger.

Chapter 9
Behavioural Adaptations

"The tiger lies low not from fear, but for aim." – Sherrilyn Kenyon

The other is silent and smooth in his movements, indulging, and that rarely, only in a low moan or a deep drawn breath, sounding like a snarl with a growl at the tail of it.

Dr. Venkata Ramana Narendra tracked a single tiger and a tigress in the Tipeshwar WLS in Maharashtra and observed their cubs as they grew up and reproduced over seven years and portrayed different emotions such as the fierce stare, the care, feminine nature of the tigress and the playful attitude of the cubs. There are so many iconic Bengal tigers in India. The stories of some memorable Indian tiger kings and queens having unique local names are described below.

1. Collarwali, daughter of Barimada (big mother), who established her territory in her mother's core area and was fondly called 'Mataram' ('beloved mother') or 'Queen of Pench', was a prolific breeder, and contributed heavily to the gene pool of the Pench TR with 29 cubs in eight litters in her lifetime, the first in 2008 and the last in 2019. The frequency of it getting pregnant is also surprising. Cubs of its 5th litter were not even 2-years-old when she became pregnant again. She littered three beautiful cubs for the first time in May, 2008. But as an inexperienced mother, she could not protect them from the harsh climate. All three cubs died of pneumonia within 24 days. In October 2008, she produced her second litter- four cubs including three males. All survived. Experts claim she has used 13 different sites to rear her cubs and most of the times it was caves. Collarwali was the first tiger in Pench to get fitted with a radio collar, thus earning her the moniker 'Collarwali'. She breathed her last on January 15, 2022 due to complications of old age.

2. Machali was the Tiger Queen of Ranthambore, named after a mark on her left cheek that looked like the outline of a fish, who gave birth

to a litter of three female cubs. One of them, the dominant one, T-16 (code name), as she was officially numbered, was referred to as 'The Lady of the Lakes'. She successfully raised four litters till her old age. Machali died in 2016 at a ripe old age of 19 years, the longest that any tigress has ever lived in the wild. This long life was probably because, 2014 onwards, she was given live bait periodically (almost once a month) as it became difficult for her to hunt successfully.

3. Bamera overthrew his ailing father B2 to become the dominant male and the Hulking Giant of Bandhavgarh. He had gained notoriety for lifting cattle in the fringe villages. He hurt both of his forepaws in a battle with Jobi. Jobi lost one eye. Later Bamera also sustained injuries, most probably from an emerging tiger Bhim. By the middle of 2016 Bamera had passed away.

4. Backwater Female- the Star of Kabini, referred to as Kismat by some, extended her territory towards Kakanakote and the backwaters of Kabini in Nagarahole. She was often seen courting with the Mastigudi Male, a massive tiger that roamed in the same area. She had her first litter in the late summer of 2018; the three little cubs were frequently sighted during the monsoon and winter months. She had three cubs in the second litter as well.

5. Paro or Paarwali belonged to the last litter of a tigress known as Thandi Maa from Dhikala Chaur of Corbett TR. Her territory was across the Ramganga river and she gradually extended it to both sides of the river, by chasing away two older tigresses. A petite-looking tiger by Corbett standards, Paro was bold, confident and fierce when defending her territory. Paro had her first litter in 2016, but she was an inexperienced young mother and lost her three cubs to a leopard attack. She gave birth to the next litter in 2017, of which one did not make it through the monsoons. The only surviving cub was a young boisterous male. In the summer of 2018, the young cub, fondly called Nawab, was mauled by a tigress invading Paro's territory. Paro's last litter was in 2018, three females. Paro was ousted from her domain and last seen in 2020. New Paarwali who almost two years back wrestled the control of the territorial area from her legendary mother Paarwali.

6. TT2 of northern Sahyadris was first photographed in the forested valley of southern Maharashtra in 2014. Later TT2 was found with three cubs! In 2017, she was captured again with a new litter of three cubs! The sub-adults from the previous litter had mostly dispersed by then. The cubs from 2017 also dispersed the following year (two survived to become sub-adults), and one of those now resides in the Mhadei WLS of Goa, photographed in 2022. TT2 still survives in the valley.

7. Link 7, a male of Kanha, named Chota Munna after his famous father Munna, spent the initial months playing smart, avoiding the larger males, disappearing for weeks together into the hills and then returning to the lowland battlegrounds. He was chased around and injured by the other males, but he always came back stronger. Later he had emerged as the top male tiger of central India ousting (possibly) two of the other males. He disappeared in 2019-20, for reasons unknown, while he was in the prime of his life. Tigers often die or incur serious injuries while fighting for dominance.

8. T-24, popularly known as Sonam, is the dominant tigress of the famous 'Four Sisters of Telia' in Tadoba. She won the Telia Dam territory from her mother T-10 aka Madhuri, a decade ago. She had a 'S' mark on the right side of her neck. In 2020, Sonam's young cubs were a major attraction! Sonam and Bajarang were together, prior to her giving birth to her latest litter of three cubs. She fought with her sister Lara to keep the territory she had built over a decade. Sonam still holds the crown of 'Queen of Telia'.

9. There are no official or public names for the tigers in the Sundarbans unlike the mainland tigers. However, there is one exception. Big Boss 'Nantu', locally called 'Dakshin Roy' of the Sundarbans mangroves, was first seen after the onslaught of 'Aila' in 2009 on the opposite side of the Netidhopani forest in Pirkhali-7 compartment during a tiger and prey-base assessment in 2010. He had a trident mark on his cheek and a distinct spot on the nose and tail. He was a dominant male tiger who held his territory in the mangroves for more than a decade. Be it Pirkhali, Nebukhali, Netidhopani or Nawbanki, his royal movement or loud roar made him the uncrowned ruler of a huge territory in the Sajnekhali Wildlife Sanctuary (WLS). Nantu was so daring that he used to move beyond

his territory to a far-flung area, an entirely different territory. He is said to have been spotted in the Bangladesh Sundarbans too. He was a notorious man-eater in the Sundarbans as reported by the companions of dead fishermen. There are several stories of Nantu, of how he attacked a human being every single month with such incredible stealth that often people a few feet away could only notice when a person had vanished in a moment. Most common sightings in the Sundarbans are when tigers are swimming from one to another island and they have nowhere to disappear at that time. Nantu was so powerful that he used to swim with the dead bodies through saltwater and dragged them over mucky shores in unnerving silence. A few foreign tourists saw Nantu hunt and devour an old fisherman. Efforts to trap and capture him failed time and again. He knew how to snatch off a goat in a mechanically devised trap-cage without getting caught.

Nantu was, undoubtedly, the most frequently seen and also the most photographed Bengal tiger in the tourism zone of STR. His familiarity with human-presence allowed him to sport a pose whenever he came across a tourist or a forest boat so as to watch his amber coat appear and disappear through the rays of sunlight on the mangroves as he walked a few feet from the edge of the island. But he was hard to track. For a moment though, he used to turn and stare straight at the mesmerised tourists. In fact, the tiger is one of the main reasons tourists flock to the Sundarbans, where they travel on boats with a cherished hope of enjoying a glimpse of the majestic cat. It is most cheering for them when a tiger crosses a river. When Nantu swam ashore, unlike the drenched, dark golden body, his face seemed to be a beautiful flame-yellow with the big head on a short and thick muscular neck and the long canines protruding through the open mouth and strong jaw. He used to sneak in and out of the mangrove forests, finally disappearing behind a mangrove called the 'tiger palm'. Studies done on the Sundarbans tigers suggest that these big cats spend a maximum amount of time in the *Phoenix paludosa* (local name *Hental*) bushes. This species of palm is also called tiger palm because of the colouration of its leaves, a brilliant hiding place for the tigers to camouflage. Past his prime, Nantu was healthy and believed to be the monarch in his territory.

Nantu was last seen on 27th December, 2021. The average lifespan of a tiger in the saline Sundarbans is 13 to 14 years. In that case, Nantu's age is thought to have crossed that limit long ago. A young male has now been sighted at the heart of Nantu's territory. From the photograph taken the newcomer seems to be larger than Nantu in his prime (2015).

It may be seen from the above accounts that a territorial fight, mostly between two massive adult male tigers or often between two adult females, took place in different tigerlands in India including the Sundarbans mangroves. The first known fight between two adult males leading to death of the younger one in STR was primarily reported by Lahiri (1974) and thereafter an article was published by Seidensticker et al. (1976). In August 1974, a young male tiger moved into a reclaimed fringe village in Jharkhali, and killed one woman and a number of livestock. The body of the dead woman had not been eaten by the tiger, presuming that it was not a deliberate man-eater. Lahiri (1974) reported that there was a goat-shed less than a metre from where she was sitting, and, as the incident occurred in the early hours of the morning, the circumstances suggest that the woman might have fallen victim by accident. The tiger had also killed dogs, cattle and a chicken, but, in most cases, he had not been able to drag away or even feed on its kills, because he was always being driven away by the villagers by shouting and creating disturbances after each kill. The tiger did not defend its kills, nor did it return to the sites.

The incident took place in the early morning of 2nd August, 1974. The tiger was sighted repeatedly over the next few days, and people were very much alarmed. The tiger was roaming in a densely populated region of this large delta island with a mosaic of paddy fields, villages and a large central mangrove marsh used as a fishery. The area had been part of RF lands until about 1955, when it was reassigned for the resettlement of refugees. The remaining RFs on the southern end of the island were about 3 km away from the village where the tiger was observed. A narrow belt of mangroves along the River Matla on the western side and the central fishery area provided the tiger with shelter and cover but without any large mammalian prey. Further south, in the RFs and in the TR, many Axis deer (*Axis axis*) and wild boars (*Sus scrofa*) were available.

On August 7th, the State Wildlife Officer, Dr. RK Lahiri, arrived there to investigate the case. After this particular incident extreme pressure was brought to bear on the responsible officials to destroy the tiger immediately. But, as the evidence indicated that it was neither a man-eater nor an incorrigible man-killer, it was protected under the Wild Life (Protection) Act, 1972. Rather than destroying the animal, the Forest Directorate decided to capture it by darting using immobilising drugs. Once immobilised, the tiger could be examined and if found physically fit it would be released in a suitable habitat in the wild and if not fit, it could be kept and treated in captivity under controlled conditions until cured, and then released back into the wild.

The tiger's approximate location was soon established. The fresh pugmarks made early in the morning of the 17th led to one of the dry thorn-covered hummocks in the central marsh, near the spot where a cow had reportedly been killed four days before, and where next day a cow's scapula was recovered. It was decided to try to hold the tiger here away from the villages. Over the next nine days, three bullocks were tied on the hummock as bait, secured with nylon rope to prevent the tiger from dragging them away. The tiger killed and ate the first two, and also two dogs that came to feed on the carcasses. A bullock that wandered into the area on the night of the 19th was attacked but it escaped along the narrow bund. After four days of observations from the *tong*, a boat with a secure hide for observation and darting was moved into position less than 50 feet from the baiting site. It was clear that the tiger would only come to the baiting site in the dark, making a night darting necessary, so a 15-foot-high machan was built 100 feet away for fixing a spotlight. The tiger then managed to pull the remains of the second bullock free and drag it into the mangrove scrub out of sight. So the third bait was tied. The water level had been lowered, conditions were favourable for darting, but the tiger returned to feed on the hidden remains of the second bullock. He did not come to the third bait until just after midnight on the 26th, when he rushed and killed it. He was darted ten minutes later and found immobilised 60 metres away on a mud bank in the mangroves.

He was a young male not fully grown. He was thoroughly examined and found without any physical defects. As in the previous days, he was observed to behave as normally as could be expected, there was no apparent reason to keep him in captivity. He was therefore put into a zoo transfer cage and moved to the launch, and by daylight he was transferred to the release site in the core area of STR. After about two hours, the effects of the drug wore off and he could push against the bars. Great care was taken to avoid any undue disturbance during the journey. At the release site, the cage was carefully moved ashore, turned on its side, and the door pulled open with a rope from the launch, thus avoiding the need to immobilise the animal again. After a few minutes the tiger walked out into the deep mud and went off into the thick mangroves behind the cage. The follow-up of the operation could only be done by periodically checking the area, as radio-location was not permitted in STR. On the afternoon of the 28th, the tiger was seen back in the transfer cage, but was gone by the morning.

On September 1st when the team members arrived at the release site, they found the tiger dead in the mangroves, 20 metres from the transfer cage. There were numerous tiger tracks around both the cage and the carcass, and the freshest tracks, which had been made in the few hours before the last high tide, and many of the older ones were larger than those of the dead tiger. The carcass was already badly decomposed. The right side and abdomen had numerous maggot-infested puncture wounds, the thoracic cavity had puncture wounds, and there were multiple puncture wounds in the hips and right shoulder. Only the extremities had been fed on by the scavengers. A small monitor lizard *Varanus salvator* ran away as they approached. It was concluded that this tiger had died of injuries resulting from an encounter with another tiger.

Translocation was successfully done, but less than a week later it was found dead from wounds evidently inflicted by another tiger. Here the implications of the incident, the publicity it attracted, and the changes in public attitudes are discussed.

Ahmad (1981) wrote that the tigers in the Sundarbans do not walk very far in search of food, as pigs and deer are plentiful. The movement through mud and slosh is slow. He saw tigers lying near the deer track after the herd came out of the tree forest onto grass fields to graze. They pounce on the deer or pigs when they go back into the forest. He also observed tigers chasing the stags which cannot run very fast through the thick forest with innumerable 'shulas' (pneumatophores) on the ground. The movement of the tiger is also retarded over muddy ground and if the deer gets ahead or a little warning before it reaches one hopping distance from the tiger, it often escapes death. It is sometimes noticed that a herd of deer will stand in the open field and risk being fired at rather than take cover in the adjoining tree forest where a tiger may be crouching to jump on them. The tiger has a peculiar habit of moving to and fro on its own track. This can be easily noted from its pugmarks on soft wet ground. It has even been noted that if a pole is lying across a stream, a tiger walks over the pole to one side and uses exactly the same route on its return.

The resourcefulness, intelligence, agility, as personified in such tigers, probably turned this magical species into a legendary myth (Chaudhuri, 2007). It has a propensity for man-eating within the forest areas and its frequent straying in the villages for hunting domestic animals has provided it a unique mysterious status. Superstition and supernatural qualities attributed to the man-eaters, however fictitious they may appear, have their basis. Locating its human prey, silent stalking him in ever closed circles for hours, high degree of intelligence to baffle trap gunner and trappers, diabolical understanding of the human mind, high precision in attacking human beings and crossing rivers, great power in controlling reflexes in mounting a boat, with least rocking, have made it a mythical animal.

Food Habit of Indian Sundarbans Tiger

Prey base	Occurrence	Mean mass (kg)	Biomass (%)	Individual (%)
Spotted deer	72.27	50	58.17	26.24
Wild boar	42.57	45	32.61	16.28
Rhesus monkey	9.9	8	4.82	13.57
Monitor lizard	5.94	8	2.89	8.14
Fish	3.96	2.5	0.85	7.69
Bird	1.98	2.5	0.42	3.84
Crab	0.99	0.2	0.21	23.75

It is still a common belief that tigers in the wild will eat only creatures that they kill themselves. But a review of case histories of scavenging tigers disproves that false opinion (Neumann-Denzau, 2006). Ahmed (2002) recalled Khasru Chowdhury, an expert on the Sundarbans tigers in Bangladesh: "He said he had seen dead dogs laced with pesticides being wrapped in polythene to be used later as bait. After digging a pit and filling it up with water, the dead dogs are put on a heap of earth by the pit. When the tiger eats the poisoned dog, the ingested pesticide, usually Endrine, makes the animal thirsty and the tiger gulps down water from the pit. As the water comes into contact with the toxin inside the tiger's body, the toxin becomes active and the tiger starts to choke and finally dies." This method has the advantage of leaving the skin of the tiger undamaged and the poachers can sell it for a higher price.

The phenomenon of cannibalism and blood rush could find new interpretations. Tigers eating dying or dead tigers that have been shot by man or even after fighting among themselves should perhaps not be called cannibals, as they do not kill their own species for consumption on purpose. Possibly, after the incident they become attracted by the unforeseen meat offering that is so easily available. Cattle-lifting tigers in a blood rush frequently kill more animals than one at a time, if they are easily available. The film-makers in Bangladesh Sundarbans (1998 - 2002) used not live but dead cows for tiger baiting due to ethical reasons. The films were: "In search of man-eaters", "Swamp tigers", and "In quest of the Sundarban Bengal tiger". Tigers, equal to a success rate of 73%, ate 11 out of 15 carcasses put out as bait. The mean time of discovery of the carcass by the tiger was 3.64 days (standard deviation 3.29 days, median 3.00 days, range 1-12 days, n=11). For good success the carcass had to be placed in an undisturbed area with fresh tiger pugmarks. In contrast, as known from historical records and very recent experiences, tiger baiting inside the Sundarbans with live baits (mainly cows, goats, or pigs) is still practised to entice the tigers to the camera trap locations.

Tigers have a bimodal peak of activity, during morning (6:00 hrs to 9:00 hrs) and late evening (17:00 hrs to 19:00 hrs), with an overall circular mean of 16:29 hrs. A high degree of overlap between tigers and the principal prey species Axis axis and Sus scrofa was found in the Sundarbans.

Since the tiger is adept at exploiting habitats other than ensconced mangrove forest interiors or inviolate core zone for foraging, dispersal, recruitment and even avoiding intra-specific competition along the buffer zone, serving a corridor/multiple use area and straying into the reclaimed fringe areas with a very high forest-dependent population, this aberration has led to severe human-tiger conflicts. In fact, this facet of tiger behaviour underscores the most significant challenges and opportunities confronting the conservationists today.

The Sundarbans tigers are comparatively small, stocky, and have muscular frames. Coarse and short coat and deep reddish-brown colouration are all adaptations to survive in this hostile habitat. If a male tiger hears the mating calls of a female from the other side of the riverbank, it responds immediately and starts swimming to approach her. Witnessing a tiger swim deceptively fast across a brackish-water channel, against a rapidly receding tide or its emergence from the water and move with consummate ease through knee-deep slush, leaving behind a trail of deep pugmarks, is a wondrous sight in the wild.

Fayrer (1875) reported: "The tiger takes the water readily and is a good swimmer. The Saugur Island and Soonderbund tigers continually swim from one island to the other, to change their hunting grounds for deer…They frequently cross the Ganges, and I have traced them by their footprints to the water's edge on one bank and taken up the track again on the other side." The tigers may disperse for more than 1,000 km and swim up to a 29 km wide river in the Sundarbans or 15 km across the sea (Garga, 1948). Garga writes: "The nearest to Sundarbans one could set from within the district of Midnapur was at Khedgeree in Contai subdivision 18 miles across the river Hooghly, to what is now known as Frasergunj(!), which is at present devoid of any jungle. But even until 1922, while the deforestation in progress was incomplete, tigers crossed this 18 mile wide fast flowing river and were killed by the villagers at Khedgeree".

Now-a-days, you will never find a tiger to swim in this large river of the region because the tiger's territory has been confined up to the River Matla and the landscape between the Matla and Hooghly is practically devoid of any tiger.

However, there is no denying the fact that the tiger is a superb long-distance swimmer. It is recorded to have crossed larger (>5 km width) channels as well. It has the capability of swimming 15-20 km through the tidal rivers. Chaudhuri (2007) observed a tiger swimming continuously for about three hours crossing two mighty Rivers Mayadwip and Matla bordering the Bay of Bengal in an April afternoon (1969). The tigers were seen to cross the rivers at right angles parallel to the current once the crossing is decided and a boat coming on the way is often attacked. They also change their course, when they feel disturbed while swimming (Mallick, 2011). The radio-collared tigers were recorded to cross, on the average, ten channels (mean width 25 m and maximum width 200 m) per day (Mallick, 2011). While following a tiger on the bank of the Nabanki khal, it was observed to move a distance of 1 km in ten minutes (ibid). There is another record of a tiger to swim at a speed of 16 km/hour (Das, 2011). From the sighting records, it appears that though the tiger generally crosses the channels at night for depredation in the fringe villages, it also swims even during the daytime in an undisturbed forest area. In the water, the tiger is just as strong as on land and an extremely crafty fighter with the estuarine crocodile. It was observed that the tigers generally avoid intertidal zones, where the crocodiles bask in the sun during winter months.

In a narrow creek near the forest, a tiger can easily outswim a country boat with only one man rowing and leap aboard. Once a tiger could swim about 600 m against the current in seven minutes eighteen seconds (Montgomery, 2008). In hot weather, it immerses itself in water, head raised high and constantly shakes to get water out of its ears. On reaching the opposite bank, before departure it first stands on its hind legs and gives itself a good shake like a dog (Bahuguna and Mallick, 2010).

In 2004, a debile tiger was caught here thrice within a month. The first time it killed a nine-year-old girl, the next time it slaughtered cattle in a neighbouring village. Twice, it returned to the forest, but the third time it was packed off to Calcutta Zoo. Bahuguna and Mallick (2010) cited a strong case of such aberrant behaviour of the tiger:

An old tiger was trapped in July 2008 on the fringe of Sundarbans. It was released in the forest. Although apparently it was healthy, it had only one canine intact. It was again captured in October at the same place. This time it was released deep inside the forest. In November, it again appeared in the same area from where it was captured. This time it was released at the southernmost tip of Sundarbans. But, this time also it came back to the same place, covering a distance of about 100 km by crossing many rivers, one of which is 4-5 km wide. Finally, it was trapped in December and handed over to the Kolkata Zoo.

During the year 2009-2010, a male tiger with its broken left lower canine had strayed repeatedly in adjacent villages, viz. Anpur-Rajat Jubilee, Lahiripur and Sonagaon (within 15 km of distance) in STR. It was trapped with live bait thrice and after the 4th time, it was sent to Kolkata zoo as it was opined that it might have lost its hunting capability. The tiger was assumed to be approximately 8-9 years old, though during its course of coming back to the villages repeatedly- it was found that sometimes it could cover a distance of 120 km approx (Mayadwip-4 to Panchamukhani-2) and came back to the same village.

On 27th January 2010 at about midday, Chiranjib Chakraborty, a Project Officer with the WWF-India's Sundarbans team, as he was sailing in a rented boat near Phiri Khali in the buffer zone of STR, a tiger was swimming calmly, a few feet away. He described the incident: "The moment it saw us, it stopped swimming. And it did so for about two minutes. Then, it turned away, as if it could care less, swimming towards a nearby island. It was swimming fast, much faster than I have seen any human swim. It swam for a distance of about 200 metres to reach the shore".

As many as five tigers have been captured in the Indian Sundarbans during the four days between 20th and 23rd May, 2010. Of these, three had 'strayed' into human habitation and two were captured for radio collaring. On 20th May, a tigress was trapped in Netidhopani for radio collaring. She was, however, found to be ill and was sent to Alipore Zoo hospital for treatment. On the 22nd, a second tiger was caught in the same area and fitted with a radio collar. On the same day, a young tigress strayed into Malmelia village. She was tranquilised and released in the Khatuajhuri forests. On 23rd May villagers found a tiger in Jamespur after it killed cattle. It was trapped and released back in the forest. A second tiger was caught that evening at Shamshernagar. Meanwhile, officials claim that the uneven distribution of prey base and the reduction of *Hental* mangroves (*Phoenix paludosa*), a favourite for tigers to seek cover in, is to blame for increased straying. It has been noted that captured-and-released tigers are making their way back to human habitation repeatedly.

Following are some instances of straying that have taken place in the region after Aila since the end of 2009.

1. A tigress, that had been raiding settlements in the Tridib Nagar Island near the Herobhanga forest for many weeks, was captured using a trap and released into the Dhulibasani forest on 20th December 2009. A male had been captured and released in the same area only three months earlier.

2. A tiger strayed into Madhya Gurguria village on 3rd January. A crowd had gathered and had cornered the tiger which was hiding behind a bush. Forty-year old Jayananda Ghatak went too close to the tiger and was severely mauled. The unruly crowd made it extremely difficult for the FD to proceed with rescue operations to capture the tiger.

3. On 16th February, a father-son duo from Jharkhali village in the Sundarbans went fishing early in the morning into the Kendokhali area. They were in their country-made boat in the river when a tiger sprang on father Jiten Majumdar. The bravery of his unarmed son, Dhamu, saved Jiten's life. The boy instinctively picked up a handful of mud and hurled it into the tiger's eyes. The cat was shocked into releasing its grip on Jiten, who was rowed back to safety.

4. On 19th March a tiger that had strayed into Sudhangshunagar village was tranquilised and captured from a keora tree it had sought refuge on. It was an old and injured tiger and had killed livestock in the region. It was taken to Sajnekhali by the forest staff after being captured.

5. Following cyclonic weather on 26th April, a young tigress entered the Adibashipara village in Basirhat range of STR. She was tranquilised, captured, medically checked and released on 28th April into the Khatuajhuri forest of the same range.

6. In 2010 a tigress forayed into the Sajnekhali Tourist Lodge in STR, having scaled the lodge's compound wall, scaring its occupants. The tigress jumped into the compound twice in a span of 24 hours

on 26th and 27th June. The tigress was tranquilised on 27th morning. The barrier around the lodge is nine feet high, tall enough to have kept the tigress out but a pile of rubble dumped near it because of some construction work seems to have aided the tigress in scaling the barrier.

The tiger chooses high tide for swimming to take off and land on hard ground and is very cautious while negotiating from one island to another. First, it observes direction of movement of the floating leaves, etc., uses the claw to dig and whirl the muddy water for observing the direction of such flow, again turns and puts its tail in the water to physically determine the intensity of current as well as direction. Thereafter, it visualises a static object like a tree on the opposite bank and tries to measure the current. Accordingly, it selects the starting point often by adjustment, i.e. proceeding ahead or going back a bit and ultimately drives into the water. In the mid-stream, if it realises that the target on the opposite bank is lost, it again comes back to its original position and follows the same procedure or starts swimming de novo. The tiger has developed a technique to keep the entire body submerged in water except for the nostrils and are able to board dinghies and boats with ease, but the man-eaters rarely enter the village. All the tigers which were killed by humans were not man-eater; for example, during 1999 to 2000 six tigers were killed without strong reason (Reza *et al.*, 2002).

The tail of the Sundarbans tiger is comparatively more muscular and consistently thicker at the base than that of other tigers, which is used as a limb for swimming, in the sinuous, sweeping, side-to-side manner of the crocodile (Montgomery, 2008).

A tiger has a record of swimming a distance of >548 m at seven minutes eighteen seconds against the tide. According to Mallick (2012), a problem tiger, which was released in the southernmost limit of STR, came back to the capture area on the north by covering about 100 km and crossing a good number of channels within a month. Tigers were seen swimming at right angles parallel to the current and changing their course, when they felt disturbed. In April, a tiger swam continuously in the morning for about three hours in the rivers Mayadwip and Bidya, before entering forests at a different location. The radio-collared tigers were recorded to cross, on the average, ten channels (mean width 25 m and maximum width 200 m) per day. In STR, the tigers have been recorded to cross larger (>5 km width) channels as well.

When a tiger walks in the Sundarbans, it crosses channels and creeks. Tigers go for long walks in the forests within its territory. But those long walks are also punctuated with sit-ins under the shade or in water before resuming the walk for scanning the surroundings. Since the mangrove forest is a hostile tiger habitat compared to the mainland habitats, the conservationists have emphasised on a habitat management approach that accommodates a wider range of both prey and predator species.

Since the entire forest area is not regularly inundated by the tidal water, the rhythmic activity in the tiger behaviour pattern owing to daily tidal movement, lunar cycles, etc was observed. In the extreme high tide, when the *Phoenix* areas go under water, the tigers have to lead an amphibious life. In the Sundarbans, the tiger has become an expert climber too. At the time of crisis, it was seen to climb even a straight tree for either safety at the time of straying in a village or for hunting in the forest. At Arbesi of STR, one tiger climbed a long and straight *Xylocarpus* tree at a height of at least 4.5 m above ground. It was tranquillised and then brought down with the help of a net spread around above the ground. When a pregnant tigress strayed into Deulbari village near Kultali in South 24-Parganas and climbed a palm tree to hide, when villagers lit a fire underneath to burn it. Once forest officials trapped it, angry villagers threw stones and beat it with sticks. Four villagers helping forest officials received minor injuries. During the cyclonic flooding of the Sundarbans in 1969, many climbed trees and a few were found drowned. During the tidal bores, it clings to the low mangrove branches or is driven to the elevated part of the island. It also climbs into sleeping shelters built on the *machan* (a raised platform used in tiger hunting) and often seizes one of the inmates. On 30th March, 2022 a Bengal tiger was seen perching comfortably on a tree branch overlooking a swamp in Katka of Bangladesh Sundarbans.

The tiger has adapted to changed food habits and become omnivore (Mallick, 2015).

A tiger typically consumes at least one deer-sized animal each week for survival (Karanth *et al.*, 2004). In captivity, it has been found that an adult male Sundarban tiger requires about 12 kg meat per day and a female requires 10 kg of meat per day of which there remains wastage of 20%. This diet culminates into a one day meal off during a week. Thus the annual requirement of a male tiger amounts to 3,780 kg/year and a female tiger amounts to 3,150 kg/year. When the statistics of human killing is analysed in Sundarbans for one year it does not amount to the annual requirement of even a single tiger. This establishes the fact that the Sundarban tiger is not an obligate man-eater.

Prey species are classified in different categories based on their weight. The tiger's food chain includes both the terrestrial and aquatic prey species. They include very small prey (<10 kg) like birds (2.5 kg), fish (2.5 kg), rodents (0.2-0.3 kg), crabs (0.2 kg), monitor lizard (8 kg) and rhesus macaque (*Macaca mulatta*)(8 kg), small prey (10-50 kg) such as the Muntjac or Barking deer (*Cervus muntjak*)(18-30 kg), and medium sized prey (50-100 kg) like Chital (*Axis axis axis*)(50 kg), Hog deer (30-50 kg) and wild boar (*Sus scrofa cristatus*)(45 kg). The large sized prey (>100 kg) like the rhinos (*Rhinoceros unicornis, Rhinoceros sondaicus inermis*), wild water buffalo (*Bubalus arnee*), barasingha (*Rucervus duvaucelii*) are now extinct in the Sundarbans tiger-land. The barking deer is extinct in the Indian Sundarbans but survives in the Bangladesh Sundarbans.

Tranquillisation and translocation of the aberrant tigers is an art which is perfected by the FD staff in the Sundarban forests. The tigers actively raid human habitations are few. Some used to swim to the villages. Either they returned to the forests themselves or were tranquilised, caged, treated and released deeper inside the forest. Aged, injured, incapacitated tigers, young pregnant tigresses and young tigers which fail turf war for control over habitat come to prey on human-inhabited deltas and swim the surrounding rivers to enter the bordering villages. Here the aberrant tigers are given a second chance to freedom.

During breeding season, tigresses often come to the fringing islands to litter cubs in safe paddy fields and bushes near the villages. To cite another incident, on 14th December, 1994 a pregnant tigress strayed into Altakhali village at night; it was caught and handed over to the FD. Another pregnant tigress strayed into an adjoining village in Sundarbans at night. There was panic all around after daybreak. Thousands of people surrounded the animal and in desperation the tigress achieved the impossible feat of climbing up a date palm tree. The animal was darted but decided to go to sleep on top of the tree. In a rather fairy tale operation, one person went up the tree and tied a rope to the tail of the sleeping tigress. She was pulled down to a stretched out net and the impact brought it back to its senses. People kept them cool and in an exhibition of extreme courage the tigress was injected with a sedative. It was caged and after proper examination released back into the wild. Such acts of courage and dedication of the field staff in the Sundarbans are many.

In the forest, a tigress chases away her cubs after they grow up. The cubs look for their own territories to dominate and control. Some of such young tigers may have chosen locality-adjacent forests as their preferred territories and from time to time, people from the nearby villages see them.

The intelligent tiger frequently managed to hunt the live bait like a goat from the trap cage. In the beginning, however, it started moving all around the cage, but did not enter it nor make any attempt to take the bait away. The field staff thought that the female is comparatively more intelligent. After a few days, it had taken away the bait without stepping on the lever, which leads to closure of the cage. In another case, the tiger avoided the main cage door, but hit the cage repeatedly from the back till one of the nut bolts fell off and it was able to secure a small opening to somehow squeeze its paw inside the cage to pull out the bait in such a manner that half portion of the goat was left inside the cage. In December 2016, the forest officials in West Bengal set up two iron cages with a goat inside it as bait and trapped a six-year-old tigress that had wandered off from the Pirkhali forest and strayed into Kishorimohanpur village.

The Bengal tiger is at the tertiary trophic level of the ecological pyramid of the mangrove forest. On the mainland it feeds on a variety of food including the larger prey species like sambar (*Rusa unicolor*), gaur (*Bos gaurus*), rhino (*Rhinoceros unicornis* and *Rhinoceros sondaicus*) or elephant (*Elephas maximus*) calves along with other medium and small animals. The Bengal tiger in the Sundarbans feeds on a large variety of active prey species, mostly medium to smaller animals. The tiger has adapted to a combination of terrestrial, arboreal and aquatic prey base. Its diet consists mostly of even-toed ungulates (Artiodactyla) like chital, followed by wild boar, but it also feeds on smaller animals like rhesus monkeys, Ganges river dolphin (*Platanista gangetica*), water monitors (*Varanus salvator*), shellfish, mud crabs (*Scylla serrata*), fish, birds [like Lesser adjutant (*Leptoptilos javanicus*), Red jungle-fowl (*Gallus gallus murghi*), etc.], snakes [a poisonous King Cobra *Ophiophagus hannah* along with a Monocled cobra (*Naja kaouthia*) were found during post-mortem of an adult female tiger during the food crisis taking place months after the cyclone *Aila* in 2009], turtles, otters, fishing cat (*Prionailurus viverrinus*) (rare), rats and frogs (*Rana* and *Bufo* spp.), and even grasshoppers. They often drink honey or liquid gold by breaking the low-hanging combs.

Though the tiger consumes a relatively low proportion of small prey species under stress, it is important to meet its dietary shortfall due to decreased availability of the preferred mammalian prey species. Often it has to take a larger number of smaller prey species. This diet shift or aberration is, of course, temporary in response to changes in prey biomass or availability. Easy availability of these substitutes might cause the tigers to eat them. Its food may occasionally include human beings who enter the forests to fish or collection of non-timber forest produces, as well as carcasses of wild and domestic animals. The tiger normally feeds on such animals which are easily recognisable in their scats collected from the wild. These samples consisted of partly digested material and undigested parts of such animals like bones, teeth, claws, scales, feathers, and occasionally plants like leaves of *Phoenix paludosa* and grass.

The density of tigers depends on several parameters like availability of prey, human interference, and male-female ratio of the tigers. In fact, the Sundarbans tiger exploits a wider range of available prey sizes to suit and fulfil its dietary needs. However, their behaviour is largely individual specific, variable and cannot be generalised like the tigers in the mainland. Above all, absence of any other large sympatric carnivores like leopard and low density of tigers in the Sundarbans in comparison to other landscapes reduces inter- as well as intra-specific competition for food.

The tigers are solitary carnivores with a polygamous mating system. Dr. A.K. Das, the then Director, Alipore Zoological Garden, observed:

> "The duration of gestation period…has been estimated as the period from the last day of mating to birth, and the mean period was 103;

and the maximum and minimum were 107 and 101 days respectively. In Calcutta Zoo, seven tigresses mated with the males in 48 heat periods from 1965 to 1977. Out of 48 mating or heat periods, pregnancy occurred in 22 cases and did not occur in 26 cases. The percentage of pregnancy is, therefore, 45.8% only and non-pregnancy is 54.2%. It was observed that the number of heat periods and percentage of pregnancy were less when both sexes lived together except at night".

The male Bengal tiger leads solitary life and congregates only temporarily for sharing plentiful supplies of food with its family. The basic social unit of the tiger is composed of a female and her offspring. Resident adults of either sex maintain home ranges, confining their movements to definite habitats within which they satisfy their biological needs and those of their cubs, which include prey, water, and shelter. The home range varies from island to island depending on the terrain and prey density as well as biomass. Male tigers in the mangroves change their territory a number of times, but females hold a breeding territory for longer periods, especially for rearing cubs. A tigress bringing up three cubs in the Sundarbans is a rare phenomenon. The number of female territories in a population determines that population's recruitment potential and ultimately its viability and resilience. If the cubs are added every two years, there are four possibilities that these new recruits can be absorbed in the current tiger population:

1. By replacing the area occupied by the old individuals

Old aged tigers are eventually replaced by the new recruits; this happens either by natural death of an old male or displacement of a resident tiger by another tiger.

2. By re-adjusting the territories and spread into buffer areas

The buffer areas are already being used by the tigers but if there are unoccupied forests, these may be filled up by the new recruits.

3. By dispersing in the landscape

There is evidence of tigers moving out of Indian Sundarbans to the Bangladesh Sundarbans and vice versa.

4. By deaths due to infighting

Being territorial animals, tigers protect their territories aggressively and intruders are not tolerated generally. The territory of the female tiger is smaller than a male tiger and a male tiger generally covers territory of 2-4 tigresses. Increasing the number of tigers will lead to more territorial fights which may cause death of tigers. Core areas may sustain a few more tigers than current numbers if habitat at crucial places is suitably amended and carefully managed. Buffer areas have large potential to accommodate more tigers but require immediate management intervention in terms of improving habitat. Landscape connectivity is an issue at present and requires further study/research. Connectivity with nearby forest areas, NP and WLSs will take much longer time compared to habitat improvement in Core and Buffer areas. Increasing tiger population in the Sundarbans beyond the carrying capacity will put more pressure on the park managers in terms of deaths of tigers due to infighting, abandoning or early weaning of cubs and more human-tiger interaction and conflict.

Vocalisations

Tigers maintain their territory and avoid intraspecific conflict by means of a range of communication methods which includes vocalisations, scent and scratch marks. The different vocalisations by tiger include moaning and roaring, coughing roar, prusten (a friendly sound mostly emitted by a mother approaching her cub), pooking (a sound believed to advertise tiger's presence), grunting (emitted by female to command the cubs to follow her), miaowing (made by cub, apparently indicate stress), woofing (an expression of surprise), growling, snarling and hissing (threatening sounds produced when the animal is annoyed) and coughing roar (indicate anger emitted during attack). Tigers vocalise a great deal during estrus and mating. One Bengal tiger in estrus roared 69 times in 15 minutes (Sunquist, 1981). Moans, groans, grunts, miaows, growls, and hisses are common during mating (Mazák, 1981).

Roar

The tiger's roar is the ultimate form of communication, which is used infrequently, primarily for long-distance communication -

(i) To draw attention and make other tigers aware of an individual's presence and location;

(ii) To announce a large kill;

(iii) By a mother to summon her cubs; and

(iv) To attract the opposite sex (also during mating).

Roaring is, however, one of numerous calls in the tiger's vocal repertoire. Some calls, like the full-throated confrontational roar, are impressively loud, while others, like chuffing, are just audible within a few feet of the source. This wide dynamic range is largely a manifestation of the tiger's larynx; the flat and broad medial surface of its relatively massive vocal folds enables the big cat to produce surprisingly low phonation thresholds and extraordinary output.

During a roar the tiger's ears are laid back and rotated so the backs are visible, the nose wrinkles and the eyes narrow. The structure of bones supporting the larynx, and anchoring the throat and tongue (called the hyoid) is connected with an elastic ligament as against more bone. This can be stretched by up to a third (20-23 cm), opening the air passages and allowing the roar to be produced.

The tiger-roars were heard mostly between the pre-monsoon and the pre-winter, but rarely thereafter. These calls are generally related to the hunting, mating and sometimes while the mother moving with her cubs encounters an enemy. Utilised in long-range communication, the roar is infrequently used. Roars, though rarely heard, are used for keeping away other tigers; it along with moaning serves as a communication between prospective mates when a female is receptive. During mating roaring is heard most often; tigers are at their noisiest when on heat and breeding. Easily heard for over 3 km, it advertises location and warns away other tigers, or attracts them when the search for a mate is on. Sometimes it is sounded after a successful kill, but never during the actual attack which is carried out in silence.

Coughing roar

Coughing roar is heard in response to another animal approaching a kill, or in challenge to a human (Schaller, 1967).

Moans

The subdued form of a roar may be preceded before the roar with a long, low moaning noise. Softer versions of the moan are used in communication between mother and cubs. The sound can be heard up to 400 m away.

Growls and Snarls

Growls and snarls are easily the most common form of communication, with the growl being aggressive, while the snarl is defensive. Growls can be turned into hissing or spitting.

Purring

This is something of a dilemma. Some people claim to have heard it in the tiger, but many experts feel it is almost certain that tigers do not, and many believe that the tiger's flexible hyoid bone does not allow it to generate a purr in the same way as the small domestic cat.

Chuffing or Prusten

The 'chuff' or 'prusten' is considered to be a friendly and non-threatening alternative to purring. In German, prusten means to sneeze, snort, or suddenly burst out laughing. This low-intensity sound is produced by holding the mouth closed and snorting through the nostrils. In the wild prusten is used when two tigers meet on neutral territory, while captive tigers sometimes start to use it with keepers.

Pook

Another unusual sound is the 'pook'. Because of this fact some experts think it may be a form of mimicry. By imitating the deer, the tiger could elicit a reply, locating the precise whereabouts of prey.

Other sounds

The tiger's range of sound is quite wide and they also meow, grunt and make a type of woofing noise.

Discrimination of individual tigers (*Panthera tigris*) from Long Distance Roars (Ji *et al.*, 2013)

The long distance roar (LDR) is sometimes referred to as a territorial roar, an estrus roar, an intense mew, or a moan. The call appears to operate in a variety of settings and is clearly intended to advertise an individual's presence. The call, a deep throated *ahh-rooom* typically lasting between 1 and 2 s, has all of the acoustic properties necessary to propagate through the environment for long distances, and field biologists and naturalists refer to the call as "one of the most thrilling noises one can hear in the jungle (Powell, 1957).

A tiger's roar might be as unique as a human's voice, i.e. no two tigers sound alike. With Raven Pro-sound analysis software developed by the Cornell Lab of Ornithology in Ithaca, New York- each tiger's call was broken down by minimum and maximum frequency, duration and other characteristics. It turned out that individual tigers varied significantly from one another: each had a unique vocal fingerprint that could be used to pick them out of a group. In addition, since females tended to roar at higher frequencies, the scientists could determine with 93% accuracy whether the recorded tiger was a male or female. Being able to identify tigers by their roars could allow for remote population monitoring, meaning researchers could take stock of endangered tigers hidden in the thick jungle without having to visually spot them.

Body postures and facial gestures

Body postures play a significant role in the tiger's communication. To look larger, a submissive tiger would bend down and flatten its ears against its head, but a dominant tiger would stand erect and arch its back. Tigers also employ facial gestures to communicate their feelings to other tigers, such as baring their teeth and wrinkling their nostrils.

An evening hunting

The sun was just beginning to set over Sundarbans, India's vast mangrove swamp, when the four woodcutters began winding up a day's work. Everything had gone better than expected. Forest guards hadn't spotted them working illegally on two muddy islands. And they hadn't encountered any of the crocodiles, poisonous snakes, sharks and tigers that roam the tidal delta.

It was low tide, and the men began wading across the channel between the two densely wooded islands, collecting fish they had caught in nets for dinner.

They joked with Bablu, their leader, as they put the fish in the pot he was carrying.

Suddenly, a flame-coloured Royal Bengal tiger leaped out of the forest. Although some experts believe tigers strike only from the rear of their prey, it attacked Bablu from the front, grabbing his shoulders with its paws and trying to bite his neck.

Bablu and another man fainted, but the other two drove off the tiger by hitting it with sticks.

Moments later, as the men desperately pulled Bablu toward their boat by his feet, the tiger returned, grabbed Bablu by the neck and dragged him away as the others fled to their village outside the Sundarbans reserve.

Elsewhere [India], tigers seldom kill people unless they are defending their food or cubs, or recovering from wounds inflicted by a hunter. But in Sundarbans, tigers behave like no others in the world. They have hunted people for centuries. Scientists don't know why. They are a mystery.

Even if the tiger disappears from everywhere else, it will survive in Sundarbans because it's very inhospitable terrain to poachers, once said Arin Ghosh when he was the Director of Project Tiger, Government of India. Sundarban is also the area where the poachers are also most likely to be killed by tigers.

Conflict and cooperation

While the tigers interact, it often happens when they are looking for mates, which may lead to conflict. When two or more tigers fight for the same prey, territory or mates, physical altercations that emerge from conflict can be fatal for one or both the tigers involved. On the other hand, tigers may occasionally hunt together to improve their success rate in capturing the prey animals, particularly when the prey is in short supply. The life of the mother and cubs is also based on cooperation.

Radio collaring

The tigers are extremely difficult to study in the mangrove forests of the Sundarbans in the absence of open roads and trails inside the forest and a high risk of potentially dangerous encounters with tigers. Moreover, the standard sampling methods that work well in other tiger habitats do not work well in this habitat. Therefore, little is known about the ecology of tigers in the Sundarbans mangroves. Two long-term studies have been undertaken on the Indian and Bangladesh side of the Sundarbans to address this gap.

The pilot study of tiger ecology in the Indian Sundarbans was conducted at three different locations each in two distinct zones, separated by a minimum aerial distance of 25 km-

(1) the core zone, which is closed to human use and

(2) the buffer zone in Sajnekhali WLS, which is open to tourism and regulated fishing and honey collection by the local population, by WII in collaboration with NTCA and the West Bengal FD.

Traps were placed 100-200 m inside the forest (from the edge of a river channel) mostly near the confluence of two or more river channels or near the rain-fed fresh water ponds created by the STR authorities because relatively high frequency of tiger pugmarks were observed in those sites during the reconnaissance survey. The traps were baited with live goats but kept non-operational so that tigers could enter and leave the cage freely. Camera traps were also deployed at each cage trap to monitor the visitation rates and the behaviour of individual tigers at cage traps. Three cage traps were deployed in each zone for a period of two months resulting in a trapping effort of 180 days in each zone. The traps were kept at least five km apart to maximise the chance of capturing different individuals in each of the traps. Traps were monitored daily to check whether the bait was taken and to monitor signs of tiger movement around the traps and calculate the time between visits and visitation rates of individual tigers to different traps. The results varied in the core and buffer areas. The visit rate of tigers to traps in the core zone was low because tiger visits were recorded only at two of the three sampled locations and that too only once in the first site and thrice in the second site at an interval of 25-30 days. Tigers in the core zone did not take the baits even after prolonged baiting outside the traps, except on one occasion. Fresh pugmarks in the mud within a 200 m radius of cage traps indicated active tigers near all the trap locations, but either did not come close to the trap or left without taking the bait. On the contrary, individual tigers visited all three trap locations in Sajnekhali WLS 5, 7 and 12 times respectively, following a gap of 9 to 15 days. Fresh pugmarks and camera trap photographs confirmed that the tigers picked up the bait from outside the cage; bait was kept inside the cage traps to lure the tigers.

For an authentic scientific study of the behaviour of the Sundarban tigers, the scientists of WII collared six tigers using vectronics IRIDIUM collars. Before that a trap was laid with bait at Chamta block(core area) because earlier a number of aberrant captured tigers were released here for rehabilitation, but it was observed that they did not enter the trap. But the pugmarks were seen within 200 m of the trap. The tiger came to the spot but became suspicious and did not take the risk, rather after making a round, it left the place. However, there are a few instances when surprisingly the Sundarban tiger took away the bait from within the cage without being trapped. GPS locations from satellite-collared tigers were analysed with ArcView 3.3.software (ESRI, Redlands, California) and Animal Movement extension v1.1 to construct Minimum Convex Polygon (MCP) and Fixed Kernel (FK) home ranges. Sign deposition rate was evaluated as the number of creeks/channels crossed by radio collared tigers per day using Google Earth.

The size of home ranges show an inverse relationship with abundance of prey- smaller the size of the home range, bigger the abundance of prey. The smallest home ranges are reported from tropical deciduous forests and resource-rich grasslands of India, like Nagarhole and Kanha, where a female tiger's home range is as low as 10 km². The largest home ranges have been reported from the Russian Far East, a result of low prey abundance. Range size of Indian Sundarbans tigers is between these extremes and suggests higher prey availability compared to Russian Far East but lower than that of deciduous forests.

Naha (2015) put up a comparative account of different subspecies of the tiger's mean home ranges arrived at on the basis of previous radio collaring experiments in their different habitats from Nepal, Bangladesh, India, Indonesia and Russia. They were effectively estimated, but it is possible that the full home ranges were not ascertained due to the limited longevity of the collars.

Mean home range size of different sub-species of tiger

Method: RT= Radio-telemetry, MCP= Minimum Convex Polygon

Place	HR adult male (n) km²	HR adult female (n) km²
Chitwan, Nepal	54.1 (2)	16 (3)
Panna, India	188.6 (2)	42.2 (4)
Nagarhole, India	25.7 (1)	16.5 (1)
Sikhote-Alin, Russia	1385 (6)	390 (14)
Kanha, India	102 (2)	10.5 (4)
Sariska, India	168.6 (1)	202.4 (2)
Ranthambhore, India	46.1 (4)	20.1 (2)
Pench, India	55.1 (1)	44 (1)
Nagpur, India	NA	726 (1)
Indonesia	236 (4)	NA
Sundarban, Bangladesh	NA	12.3 (2)
Sundarban, India	NA	44 (1)
Sundarban, India; study conducted in 2014 (Naha, 2015)	141.7 (3)	336.8 (1)

Tiger range sizes are much larger than reported from Bangladesh Sundarban (Barlow et al. 2011) suggestive of a low tiger density in the Indian side. The radio-collared tigress in Sundarban, during the present study, was a dispersing individual captured on the edge of the mangrove forest near a village. She had a home range of 329.9 km^2 (95% MCP) which is comparable to home range of a collared tigress 726 km^2 (95% MCP) in a human-dominated landscape near Nagpur, Maharashtra. The Sundarban tigress travelled a distance of 120 kilometres in 27 days from the day of her release till she dropped her collar in early April 2010. Earlier telemetry studies have shown that breeding tigers are territorial with an average home range size of 12.3 km^2 in Bangladesh Sundarban (Barlow *et al.*, 2011), 40 km^2 in Indian Sundarban (Sharma *et al.*, 2011). The male 7825 captured from one of the peripheral villages on the northern part of the TR during the present study crossed a kilometre wide River Harinbhanga in early June, 2010 and settled in Talpatti island of Bangladesh, where he remained for two months till the collar remained operational, suggesting that tigers preferred not to cross wide water channels yet these were not barrier to dispersal. The tiger population of India and Bangladesh Sundarbans is a single continuous population and trans-boundary conservation between the two countries is vital for long term population persistence.

The tigers in a part of the Indian Sundarbans were successfully habituated to cage traps and this technique was used to radio-collar (GPS/ARGOS/VHF) a healthy adult female in December 2007. Although the collar had worked well during testing, the satellite link stopped working within a day of radio-collaring the tigress. The GPS also failed within a week. However, the VHF beacon continued to work, providing a range of over 2-3 km that allowed the researchers to monitor the tigress from a boat. Based on data collected over 45 days, the home range of the tigress was estimated at 40 and 28 km^2 using the 100% and 95% Minimum Convex Polygon method, respectively (Sharma *et al.*, 2011). The weeklong continuous monitoring also revealed that the tigress usually rested in the day and was most active at dusk and dawn. Additionally, it was once observed to hunt a chital (*Axis axis*) during the continuous monitoring period. Thus, tigers in the Sanctuary buffer appeared to be bolder and less wary of disturbance than those in the core zone. Moreover, female home ranges reported for two radio-collared tigresses (F1 and F2) from the

Bangladesh part of the Sundarbans between 2004 and 2006, namely 12.3 and 14.2 km², with a monitoring effort of 7 and 3 months, respectively (Barlow, 2009; Barlow *et al.,* 2011). The tigers were immobilised with 6–8 mg kg⁻¹ of Telazol, administered using a projector and dart (Palmer Cap-Chur Inc., Powder Springs, USA). Tiger F1 weighed 75 kg and was estimated to be 12–14 years old (based on discoloration, damage and general wear of teeth). Tiger F2 weighed 80 kg and was estimated to be 10–14 years old (based on teeth condition). There was no evidence to suggest that either female had any dependent offspring. Both tigers were fitted with GPS collars (Advanced Telemetry Systems, Isanti, USA). F1 was tracked for 5.5 months until she died of unknown causes. F2 was tracked for 2.5 months until the GPS collar batteries expired. F2 was then recaptured and released at the capture site after the collar was removed. The GPS collars recorded 679 locations for F1 during April-October 2004 and 1,528 for F2 during March-May 2006. After 4 months of monitoring, F1 made a foray to the east of her normal home range, returning after 3 days. She moved to the same area 6 weeks later, and died c. 9 km from her normal home range. Tracks, judged by their size to be that of a new female, were observed in F1's former home range within days of F1 moving out. Female tracks together with those of a large cub were observed in the same area 1 year later. The poor condition of F1's teeth, her movement pattern and the appearance of a new female, suggests that F1 may have been unable to defend her territory from a rival. Therefore the forays to the east were discounted from the calculation of home range size, as they were not representative of F1's normal movement pattern.

However, these females were from a higher than average productive area, hence extrapolating these areas to be the average female home range sizes in Sundarban would be misleading. Information on female home ranges in low to medium prey density areas is lacking. Areas like Royal Chitwan, Nepal and Nagarahole with high prey density have recorded female home range sizes to be about 16 km² (Sunquist, 1981) and 16.5 km² (Karanth and Sunquist, 1995) respectively. Variation in the home range of tigers between India and Bangladesh may thus reflect differences in prey availability and abundance between these two different tracts of the Sundarbans (Sharma *et al.,* 2011).

The second radio collaring in the Indian Sundarbans was done in February 2010. On 21 February, a tigress strayed into Sonagaon village from Pirkhali block by crossing the River Bidya and injured a housewife. Six months ago, she reportedly entered this village. She was tranquilised the next day at about 13:30 h and radio collared. She was released in Storekhali forest of Netidhopani block on 24 February. From the incoming signals it was learnt that her movement was confined to about 3 km radius. But for five continuous days, the signal was being received from the same location. It was found that the collar had somehow fallen off. Pug marks of a tiger were also traced in that area.

The third tigress was radio collared on 28 February after trapping it at 5:00 h in the Pirkhali forest. She was released in the same forest on 1 March at 1:30 h. Initially, the signals were received for a few days, but stopped later. Her detached radio collar was found in another compartment of the same block on 6 March.

The fourth radio collar was fitted on a male tiger trapped on 19 March at about 17:00 h in Netidhopani forest. It was released in Pirkhali forest on 21 March at the same time. His radio collar was also stopped signaling. During 2010-2011, three tigers were radio collared in STR. On 19 May a male tiger, which strayed out to Kalidaspur on the same day from Jhilla-2 was immobilised, radio collared and released at Harinbhanga-1. On 22 May, another male was trapped at Netidhopani camp (Netidhopani-1), immobilised, radio collared and released at Netidhopani-1 on the same day. On 2 October, again the same tiger was trapped, immobilised, the radio collar changed and released at the same location.

The data recorded during monitoring of the radio-collared tiger or tigress has showed that the home range of male Sundarbans tiger was 110 km^2 as against 56.4 km^2 in case of the females, whereas in Pench TR, Madhya Pradesh, Central India, the estimated home ranges were 43 km^2 (adult female), 55.1 km^2 (adult male) and 52.2 km^2 (sub-adult male). A minimum of 25-30 km^2 undisturbed core area is required for a breeding female. Naha (2015) observed that the tigers move more during the early hours of dawn, day and afternoon with a distinct heightened movement between 7 am and 10 am. For an hourly interval, the highest distance moved was 595.25 m during 7 am and the lowest being 31.58 m during 7 pm per day with a peak around 4 pm in the afternoon. Tigers travelled an average of 5.06 km (SE 0.47) during neap tide phase and 4.43 km (SE 0.40) during spring tide phase per day with no significant difference observed between these two tide phases (Paired t test t = -1.193, df = 3, P = 0.319).

Mother tigers may share their home ranges with their daughters to increase the reproductive success of the latter. Sub-adult male tigers generally start dispersing between 19 and 28 months, to longer distances from their natal areas. Vacated home ranges of the deceased male or female are usually filled by related or unrelated neighbouring tigers. The dominant male, known by the name Big Boss, was around 6-7 years old in December, 2015 when he had a huge territory of 360 km² in Sajnekhali WLS, but now after eight years, he seemed to have lost quite a bit of his area to a new male, who most likely is his son. Another rare incident was observed by the FD personnel, while patrolling the forests of Jhinghakhali, Basirhat Range, India, on 5th September 2013 that a male tiger approached his mate and cub out in the clearing near a sweet water pond, where he stayed together till dusk.

Bangladesh FD's Sundarbans Tiger Project activities between February 2004 and June 2006

Tiger capture and immobilisation

Live cows were used as bait, set out in rudimentary enclosures made of local materials. Two to three entrances were left open, with snares set at these points to catch any tiger attracted by the bait. All cows were checked each morning and evening. The snares were designed especially for trapping tigers. The snares were designed to minimise potential injury to the tiger by not tightening around their wrist as they pull on the snare while attempting to escape. Snared tigers were immobilised with Telazol (tiletamine hydrochloride and diazepam hydrochloride) using a dosage rate of 4 mg/kg.

Telazol is administered as an intramuscular injection so, when darting the tiger, the dart is aimed to hit either the rump or the shoulder. Like other dissociative anaesthetics, Telazol can reduce the normal thermoregulatory capabilities of an immobilised animal, potentially resulting in overheating. However, this can be avoided by constant monitoring of the body temperature. If the tiger's body temperature rises to 102 F or above then water is applied to the animal's flank. Fanning of the damp body facilitates evaporation of the liquid and cooling of the animal.

Once the tiger appears to be unconscious, it is approached cautiously from behind and prodded with a long stick. Once confirmed as asleep, a core team of 5-7 people moves the tiger to a shady area where it will have a comfortable recovery.

The tiger is then measured, weighed and fitted with a GPS radio collar. As the tiger starts to show signs of recovery, such as ear, tail and head moving, the capture team moves back to observe from a safe distance. Once the tiger has moved out of site, it is tracked using radio telemetry to monitor its progress.

Two radio-collared tigers- F1 named "Jamtola Rani" (JR), captured at the Jamtola area on the 22nd of April 2005, and F2 "Chaprakhali Rani" (CR) died in the Bangladesh Sundarbans during research of a joint venture between the FD of Bangladesh and the University of Minnesota by anaesthesia and radio-collaring (Barlow *et al.,* 2006). The first tigress was darted at 6:58 pm with 600 mg of Zolatil and died six months later having the collar on. When captured, all its upper incisors were either worn down to stumps or totally broken off. Two lower incisors and the right lower canine were also missing. Although thin, the rest of the body seemed to be in relatively good condition. All claws were intact and, apart from three small kinks in the tail, the tiger showed no signs of previous injuries. The tiger weighed approximately 75 kg and measured 234 cm from nose to tail along the curves. a conservative estimate for the captured tiger would be around 14 years or older. She recovered slowly but steadily through the night. The first head up was recorded at about 1:30 p.m. She then went through intermittent bouts of sleep until she walked away from the capture site at 10:00 am.

JR's territory was defined using the 648 GPS locations recorded in the collar. Her estimated home range was 17.38 km² (95% MCP estimator). JR's territory encompassed some grass meadows that stretched from Jamtola to Kochikali, together with a mixture of keora (*Sonneratia apetala*), gewa-goran (*Excoecaria agallocha - Ceriops decandra*), and sundari (*Heritiera fomes*). JR stayed in its home range from April 22nd to September 29th (5 months), apart from a brief exploratory foray between the 23rd and 5th of August. On the 29th of September it made a sudden shift to the east, where it stayed until its death. The last location recorded for the tiger when it was still alive was on 15th of October. Her remains were discovered on the 28th of October, just north of Chandeshwar patrol post and approximately 8 km from her former home range. The old age of the tigress, indicated by teeth's condition, its movement patterns, and the recorded presence of a new tigress in her territory, all strongly suggested that JR was displaced by a younger and stronger female.

The second tigress was captured on the night of the 2nd of March, 2006. Judging by teeth wear it was between 10-12 years old. She weighed an estimated 110-125 kg, and measured 228 cm from nose to tail tip along the curves. Her body was in good condition; all claws were intact and the only sign of previous injury was an old 15 cm long scar on her right shoulder, and a missing upper canine. The tiger recovered well from the immobilisation, leaving the capture site at 8:30 am. She was radio tracked for 62 days between 3rd March and 16th May, 2006. During that period, the tiger was located for 57 days (91% success rate). CR's home range extended east and west of Chaprakhali village and totaled 12.1 km^2 (95% MCP estimator). Her territory had large stands of gewa-garan (*Excoecaria agallocha - Ceriops decandra*), sundari (*Heritiera fomes*) and keora (*Sonneratia apetala*) similar to JR, but no substantial patches of grassland. Her boundaries were well defined and at no time during the radio telemetry period (March 2nd-May 16th) there was no sign of another female tiger being inside CR's home range. A male tiger was, however, infrequently detected (differentiated by the larger track size). Both CR and the male tiger visited the fresh water hole maintained by the villagers at Chaprakhali.

It was for the second time tranquilised in December 2006 to remove the collar. However, both tigers made full recoveries to their pre-tranquilised state and were released to the wild and continued their normal life in the jungle.

Tigers in the wild rarely live to more than 14 years old before they grow weak, unable to catch prey, and are forced out of their territories by younger stronger tigers. Both collared animals in question were old, approximately 12-14 years of age, when they were caught. Because of their age both tigers had broken and discoloured teeth. Jamtola Rani left her home range 5 1⁄2 months after her collaring, presumably in response to another female tiger that took over her territory and she later died; a normal event for an old female. Chaprakhali Rani was recaptured nine months after her collaring. She was weak because her bad teeth did not allow her to feed efficiently. She was released alive back into the jungle and observed for three consecutive days as she came back to feed off the same bait with which she was captured.

Tiger Kills

JR: The difficult weather conditions of the monsoon and the damage of a receiver prevented kills being identified during the course of the radio telemetry. Kills were instead inferred by the movement patterns recorded by the GPS collar. The tiger was considered to have made a kill if it either spent more than 12 hours in a 100 m radius during the day or more than 8 hours during the night. Using this criteria, a total of 39 hypothetical kills were recorded, ranging from 2 to 9 months.

Number of hypothetical kills made by tiger JR

Month	Hypothetical kills
May	6
June	9
July	9
August	2
September	9
October	4
Total	39

Note: Kills are based on time spent within a 100 m radius of a location. October data is for 15 days only.

CR: Tracking CR every day using radio telemetry enabled the discovery and identification of some of her kills. If she stayed in an area for more than a day it was nearly always indicative of her having made a kill. An attempt was made to find all of her kills but undoubtedly some went undiscovered. Nine carcasses were recovered and 2 other kills were confirmed by kill sites and drags, giving a total of 11 confirmed kills. All kills were of chital, and of the 9 animals whose remains were found 4 were large stags, 3 were adult females, 1 was a small fawn and one was a medium sized animal of unknown sex. Once killed, CR typically dragged the carcass into thick goran (*Ceriops decandra*) to continue feeding.

Pilot study of prey assessment methodology

A total of six 100 m² plots and two transects of fifteen 20 m² circular plots were set out. Counts were carried out 10-20 days after the plots were initially cleared of pellets. Pellets from chital, wild boar (*Sus scrofa*) and monitor lizard were detected. Four of the circular plot transects were washed out by either high tides or storms.

Results of pellet count pilot study, 2005

Plot	Area m²	No. of plots	Habitat type	Date	Prey species	Days before count 10	Days before count 20
1	100	1	Gewa/Garan (*Excoecaria/ Ceriops*)	7 March	Chital	1	4
					Wild Boar	1	1
					Other	0	0
2	100	1	Gewa/Garan /Keora (*Excoecaria/ Ceriops/Sonn eratia*)	7 March	Chital	3	9
					Wild Boar	0	0
					Other	0	0
3	100	1	Gewa/Garan	7	Chital	5	10

Plot	Area m²	No. of plots	Habitat type	Date	Prey species	Days before count 10	Days before count 20
			Keora/Sun-dari (*Excoecaria/ Ceriops/ Sonneratia/ Heritiera*)	March	Wild Boar	0	0
					Other	0	0
4	100	1	Keora (*Sonneratia*)	7 March	Chital	10	-
					Wild Boar	1	-
					Other	0	-
5	100	1	Keora (*Sonneratia*)	7 March	Chital	6	-
					Wild Boar	0	-
					Other	0	-
6	100	1	Keora (*Sonneratia*)	7 March	Chital	9	-
					Wild Boar	0	-
					Other	0	-
7	255	13	Gewa/Garan /Keora (*Excoecaria/ Ceriops/Sonn eratia*)	26 March	Chital	2	-
					Wild Boar	0	-
					Other	0	-

Plot	Area m²	No. of plots	Habitat type	Date	Prey species	Days before count 10	Days before count 20
8	236	12	Gewa/Garan (*Excoecaria/ Ceriops*)	27 March	Chital	41	-
					Wild Boar	0	-
					Other	1	-

Ear Tag and Micro-chipping in Indian Sundarbans

Since July 2009, some of the straying tigers in STR were ear-tagged with microchips. Electronic microchip was introduced to track those tigers that repeatedly stray into human settlements. In the first case, a pregnant tigress, who was unable to catch prey in the forests, had entered Adivasipara on Kumirmari Island and, opportunistically, killed some livestock. It was caught, examined and tagged with a microchip placed at the base of her tail before release. Her microchip code was 001 and she returned. Then she was captured for the second time and then released. Her first straying took place in July and barely two months later, she returned. After killing a cow and a pig over a week she walked into a trap in Rajat Jubili village on Satjelia Island. To ensure that she didn't stage a repeat comeback, she was released in Kalash Island in Dhulibhasani at the estuary of the River Matla, about 74.08 km away from where she had been trapped. But hunger might have driven her back to human habitation again- as did a tiger a few years ago, three times over, before he was permanently placed behind the bars in the Alipur Zoo.

While male occupants of a territory change more, females tend to hold a territory for longer periods. The territory of a male also overlaps the territories of more than one female. The cubs are reared exclusively by the mother and remain attached to her for up to two and half years. The female cubs often occupy territories or home ranges adjacent to that of their mother. During the daily activity, every tiger records its presence while walking on soil amenable to receiving its pugmarks.

In their individual territory, the tigers maintain contact with other tigers, especially those of the opposite sex and regularly monitor their movements and activities. A male tiger keeps a large territory in order to include the home ranges of several females within its bounds for breeding, but defends its territory against intrusion by any other male. Territorial disputes are usually solved by intimidation rather than outright violence. Once dominance has been established, a male may tolerate a subordinate within his range, as long as they do not live in too close quarters. Sub-adults are killed by adult tigers at the time of territorial fights. Hence territory acquisition is important, especially for sub-adult or transient tigers in core habitats where competition is high. Territorial fight between a dominant and a young tigress is a rare incident in the mangroves. Ultimately the young submitted.

The tigers are mainly nocturnal predators, but they also hunt in daylight, particularly the morning slot. They generally hunt alone and ambush their prey. In most cases, tigers approach their prey from the side or behind at a close distance and grasp the prey's throat to kill it. Then they drag the carcass into cover, occasionally over several hundred metres, to consume it.

Initially the project was called "Tiger Study Project in the Sundarbans" but later the project was addressed as "Sundarban Tiger Project" or STP. Initially the main goals of the project were:

A. To find out the home range of tigers.

B. Tiger density and prey abundance in relation to forest cover.

C. Tiger-human interaction.

D. Change of behaviour in response to tidal, diurnal and seasonal fluctuations.

E. Behaviour change with cubs and last kill etc.

The initial objective of the project was purely scientific but after the death of the first tiger and related public reaction, the project added two more goals-

F. Conservation capacity building and

G. Creating public awareness.

The project's working manpower was composed of a PhD student Adam Barlow, one forest guard, one speedboat driver and three helpers. Barlow was going to radio-collar eight to nine tigers, six female and two or three male by baiting cows, trapping them using snare and then tranquillise by using a controversial drug Telazol (Tiletamine hydrochloride and Diazepam hydrochloride) using a dosage rate of 4 mg/kg. However, it is argued that Telazol has a fast knockdown (2-8 minutes) and slow recovery (normally 2-6 hours, depending on dosage and condition of the animal). The fast knockdown reduces stress to the animal and the slow recovery allows time for researchers to carry out the necessary measurements and collar fitting without concern that the tiger will regain consciousness unexpectedly. Instead of Telazol, some experts are suggesting combined injection of Xylazine Hydrochloride (dosed at 1.0 mg/kg bwt, IM) and Ketamine Hydrochloride (ketamine hcl)(dosed at 3.5 mg/kg bwt, IM). However, the protocol is more complicated for Ketamine than Telazol.

Hunting records of JR and CR were also available (Barlow *et al.*, 2006). JR was considered to have made a kill if it either spent more than 12 hours in a 100 m radius during the day or more than 8 hours during the night. Using this criteria, a total of 39 hypothetical kills were recorded, ranging from 2-9/month (May 6, June 9, July 9, August 2, September 9, and October 4).

In spite of the fact that the Sundarbans tiger is an excellent killing machine, often immobility takes a toll on him. Rough surface is one of the reasons behind this ailment. During the 21st century, the Alipore zoo doctors have treated the Sundarbans tigers for various ailments - from posterior weakness and infected osteoarthritis to corneal opacity. Rough surface, lack of exercise and excess body weight are reasons behind an ailment triggered by hygroma, which means accumulation of fluid in the joints.

Treatment details of the Sundarbans tigers in the Alipore Zoological Garden

In the incidents of straying in the Sundarbans during 1980s two tigers and one tigress have been tranquilised and transferred to the zoo in Kolkata by the forest officials. These tigers have since been living happily in the zoo. In the last of the three incidents the tigress was immobilised in the Uttardanga village near Gosaba on 4th January, 1982. The animal had entered a cowshed in the village. Around eight in the evening the forest officials reached the disembarkation point in their launch. After walking about a kilometre on the embankments, they arrived at the village. The villagers had already shut the door of the cowshed and secured it with strong fishing nets to prevent the animal's escape. In the presence of about a thousand people, the darting of the tigress began. As many as three darts were used for immobilising the animal. She was tied to bamboo posts and carried to the embarkation point. She regained her senses only after three hours. Now, she is known as Sundari to the visitors of the Kolkata zoo and has recently given birth to two cubs.

2008: In February, a ten years old male big cat had strayed into Jharkhali village. The tiger had multiple injuries on left hind leg and right foreleg was captured and sent to the zoo for treatment. The animal had ten wounds on its body, including a gash on the left hind leg and one of its claws was missing. It was observed that there was no immediate possibility to release it into the wild till the wounds were healed. It was later sent to North Bengal's Khayerbari.

2009: A female tiger with posterior weakness brought to the zoo. It was shifted to Jharkhali rescue centre in 2015.

On 13th March a nine years old tiger crossed over the forests and strayed into Shamshernagar village in search of food because it was injured in its hind leg and probably could not hunt. It was found in a cattle shed and the villagers raised hue and cry. The FD staff rushed in and tranquilised the tiger. On 16th March it was brought to the Alipore zoo where treatment began. The tiger recovered rapidly. On 21st July the tiger was sent to Harikhali by a caged truck and released in the wild from a boat on 22nd on the riverbank, 100 feet away from the forest. The tiger sat at ease in the cage, now and then licking its previous wound area. Around 2 pm the cage door was opened. For a minute, the tiger remained motionless. Then it hunched its back and yawned, looking out through the opened door towards its new home. It made awkward steps ahead and looked around. It lowered one of its paws into the river water and jumped below at a lightning speed to move forward through the muddy bank. It shocked itself dry and after a few hesitant steps ran away into the deep mangrove forests. Setting this tiger into its natural habitat was important because it was the season for mating in the Sundarbans which means that back into the wild it would search for a mate to sire.

2011: A male tiger with injuries on hind legs, triggered by a chronic wound, treated at the zoo. STR officials patrolling the stretch near Pirkhali-6 in the morning first spotted the tiger sitting on the edge of the island. They kept an eye on it for 24 hours. On finding that there was not much movement, they decided to tranquilise it the next morning so that it could be brought to the zoo for emergency treatment. The tiger was about eight years old. They tranquillise the tiger around 10.30 and brought it to the Alipore zoo for treatment. It was treated and later released in the wild. It may be noted that in the last three years, this is the fifth instance of a tiger being captured from the forest and brought to Kolkata for treatment. Of the four tigers captured before, two were released back to the wild after treatment. Two were still undergoing treatment in the zoo hospital.

2012: On 26th July, 2012 at 06.50 pm, a male tiger was received by the zoo in drowsy state due to chemical immobilisation by intramuscular ketamine (800 mg) with a history of hindquarter weakness and debility. After prolonged treatment the animal became healthy and finally released in the Sundarbans again on 29th July 2013 (Majie *et al.,* 2013).

The said animal was spotted without any movement for nearly 24 hours in the Pirkhali-7 area of the Sundarbans. Then the STR authority decided to tranquillise and shift the animal to Alipore Zoo Veterinary Hospital on an emergency basis for proper treatment. From haematological, biochemical and radiographic examination it was diagnosed that the tiger was suffering from infective arthritis. All the reflexes of the animal were less with fixed eye ball and one major penetrating wound near left coxofemoral joint with inward tunnel surrounding the joint and three other wounds in right forelimb and right thoracic region with signs of dehydration as noticed during immediate clinical examination after receiving the animal. All the canines of the animal were intact with sharp tips indicating the animal was aged about 6-7 years. Treatment was started immediately with intravenous fluid (Ringers Lactate) 1000 ml, Inj Dexamethasone (4 mg/ml) 4 ml to counter residual effect of ketamine. The tiger responded gradually to the treatment on the same night. The wound and hind quarter weakness gradually improved and overall health condition including aggression and agileness enhanced to the desired level. The animal was handed over to the STR authority to find a better home for the tiger to reduce the adverse effect of captivity. Then STR authorities decided to release the tiger in the wild again and the tiger was released in the Pirkhali-7 area.

2014: An old male tiger, which was moving very close to the nylon net separating the Sundarbans village and the mangrove forests for a few days and earlier it had charged at a forest guard and was captured by the foresters. An initial health check-up indicated that it had developed cataract in the right eye. He was about 10 years old and all its canines were intact. However, it has an injury on its left front leg, triggered by a territorial fight. and was treated and shifted to Jharkhali rescue centre in 2015.

2015: An adult male tiger was caught from the Kultali village in the Sundarbans in December, 2015. It was kept in the Jharkhali rescue centre for treatment where it developed a problem similar to 'capped knee' in its front legs triggered by hygroma which means accumulation of fluid in the joints. The Sundarbans vets were initially taking out the fluid accumulated in the knee joints. But it turned out to be a recurring problem and hence it has been brought to the Alipore Zoo for better treatment.

2017: One male tiger was rescued on 29.06.2017 from Herobhanga-9 compartment and brought to Sundarban Wild Animal Park at Jharkhali. It was examined by the veterinary doctors in the early morning on 29.06.2017. The veterinary doctor advised the authorities to keep the animal in shelter for treatment till it is fit for release in its natural habitat. Accordingly the tiger was kept under medical treatment till 25.10.2017. The said tiger was released at Ajmalmari-l0 on 26.10.2017.

Man-eating propensity

Bengal tigers do not under normal circumstances kill or eat humans. They are by nature semi-nocturnal, deep-forest predators with a seemingly ingrained fear of all things bipedal; they are animals that will generally change direction at the first sign of a human rather than seek an aggressive confrontation. Yet at the turn of the twentieth century, a change so profound and upsetting to the natural order was occurring in the Indian subcontinent as to cause one such tiger to not only lose its inborn fear of humans altogether, but to begin hunting them in their homes- a tragedy for hundreds of people who would eventually fall victim to its teeth and claws.

Corbett's hunting stories present many such examples from the early part of the 20th century. He, however, put forward a theory in defence of the man-eaters- they are either injured (wounded paw or legs or broken canines) or were old or weak tigers. Most of the man-eaters killed fall into this category. But, according to late Professor Ratanlal Brahmachari there are about 20-22 cases where healthy tigers regularly made humans its prey. Hence, the question remains to be answered: why do the tigers kill human beings? Driven by hunger is what J E Carrington Turner, an IFS officer of British India posted in various parts of India, believed. With habitats shrinking every passing day, the usual prey base (various deer, wild boar, buffalo, gaur) diminishing steadily, tigers do move continuously, bringing them in close proximity to human beings. Possibly out of this struggle for existence there emerge some tigers that suddenly start killing humans for food. There are places in India like Chandrapur, Pilibhit, and the Sundarbans where tigers attack and kill people. Tigers do go near villages to lift cattle in most cases and not human beings, and are poisoned or killed in many instances.

However, not all killings can be attributed to "man-eaters". At Bandipur for instance, a few tigers were labelled "man-eaters" a few years back during separate instances after they repeatedly killed and dragged victims. One of them was killed during such an operation while the other was captured. The tiger captured had a wounded paw and was hungry.

In 2017 at Tadoba TR, when Matkasur killed a forest guard (but did not eat his body) he was not labelled a man-eater. He was still seen in tourist zones and hadn't reportedly killed anyone thereafter. Even at the Sundarbans, which account for nearly 25% of these fatalities, people who venture into forests for fishing, crab hunting or honey collecting, are in turn taken away by a tiger at times. Some experts attribute this to excessive salinity of the water that tigers drink, making them irritable but these arguments remain largely inconclusive. There is absence of concrete evidence to prove the theory that tigers are man-eaters by nature and it may be a behavioural characteristic manifested in some tigers at Sundarbans.

In the estuarine tract in Sundarbans, physical, chemical, mechanical and animate environments create a series of specialised niches where aggression, tenacity, battle for survival are relentless and supreme (Chaudhuri and Chakrabarti, 1980). The Sundarbans tigers are generally branded as maneaters. Tigers are known to be man-eaters elsewhere in India, too. In the colonial imagination, the 'man-eater' morphed from a metaphor of Oriental despotism into a metonym for the alleged criminality of big cats. By casting tigers as man-eaters, colonial officials sought to legitimise their control over the subcontinent's wilderness. The British erased the traditional tiger symbolism and turned the tiger into an exemplar of depravity.

In fact, the implicit correlation of tigers with man-eating laid the groundwork for colonial hunting policies. The British effort to exterminate tigers was driven partly by a desire to push the forest back and enhance agrarian revenues. Mahesh Rangarajan has argued that after the Bengal famine (1770), when large tracts of land lay uncultivated, "fewer tigers meant more cultivation and more revenue, their elimination a blessing of imperium after the elimination of an Oriental despot."

Schaller mentioned the jungles of Mandla, the Kumaon hills and of the Sundarbans as infested by man-eating tigers. He agreed with Champion that in general tigers are harmless and tend to avoid man and criticised the early British hunters who took the tiger to be a devilish beast. Like Corbett, Schaller found the tiger to be a good natured beast if undisturbed.

Age and damage may cause some teeth to be lost and this is considered a prime reason for tigers turning man-eater. Bad teeth make catching and holding prey extremely difficult. It takes more than one or two lost teeth to turn a tiger into a man-eater. In one case, a post-mortem examination revealed a tigress with two broken canine teeth (only about a third remained of each tooth), four missing incisors and a loose upper molar. Only once reaching this stage did she attack a workman.

Attitude towards the man-eating tigers differs extremely. On the one hand, there are people who accept the sensational accounts of the Sundarbans which depict the tiger as a man-eater, ever-ready to attack humans. For the locals, the tigers are not just apex predators, they are the ghosts of ancestors who have returned for revenge, and demon gods who need to be appeased by worshipping Dakshin Rai, the tiger demon, and Bonbibi, 'lady of the forest', who is a guardian spirit of the Sundarbans forests. The common people and forest-goers, Hindu or Muslim, believe that it is her protection that keeps the tiger at bay. On the other hand, there are conservationists, who try to make out that human killings are mostly accidents caused when people enter the tigerland, become easy prey of the tiger and get mauled.

The adaptation of the Sundarbans tiger to the mangrove ecosystem is every bit as remarkable as that of the mangrove system to tidal ecology. The Sundarbans tiger is unlike its counterpart on the mainland. In the marshy land and brackish channels caused by encroaching tides, the huge terrestrial animals took to the water. It drinks saline water in crisis, is an expert long-distance swimmer with an estimated speed of 13 km per hour. The tiger is also a moderate climber, and hunts with as much finesse in tidal currents as it does on land. In the difficult impenetrable swampy forest, interspersed with rivers, rivulets and creeks, and flashed by tides twice a day at intervals of 6.2 hours on an average. The tidal rhythms of the Bengal tigers are simply direct responses to the rhythmic ebbing and flooding of the tides.

The mangrove tiger, which has adapted to this environment, is not just an apex predator, but also an enigma. The particular niche of any "Apex Predator" is not properly defined in the Sundarbans, rather a partial overlapping between terrestrial and aquatic Food chain discloses an established aberration in the interest of survival of the fittest (Pandit and Guha, 2015).

Some of the Sundarbans tigers become so fearless from the repeated lifting of man that they do not hesitate to board small boats at night and to climb stilt huts erected by wood cutters on the bank of rivers (Gupta, 1966). In this context, an extract from Trafford's report of 1909-1910 is reproduced below.

"Man-eating tigers were more aggressive in the Sundarbans this year than they have ever been known to be previously. An alarming feature in two different (timber felling) coupes was the climbing by tigers into sleeping shelters built on *machans* (a platform erected in a tree, used originally for hunting large animals and now for watching animals in wildlife reserves), and the seizure of one of the inmates on each occasion. The further development of this new propensity, however, was fortunately arrested as both these tigers succumbed to the temptation of traps in which they were caught and shot. This aggressiveness is attributed to the fact that enormous numbers of deer were killed or drowned in the storm wave which accompanied the cyclone of October, 1909, which upset the balance between tigers and their natural prey, and induced them to attack men".

A question is sometimes asked as to what deters the tiger, which is a devil to reckon with in the Sundarbans jungle, from carrying out its depredations in the adjacent reclaimed areas with equal audacity. The answer is that the tiger, like most other animals, is a slave of habits and practices, and while he would not hesitate to attack a country-boat or a stilt-hut within the forested area as generations of tiger have done before him, he instinctively shrinks from launching upon an adventure of swimming across a boundary river to hunt in an unknown domain devoid of any cover. In other parts of the country, where the jungle area is fairly small or where there exist cultivations in enclaves within the jungle, the tiger gets an urge to step across the border and explore the cleared areas outside. With the development of familiarity in the gradual course this becomes a habit, and he extends his hunting ground. But here in the Sundarbans the forested area is immense with no enclaves of cultivation anywhere, and there has, therefore, never been any good and sufficient cause for the tiger to leave what he believes to be his homeland, and venture out into the open domain of man in search of prey. There was thus no occasion for him to develop a habit in that direction. There were a few stray cases where a tiger crossed over to the cultivation under the cover of darkness, and killed cattle, but these were of the nature of a limited adventure, the extent of penetration being small, and besides, such cases are by no means common.

Post-independence period

India

Sometimes a few people opine that "man-eater" may not be the right word to describe a tiger after all. A ferocious beast that sometimes kills its own kind and doesn't show any emotions- are usual beliefs associated with a tiger. But at Tadoba we heard and witnessed the remarkable struggle of a tiger mother, Maya, desperately trying to save its young from bigger male tigers. Same behavioural pattern is observed at other forests like Ranthambore, and Sundarbans where a tigress lures away the intruding male to protect the young. There are male tigers which kill others' cubs to force the mother to mate again, or fight with other males to establish territory. Is this not enough to make a big cat 'cruel'? What can we say when we hear about Zaalim (T25, and a misnomer for sure) taking care of the cubs after their mother died at Ranthambore? In the same forest, we have witnessed Aurangazeb (T57) being tolerant of the cubs of his partner Noor (T39)- the cubs belonged to Ustaad (T24) who was moved away after being labelled "man-eater". At Bandhavgarh, a tiger behaved as a babysitter for his siblings (who belonged to a separate litter from his mother) when his mother left for hunting.

Emotions or feelings are prevalent in tiger 'society' but are probably invisible to the human eye. Bandhavgarh guides or old-timers will tell you about Charger calling for his mate Sita after she disappeared one day; or at Pench you would hear about the story of Collarwali and Raiyakassa being together for more than four years now building the foundation of future generations of tigers. The tiger 'society' probably is more than just being synonymous with ferocity or cruelty. The story of Telia sisters from Tadoba belies the usual theory that tigers kill and eat individually. Maybe they still do most of the time but then Valmik Thapar did record an incident of nine tigers on a nilgai (*Boselaphus tragocamelus*) kill (led by the famed Padmini tigress) at Ranthambore and not all were related to each other.

Indian Sundarbans

There are various conceptions regarding the man-eaters (Chaudhuri, 2007). So many studies have been conducted on the man-tiger interaction in the Sundarbans. Tigers were present in all the fifteen (15) blocks in STR (2,585 km^2) and casualties were recorded in all the compartments (65 nos.) of these 15 blocks; in fact 21 blocks of the entire Sundarban forests recorded the incidence of maneaters (Chaudhuri and Chakrabarti, 1980).

Total casualty in 'pre-Project Tiger' and 'Project Tiger' Periods

Forest Block	Pre-Tiger Project period (14 years)	Project Tiger period (5 years)

	1960-1969	1969-1973	Total	1974-1978 up to September
Chamta	80	39	119	17
Matla	41	36	77	25
Arbesi	8	20	28	16
Gosaba	37	12	49	1
Netidhopani	17	13	30	27
Chandkhali	16	15	31	52
Pirkhali	22	8	30	37
Bagmara	22	-	22	-
Khatuajhuri	12	8	20	17
Others	81	21	102	Panchamukhani 32
Total	336	172	508	224

The table shows casualty figures in nine blocks out of 15; the last column shows total casualty in six blocks (except Panchamukhani). Of the nine blocks, Chamta, Matla, Gosaba and Bagmara are in the core area of STR where no permit-holders were allowed to enter; obviously the casualty figure was very low in these four blocks. In columns 6, 7 and 10, i.e. the blocks Chandkhali, Pirkhali and Panchamukhani recorded sudden bursts in high casualty figures. It shows that Pirkhali-7 and Netidhopani 1, 2 and 3 registered 64 casualties, Chandkhali 1, 2 and 3 adjacent to Chamta registered 52 casualties.

Fluctuations of casualty in different blocks

Year	Chamta block	Gosaba block	Matla block
1961-1963	9	5	4
1963-1964	13	5	3
1964-1965	7	4	2
1965-1966	15	7	-
1966-1967	5	1	3
1967-1968	10	3	5
1968-1969	17	4	2
1969-1970	2	5	5
1970-1971	4	4	14
1971-1972	-	-	10
1972-1973	1	-	2
1973-1974	5	3	2
1974-1975	9	-	1
1975-1976	8	-	1
1976-1977	1	-	-
1977-1978	4	-	-
1978-1979 (Sept.)	3	-	-

The following are the analyses of data and findings:

1. The wave of human casualty maintained the same rate of about 40 persons a year. It is significant that eight blocks out of a total of 15 in STR were closed to the permit-holders from 1974 (September); yet the rate of death was the same. This positively indicated:

(a) Migration of tigers from the blocks which were made out of bound for the permit-holders to the areas where the permit holders were now working. It is evident that the concentrated zone of the heaviest kill in the Chamta and Matla blocks got diluted owing to migration of tigers to the adjoining Chandkhali, Netidhopani, Panchamukhani and Pirkhali-7 blocks.

Prior to this period, the incidence of casualty, especially in Chandkhali and Pirkhali-7, was very low, and the maximum was in Chamta, Matla and Gosaba blocks.

(b) The rhythmic activity in the tiger behaviour pattern is common with many micro- and macro-animals owing to daily tidal movement, lunar cycles, etc. In the past 17 years it has been found that the zone of the maximum kill frequently shifted, and that a particular area did not maintain a high incidence of casualty for a long period. In the present case the inactive tiger of Netidhopani, Chandkhali and Pirkhali-7 might have burst into activity owing to lunar periodicities or some endogenous oscillations; or the inherent man-eaters migrated from Chamta and Matla blocks.

(c) Increase in tiger population is also indicated, if the aforesaid analyses of migration and rhythmic activity were not accepted.

(d) The entire land area of all these blocks was not regularly inundated by tidal water.

(e) Salinity and tiger activity might have some positive correlations, but this might be a controversial issue unless proved with sufficient data.

2. It was a conspicuous factor that the Chamta and Netidhopani blocks of forests with an open crop of low height and patches of *Phoenix* (Hental) formation which were above daily tidal inundation level, have recorded high concentration of the tiger.

3. Another very conspicuous fact was that the three blocks- Netidhopani, Chamta and Pirkhali-7, which flank the rivers Netidhopani, Naubanki and Panchamukhani, the entry and exit water route to central and southern Sundarban blocks, recorded the maximum human casualty.

4. These factors also proved high degrees of intelligence and diabolical understanding of human behaviour by the man-eater of the area. The salinity concentration, therefore, may not be a deciding factor in high human mortality although the Gosaba river flanking Chamta, Netidhopani and Pirkhali has the maximum salinity. Strict protection of the 'core' area of the 'Project Tiger' from the permit-holders was sure to reveal more facts.

5. It should be possible to isolate the inherent man-eaters and deal with them separately to save loss of precious human lives.

6. Aerial photographs of the tiger habitat showed limitation of suitable territory, and a smaller number of saline blanks (such blanks are many and big sized in Namkhana Range west of the TR). There were too many creeks which made the movement of tigers extremely difficult.

7. There was adequate cover all over the forests.

The remarkable success of the Project Tiger in the Sundarbans in protecting the Bengal tiger as also the rare and beautiful mangrove ecosystem has been acclaimed by wildlife-lovers all over the world.

Chaudhuri and Chakrabarti (1980) analysed the problem of human-tiger conflict in detail.

1. Predatory aberration of Sundarbans tiger must be looked into carefully and steps taken to mitigate the circumstances leading to acquisition of such behaviour. The genesis and the tragic death of human beings due to predatory aberration of estuarine tigers is perhaps due to the combined effect of environmental attributes and human activity pattern. The saline water, muddy terrain, recurrent high tide, dense vegetation etc. coupled with difficulty in preying upon animals may bring about a psychological change in the tiger behaviour. This change leads to a cyclic and chronic pattern of the tiger's predatory proficiency. Habitat amelioration is not possible in this tract but the human activity pattern has to be changed. It may be a necessity to isolate the man-eaters.

2. It follows, therefore, that the study of the habitat and behaviour pattern of Sundarbans tiger is a subject of research and deep study. It is a difficult task. It should be possible to isolate the man-eaters and study their nature. Creation of large enclosures (if at all possible) in the forest habitat and releasing captured man-eaters to study their behaviour should be

preferred. Investigation on the blood, urine, kidney, liver, etc. of the problem animal is an immediate necessity. It is the research, and not management, on tigers which is of prime necessity in this estuarine tract. Fantastically high death rate in Chandkhali, Pirkhali-7 and Netidhopani-5 leads one to accept either of the three theories- (i) migration of inherent man-eaters or (ii) increase or concentration of tiger population or (iii) man-eating tendency revitalised in the dormant ones showing a rhythmic pattern of activity.

Relatively high negative perceptions of tigers in the human communities around the Sundarbans are exacerbated by other locally experienced poverty related problems as well as cyclones, floods, and soil erosion. Namita Singh of The Independent/Getty (26 May 2023) described an incident of surviving a tiger attack and how hunters became the hunted in STR in December, 2021. 52-year-old Mihir Sardar went fishing with his friend Babloo, taking his boat out at dawn and laying nets across a quiet stretch of brackish water. With the bulk of the work done by 8 am, the pair planned to stay through the day and overnight before returning home with a full net around the same time the next morning. "It was late evening. I was tired, I slept in the boat," Sardar recalls. "There were two doors on that boat, in front and at the back. So, while I had locked the front door, I forgot to lock the back door." It was an oversight that almost cost him his life, he explains. "That's where the tiger came from and attacked me." "The tiger caught me by my left leg and dragged me into the jungle," he recalls. "It was my friend Babloo who fought the tiger to save my life. While I held the tiger with his two front claws as it attacked me, Babloo managed to sit on the tiger's back and tried to hold him by his head, trying to prevent it from eating me. Then he fell to the ground." Sardar explains that tigers typically only target one person that they have identified as prey during an attack, ignoring others in the process. "So, as Babloo tackled the tiger, I tried to slowly crawl back in the boat. Babloo threw mud in the tiger's eyes." The tiger finally fled, and Sardar escaped by whisking what would have been a deadly bite to the head from the cat's jaws. But his injuries were still so severe that his life hung by a thread: the attack cost him a leg, an ear, an eye and part of the left side of his skull. His friend Babloo saved his life a second time over, Sardar explains, by using mud to cover his wounds "in a bid to stop the bleeding". "If I had lost more blood, I could have died there," he says. Since the attack none of his family members have stepped foot in the forest. "My son has taken a loan to buy an auto (rickshaw), so we are no longer dependent on the forest as our source of income," he says. "Now that I have survived the tiger attack, I cannot let another member of my family go through the same."

Bangladesh Sundarbans

A few studies on tigers and their prey have been conducted on the Bangladesh side of Sundarbans. In 1971, Hubert Hendrichs conducted a three month's study to identify reasons for man-eating by Sundarbans tigers. However, the project could not be completed but the initial data indicated an association between man-eating behaviour amongst tigers with increasing salinity levels. In more recent times, a long term study was initiated in February 2005 by the Bangladesh Wildlife Department from funding by Save the Tiger Fund and the US Fish and Wildlife Service to study tiger ecology and prey availability. Some other studies to assess prey density have also been conducted in this landscape by Reza *et al.* (2002). However, the most important contribution to information on tiger ecology in this region is an outcome of studies conducted by Adam Barlow in Bangladesh Sundarbans, which includes monitoring tiger populations in mangrove landscapes (Barlow *et al.*, 2008), designing conservation framework to reduce human-tiger conflict (Barlow *et al.* 2010).

Hendrichs (1975) found 392 tiger casualties between 1956-1957 and 1970-1971. Siddiqi and Choudhury (1987) recorded 554 human casualties during the period 1956 to 1983. Reza *et al.* (2002) recorded a total 401 people (about 24 people per year) were killed by the tigers whereas 4l tigers (about 3 tigers per year) were killed by humans during 1984 to 2000 in the Bangladesh Sundarbans. From January 1999 to March 2000, a total of 6 tigers were killed by humans. Most people were killed in the Burigoalini (45%) and Sarankhola (24%) ranges. Fishermen (44%), woodcutters (36%) and honey collectors (18%) comprise the bulk of tiger victims. Age groups of 26-35 years (38%) and 36-45 years (30%) were most vulnerable for tiger attacks in the Sundarban.

Neumann-Denzau and Denzau (2010) found the annual average number of tiger victims in the Bangladesh Sundarbans to be 168 for the years 2003-2005, with a high percentage being illegal entrants.

However, on the Indian side, while several books have been published on this region and man-eating tigers, scientific studies on the tiger are lacking. The inaccessible terrain of these habitats makes scientific research a challenge thus few such endeavours have been attempted in this zone. The first effort to assess tigers and their prey numbers in this region using more reliable scientific methods was made by Ullas Karanth and Nichols in the mid-1990s followed by a more recent attempt at understanding tiger ecology using radio-telemetry.

The methodology used across India (Phase I) for monitoring tigers and their prey could not be applied in the Sundarbans. A methodology to suit the environment of the Sundarbans was applied. Tidal channel searches were done across the Sundarbans to record sign and animal encounter rates. One hundred and twenty-six boat transects with an effort of 1,163 km were sampled across the entire TR. A similar approach has also been used in the Bangladesh Sundarbans as well (Barlow *et al.* 2008). The sign intensity data across the Sundarbans constituted the Phase I data set. Then a combination of satellite-telemetry and camera traps was used to estimate home range size, population and density of tigers (Phase III). The camera traps could not be deployed in a systematic grid based approach used across India. Instead, the camera traps were set up at strategic locations, near fresh and brackish water ponds, using attractants to lure tigers to the camera stations.

Mallick (2013) has divided the man-eating tigers in the Sundarban into three broad categories according to their behaviour pattern as shown below.

Category I: Inherent and designed maneaters (25%):

This subcategory is again subdivided into two micro-groups:

a)   Aggressive maneater (80%) and
b)   Lusty and adventurous (20%).

Category II: Undesignated man-eaters (15%); and

Category III: Circumstantial man-eaters (60%).

Whereas the man-eating propensity is attributed to the old age, disease, and injury of the mainland tigers, in the Sundarbans, the man-eaters are often young, strong, and healthy. Adult tigers enter the villages due to old age, sickness and injury (Rishi, 1992), when they are unable to hunt prey. Out of seven tigers autopsied, four (one subadult and three adults) had huge nematode colonies in their stomachs and the veterinary surgeons also reported that two tigers were infested with ticks and had liver problems (Gani, 2002).

Chakrabarti (1992) believes that salinity of water is probably the most important factor responsible for a good percentage (25%) of tigers turning man-eaters. The salt water might have affected the kidneys and liver of the animals and hence changed their physiology (Chakrabarti 1987). But analysis of the data shows that although water salinity tends to increase as one proceeds from the Northern zone to the Central zone and then to the Southern zone of the Reserve, a similar pattern was also noticed so far as soil salinity index is concerned, but the frequency of human casualties does not vary according to such zonations (Chaudhuri, 2007). This indicates that the man-eaters of the Sundarbans are concentrated to some habitat formations only, which also incidentally records significantly more salinity. Chaudhuri (2007) opined that the man-eating urge, a peculiar trait marked in a percentage of the Sundarbans tigers, is governed by cumulative action of physical and chemical environments of the Sundarbans. He also stated that the abundance or scarcity of prey animals are not directly related to the human casualty figures of tigers.

One Indian expert calculated that if people comprised a major item of the tigers' diet, Sundarbans tigers would kill 24,090 people every year. The man-eating nature of the tigers was studied many times in Bangladesh Sundarbans. 392 casualties were reported between 1956 and 1970, out of which 365 were known by place of incidence with 198 (54.2%) occurring in Satkhira Range. The possible reason behind such occurrences, among other factors, was the saline estuarine water.

Subsequent data (2002, 2003 and 2007) proved that the portion of casualties in the Satkhira Range had always been high and increased slowly. During the period 1974-1983, the differences in casualties between the low and high salinity zones became insignificant and the casualties in the medium salinity zone were significantly higher than those of the other two zones. This trend does not justify the hypothesis that the salinity of the water causes tigers to develop man-eating behaviour. In fact, the frequency of man-killing is highest in areas and at times of heaviest concentration of people representing various occupations, suggesting that the man-killing and the frequency of man-tiger contacts are directly correlated. But there are some uncertainties in calculating the number of people entering forests for the purpose of extraction including the illegal entrants to establish this interrelationship statistically because-

(a) It remains unknown how many members of the primary dependent households enter the forest for how many days per year;

(b) The group of secondary resource users, whose main income is generated outside the forest but who enter the forest occasionally, was not analysed in the SBCP Baseline Study (it might be the bulk);

(c) The number of illegal entries is not available; and

(d) The number of persons who enter the forest from outside the Impact Zone is not recorded.

It is assumed that this region is under intense human pressure with around 3.5 million people living within 20 km of its northern and eastern borders and depending upon the forests for livelihood resources. Annually around 35,330 people enter the forests of Sundarbans to collect timber, fish, honey and other products. This is an estimation based on the permits issued by the FD. The actual figure may multiply but it is beyond calculation.

In the Indian Sundarbans, on-ground verification does not support the statement that all the Sundarbans tigers are man-eaters. In the last 35 years, humans were killed inside the forest only when they entered illegally for NTFP collection. There was only one incident of a tiger straying into the fringe areas and killing a human, and that also was accidental. Based on these evidences it is assumed that the Sundarbans tigers carry a dual characteristic. Inside the forest they have no mercy, but outside their territory they understand their limits.

In fact, the Sundarbans tigers have been branded as man-eaters since the early colonial period but before that human-tiger confrontation was bare minimum. But during and after reclamation till present time entry of the human beings into the forests has not been diminished. Men and tigers are living together through ages despite cases of human fatalities. The problem of man-eating in the Sundarbans does not restrain the people from seeking a forest-linked livelihood, as these people accept these incidents in their mode of life because of dire poverty. The problem of man-eating tigers in the mainland can hardly be compared to this tract.

When a human has been killed by a tiger, the warning is a stick flying a fragment of cloth, often red, placed in a prominent creek-side position. People are reluctant to go ashore. They tie up in mid-river, anchored to a pole driven deep into the mud. But tigers are not averse to swimming out to moored boats to secure their human victims, even in the dead of night. Isolated in the mangroves, people are both vulnerable and available, 24 hours a day. Sundarbans tigers know this and they have acquired a reputation for being more nocturnal than tigers elsewhere. They come and go as they please, like phantoms of the night.

Chapter 10
Intraspecific Interactions and Relationships

"No one can walk beneath palm trees with impunity, and ideas are sure to change in a land where tigers are at home." – Johann Wolfgang

The tigers of mythology, of legend, of stories from childhood are all embodiments of raw power - the quintessential tiger is a creature of great strength, unmatched ferocity and savage instinct. The mangrove tigers are phenotypically unique and different from their mainland counterparts because they are specifically adapted to life in the mangrove forests. Tigers dispersing out of this habitat need to be restored either in the Sundarbans or in zoos. No tigers from any other cluster should be introduced into Sundarbans nor should tigers from Sundarbans be introduced elsewhere.

Personality types

To thrive in some of the harshest environments on Earth, the mangrove tigers rely on their morphological adaptations. But another critical survival tool is not so easily glimpsed: their personalities. The psychological makeup of the mangrove tigers may affect their hunting, mating, and even social standing among their peers. Insights into how they interact with their environment are vital. The adjectives that describe the tiger's personality traits range from "savage" and "imposing" to "dignified" and "friendly."

Two distinct personality types emerge that account for most of the tigers' behaviours. Tigers that score higher on words such as confident, competitive, and ambitious fell under what is described as the "majesty" mindset. Those that exhibited traits such as obedience, tolerance, and gentleness are grouped together under the "steadiness" mindset. Together, these two personalities explain most of the behavioural differences displayed by the tigers. Further analysis reveals tigers that score highly for "majesty" are healthier, prey more on live animals, consume more, mate more and rank higher on group status.

It is found that the most common personality types across felines are sociable, dominant, and curious. It would seem that 'majesty' aligns quite closely with a 'dominant' personality component and their 'steadiness' component aligns with components such as 'calm.' The researchers stress that more evidence is needed to determine why tigers with majestic mindsets appear to fare better than their steadier peers. It could be that majesty is the dimension that gives animals the extra edge to take that extra risk. That is to say, if food resources are scarce, the tiger that is the higher risk-taker will get one extra hunt each month.

Tiger is, therefore, a majestic predator as it plays a vital role in terms of preserving the ecosystem, environment and climate change. Tiger is perfectly designed to handle its prey animals and no other animal has such a feature. Small crabs in the mangrove habitat to big rhino or wild buffalo in the mainlands are the prey for them. Many think that tigers hunt successfully, but their success rate is only from 5 to 10%. For instance, many tigers die due to starvation. Tigers can be severely injured by spotted deer and there are such examples. The massive male wild boar chases away a tiger. A tiger consumes 3,500 to 4,000 kg of meat per year. The big cat kills an average of 60 to 70 spotted deer a year. The tiger has a unique communication system and one tiger communicates to another for a few kilometres. In many cases, male tigers go out of the natal forests to establish their territory. Male tigers look for territories which can be acquired easily. Young male tigers are not capable of fighting with adults to establish their territory. The tigers avoid straying out of forests into the fringe areas during the day, due to fear of human beings. But, one can see tigers hunting and dragging prey during daytime in Sundarban mangroves.

Tiger's day and night

In Nagarhole NP, India, the rates of movements of radio-collared Tigers were lower during the daytime (mean= 0-07 km/h) than at night (mean= 0-21 km/h). Radio-collared tigresses in Chitwan NP, Nepal, travelled about 0-7 km/h (range 0-2-1-2 km/h); estimates of the distances travelled in a night by females varied from 3-8 to 9-6 km. These estimates are based on straight-line distances between hourly nighttime locations, so the actual distances travelled would be somewhat greater. On some occasions tigresses crossed their ranges in a single night, while at other times they remained in a portion of it for several days. However, except when they had a kill or young, tigresses in Chitwan normally choose a different rest site each day. Hereinbelow I am describing a thrilling episode in the Sundarbans mangroves.

Adventures of the first satellite-tracked tigress in STR

To reduce the human-tiger conflict, a series of operation from trapping of the problem tiger/tigress followed by darting, sedation, radio-collaring, and release in the wild, to intermittently or continuously monitoring of its mobility and activities by radio telemetry is being done for the last seven years. Since both the young and old adventurous or aberrant tigers in Sundarbans are accustomed to the conventional cage traps, this technique was used for the first time to radio-collar an adult female in December 2007 for conducting a study through satellite tracking, when she was fitted with a customised (species-specific, i.e. a device for the tiger does not fit a penguin) imported US-made radio-collar (GPS/ARGOS/VHF) and tiny radio transmitter attached to a little harness or strap on the neck (the device well-tested before by placing it at different mangrove forest locations with variable canopy density). There were a number of receivers high above ground level, which picked up the signal of the transmitter carrier via satellites and conveyed this information to the researcher stationed elsewhere. When the information is being conveyed from the animal to the receiver, it is called an uplink. When it is sent from the satellite back to the researcher, it is called a downlink.

Before communicating the very interesting firsthand knowledge of this start-up case involving the resident tigress, named 'Priyadarshini', let me explain the objectives and methodology of this tracking in brief. GPS (Global Positioning System) tracking is a process to (i) remotely observe relatively fine-scale movement or migratory pattern of the radio-collared animal from island to island, (ii) calculate home range (normally 10-12 km in the mainland) and (iii) get an idea about habitat use and behaviour. The receiver is placed on a mechanised boat. The device records and stores the incoming data at a predetermined interval (duty-cycle) by an environmental sensor. The tiger's locations are plotted against a map or chart.

GPS collars are programmed to acquire locations at set time intervals either in advance or in the field as per requirements, whereas the radio transmitter allows monitoring of the tiger's movements on a daily basis. The battery in the collar allows for a set number of possible GPS locations. The collar's VHF signal is programmed to be transmitted for a set period of time each day, normally 8 hours. Its battery life in the field is determined primarily by the time interval between attempts to fix a GPS location. If it is set for every 2 hours, the field life would be approximately 8 months, whereas at an interval of less than 2 hours its field life would decrease and vice versa. Furthermore, battery life varies depending on the ratio of successful and unsuccessful GPS locations because it is for a longer period in case of the unsuccessful locations.

The collar locations are stored within the collar which has to be retrieved either by recapture, automatic release after the batteries run down, or by remote release using the equipment that transmits a signal to activate the collar release system. After the limit of GPS locations are acquired, the collar retains some battery life for radio telemetry and activation of the self-release mechanism. Even if the collar has fallen off subsequently, the still active radio signal helps retrieve it.

Moreover, it is vital to track the stationed or moving tiger every day using radio-telemetry. The GPS locations by themselves do not divulge information regarding the activities of a tiger at each location. By radio-tracking, it is sometimes possible to determine the tiger's behaviour, for example, when the tiger hunts or if it is associating with other tigers.

GPS collar locations are analysed to construct Minimum Convex Polygon (MCP) and Fixed Kernel (FK) home ranges. Sign deposition rate is evaluated as the number of creeks/channels crossed by the radio-collared tiger per day using Google Earth. If a tiger remains at a particular location within a periphery of only 100 m for more than 12 hours during the daytime, it is considered that it must have hunted. Similarly, if it is localised for more than 8 hours at night, the same conclusion is arrived at. When the transmission is temporarily cut off in case the tiger enters deep inside the forests beyond the signal receiving range (2-5 km), the monitoring team tries to retract the animal by to and fro movements for hours. Therefore, it needs calculative use, otherwise frequent readings will drain limited battery power of the collar more rapidly, whereas longer intervals between readings may provide lower resolution but over a longer deployment.

Before conducting the first radio-collaring experiment in STR in 2007, a preliminary ground survey was carried out in the PAs during 2006. Camouflaged cage traps were placed in six different locations, at least 5 km apart (to capture maximum number of different individuals), three each in two distinct (separated by at least 25 km aerial distance) zones NP (Core area closed to human use other than management purposes) and WLS (Buffer area open to tourism and regulated fishing and honey collection by the fringe people), 100-200 m inside the forest (from the edge of a river channel), mostly near the confluence of two or more river channels or near the rainfed freshwater ponds as these places are visited by the tigers frequently.

In addition, camera traps Trailmasters™ 1550 (Goodson Associates Inc., Lenexa, KS, USA) were deployed at each cage trap to monitor the visitation rates and the behaviour of individual tigers at cage traps. Initially, the traps were kept non-operative so that the tigers could enter and leave the cage freely so that they are habituated to such practice in future when the traps are made operative. The traps were deployed for two months (30 days x 2 months x 3 sites x 2 zones = 360 trap-days in total). The trap sites were monitored each day to check the signs of tiger's movement and whether the live bait is hunted. Tiger visits were recorded only at two of the three sampled locations in the core area. One of the camera trap locations recorded a tiger visit only once, while the second location registered three tiger visits at an interval of 25-30 days, but the visitors could not be identified because the camera trap did not trigger on the first occasion and on two subsequent visits the tiger stayed away from the infrared beam necessary to trigger the cameras. Whereas the individual tigers visited all the three trap locations in the Sanctuary 5, 7 and 12 times respectively at a 9-15 days' cycle. Fresh pugmark sets and photographs from camera traps placed at the mouth of the traps confirmed that the tigers picked up the bait from near the traps. Hence, the live baits were kept inside the cage traps and tigers did not seem wary of picking the baits from right inside the traps.

In the field, it appeared that the tiger's visit rate was low in the core but high in the buffer area. For example, a trap was laid with goat-bait at far-flung Chamta block (NP area, centrally located south of Panchamukhani, in between Chandkhali and Netidhopani) because earlier a number of aberrant tigers had been released here for rehabilitation, but it was observed that none of them was encouraged to enter the trap and lift the bait. A few pugmarks were, however, seen within 200 m radius of the trap, which indicated that a tiger came to the spot at night but it was very clever (or recollected the bitter experience when captured earlier) and, naturally, being suspicious, did not take the risk of further confinement, instead made a few rounds surrounding the spot for inspection of the trap cage and then left the place to utter dismay of the field staff. They did not attempt to lift the bait, even after prolonged baiting or when it was tied outside the trap at the same location, except on one occasion. However, fresh pugmarks were found on the mudflat to indicate that they used to pass by ignoring the lure of the easy prey.

On the contrary, in the case of the sanctuary, the tigers used to visit all the trap locations. Individual tigers followed a cycle (gap) of 9-15 days to revisit the area. Here, the visiting tigers were not at all shy or hesitant like those of the core area. Most surprisingly, there were a few instances when the intelligent tiger(s) frequently took away the bait from within the cage without being trapped. Fantastic, a jewel of the cunning hunters! You may be eager to know the innovative tactics of such bold tigers. First, it started moving all around the cage, but instinctively did not enter at the time of its first visit, not to speak of any attempt to take the bait away. It appeared that the females were comparatively more intelligent. However, after a few days' visit and observations, it had taken away the bait without stepping on the lever; lest it would lead to closure of the cage. In another trapping site, the animal avoided the main cage door, instead hitting the cage repeatedly from the back till one of the nut bolts was slackened and fallen off and it was able to secure a small opening. It then somehow squeezed its paw inside the cage through that small opening to pull out the bait in such a manner that, as a result of the tremendous tug, half of the goat's body was left inside the cage. However, all the residents are not equally cautious or innovative in securing their prey and a few of them entered through the main cage-door in lure of the live bait and could be trapped.

Considering that majority of the tigers were sighted in Pirkhali and Panchamukhani blocks of Sajnekhali WLS and, as per the annual monitoring reports of the previous years, highest frequency of sighting during the four months from January to April and the sightings started reducing from May onwards during the monsoon, recovering only in December, indicating a periodical fluctuation of tiger population in these northern blocks, Sajnekhali WLS seemed to be the most effective site for capturing and radio collaring of the tigers, particularly from December onwards. This hypothesis proved to be true, when fully operational traps were deployed in December, 2007.

On the 4th, a healthy tigress, approximately five years old, later named 'Priyadarshini', could be captured at Panchamukhani-3. It was tranquilised and radio-collared on the 5th. After recovery from drowsiness, the tigress was released in the same block, i.e. her original habitat, at Choragajikhali location next day (6th) at 8.20 am. It was being tracked for the whole day, when she was found either resting or moving intermittently and, till the evening, covered only a distance of about 2.5 km from the place of release. The continued monitoring conducted later also revealed that it usually rested during the day and became active at dusk and dawn and once observed to kill a deer.

The satellite link stopped working within a day of radio-collaring the tigress. The GPS also failed within a week. However, the VHF beacon continued to work, providing a range of over 2-3 km that allowed us to monitor the tigress from a boat. During the period from 7th to 18th, Priyadarshini remained within the surrounding areas of the Panchamukhani forests because signals received confirmed her local movement. As usual, she was mostly resting. Perhaps she might have managed to hunt in the meantime and consumed food sufficient to meet her appetite for a few days.

Continuous monitoring began on the 20th afternoon, when Priyadarshini was staying at Pirkhali forest near the Deulbharani *khal/khanri* (narrow creek), opposite the river Gosaba. She was located in almost the same habitat for the last few days. On 21st, she moved towards Panchamukhani by crossing the *khal*. But, she did not proceed for long and, ultimately, came back to her earlier refuge at Deulbharani-Gosaba junction. On 22nd, up to the afternoon, she was resting. At 5 pm she got up and started moving again towards the Deulbharani *khal*. On the 23rd too, she did not shift to any other location. It was only on the 24th evening (7 pm) that the incoming signals were being strengthened due to her quick movement. She was in dire need of flesh now. At 9 o'clock, it came within a range of only 200 m from the monitoring boat. The signal was almost fixed, i.e. she was resting in the forests. Silence prevailed during the next one and half hours. It was 10.40 p.m., when the signal showed rapid up and down movement. At about 11 p.m., she could locate and select the prey for hunting. But the targeted prey could sense the presence of its camouflaged enemy nearby. The surrounding forest was reverberated at the repeated, loud and sharp alarm calls 'ack-ack' of the *Chital* deer (*Axis axis*) and within a minute the sound of splashing water was heard ahead within a distance of 10-70 m. The predator started chasing her prey in the flooded terrain. It was already high tide then, engulfing the lower forest floor at a rapid pace. It was a full moon night too. Still neither the predator nor the prey could be seen, though a tough struggle seemed to take place between the sturdy tigress and the lanky deer, as you often see in the popular tv channels. Whereas the chaser was galloping, the forerunner was trying to escape by leaps. But Priyadarshini was in vain and the deer won the race. A great relief for the panting even-toed ungulate! Although Priyadarshini was utterly frustrated, she did not give up hope and regained sufficient energy for the second chance. At 11.20 pm, again the sound of splashing water was heard. The tigress relocated her prey and this time ran after the target in a zigzag way. The prey was really very tired and this time could not compete successfully. Suddenly, it gave out a loud cry followed by a thud. At last Priyadarshini could seize the prey (average weight 50 kg). It was exactly 11.24 pm. She dragged the kill through water towards a high dry land for the dinner. It was past midnight, 00.09 am, i.e. 25th December. Another chital cried afar. But the tigress had then concentrated on her kill. While feeding, she was relaxed and gave out low growls in pleasure. A great feast of X-mas indeed!

In the forenoon of 26th, she moved with her half-eaten kill about 1 km away for relishing during the remaining few days of 2006. However, nothing could be seen as dense fog rolled into the entire forest and over the water bodies. In the afternoon, continuous monitoring was called off.

Based on data collected over 45 days, the home range of the tigress was estimated at 40 km² and 28 km² using the 100% and 95% Minimum Convex

Polygon method, respectively in Panchamukhani and Pirkhali forest blocks, located side-by-side with similar type of vegetation and prey-base, where the tiger density was seasonally high. However, the radio-collar of the tigress was operative for four months (up to April next year) and then ceased operation, but the tigress with collar was seen physically many times afterwards in the same location, showing the sign of territoriality, i.e. within a definite area of operation in comparison to that of the strayed tigers.

Social structure

Based on the preferences of living habits, animals can be classified into solitary animals, colonial animals, or gregarious animals that tend to live alone, in colonies, or in packs or groups, respectively. Tiger is the best example of a solitary animal and not very friendly in nature. The biggest difference between two top big cats in the world- the tigers and the lions- is their social structure: lions are gregarious and live in pride, a family group composed of many females (always related) and either a single male or a coalition of males (usually the male lions are brothers but they can also be just friends who sided together to form a team or both), while tigers usually live alone with the exception of the cubs with the parents.

Breeding chart

Traits	Criteria
Age of maturity	Females sexually mature at 3-4 years and male at 4-5 years.
Mating system	Polyandrous mating system. Copulation frequent during mating, mating induced ovulation.
Seasonality	Any time of the year.
Life expectancy	20-26 years.
Oestrus cycle	About every twenty-five days
Gestation period	Average of 104-106 days.

Traits	Criteria
Litter size	Average 2-3.
Growth and development	Eyes open after 6-14 days; mother's intensive care for 3-6 months; begin to move with mother at 5-6 months; able to hunt by 11 months; males grow faster and the females stay back longer with the mother.
Dispersal	Both sexes at two to three years of age.
Inter-birth interval	33.4±3.7 months (range 24–65 months).

Tiger dating

If urine and faeces can reveal whether the tigers have the right chemistry to be paired, it is called tiger dating. Tigers are very particular when they choose a partner. If a male and a female tiger are paired, there's no guarantee that it will result in tiny small tiger cubs. The female tigers can pick a mate by sniffing samples of urine and faeces from a selection of male tigers. The method emulates the tiger's behaviour in the wild, where they normally live scattered over large areas. Here the male marks his presence and exhibits his qualities by squirting on trees or dropping stool, which the tigress can then sniff and thereby draw conclusions about each of the males. If she fancies one particular scent, she may stay in the area and allow that male to mate with her. Captive breeding is the process of breeding animals in human-controlled environments with restricted settings, such as wildlife reserves, zoos and other conservation facilities. The process is sometimes construed to include the release of individual organisms to the wild, when there is sufficient natural habitat to support new individuals or when the threat to the species in the wild is lessened. It is mainly the female who selects a mate, because she's the one running the greatest risk by carrying a cub. That requires a lot of time and effort, so the tigresses are quite picky.

Interactions

Interactions between the tigers influence a range of functional behaviours and can affect the dynamics of populations and communities. Significantly, interactions between tiger conspecifics can influence fitness outcomes and may help these animals cope with environmental and anthropogenic disturbances in Nepal's Chitwan NP with approximately 120 individuals monitored over eight years (2008–2017) (Carter *et al.*, 2023). It is expected that the conspecific interactions in solitary carnivores vary by individual characteristics, such as age, sex, life-history stage, or reproductive state. Tigers, though considered "solitary" predators, have extensive interactions with conspecifics, but that interactions are modulated by intrinsic factors such as sex, residency, and reproductive status.

In such a network, nodes are often individual animals, with individual characteristics such as age and sex. Edges represent how two nodes relate to one another, such as co-occurrence at a given site. Network analyses can then be used to assess network structure, including the presence and strength of conspecific interactions and the factors underlying those behaviours. This framework has greatly improved our understanding of behaviours ranging from mate-choice to competition, from dispersal to predator avoidance, and from social learning to disease transmission, all with important implications for survival.

Saying that tigers are solitary animals can make it appear that they do not have a social structure or that this is simplistic. Since they are loners, mainly living an isolated existence, it is hard to view their social structure on a regular basis which can change during the various periods of their life. In fact, the tigers establish systematic relationships that make up a social organisation, of course, typical of this animal species. Unlike other social carnivores, they do not move around in groups. So mostly they are not in constant touch with each other visually. But they still can communicate with each other by different means which helps maintain their society and hierarchy because they form dominance hierarchies through social bonding for survival. Hence, they create complex communities that depend on a variety of inter-individual interactions. Their ability to survive in their surroundings depends on this social interaction. How individuals interact with one another is essential for understanding their culture.

In Panna TR one female tiger with three nine to ten months old cubs was radio-collared in January 1997. The collared male has expanded his home range considerably from 55 km² to over 230 km². But 90% of his activities are restricted to 115 km², which primarily lies in the middle plateau in the

Hinauta range. This is the least disturbed habitat of the park and it is the home of the recently collared female. Her movements are restricted to this undisturbed and prey-rich area and four months of monitoring reveals her home range is 25 km². She was predating mainly on sambar at intervals of 6-8 days. So far 13 of her kills had been found, of which seven were sambar (mostly females), two nilgais and one chital. For a tigress that was raising three cubs, the kill rate observed is less than what is required. Sufficient availability of food is critical for the mother to raise the litter successfully.

In Panna, the principal prey species in order of their abundance are nilgai, sambar and chital. Though nilgai is the most abundant prey species of tiger, it occurs in the more open thorny forest which is seldom used by tigers. Tiger's principal prey chital occurs in low density and sambar is patchily distributed. Such limitations can certainly affect the reproductive success of the population. The male tiger, on the other hand, was feeding primarily on livestock, especially buffaloes and he ventures outside the park boundary frequently. His habit of livestock killing and venturing outside the park exposes him to man-caused mortality.

It is presumed that males would be more influential within networks (i.e., higher degree, eigencentrality, and betweenness) due to their larger home ranges and increased likelihood of encountering conspecifics and being more likely to associate with females due to tiger mating systems (i.e., single males overlap several female territories). The residents would be more influential in networks due to interactions involved in the maintenance of territories and may have genetic or other relationships with individuals in adjoining territories, while non-residents may avoid residents to avoid territorial conflict. Likewise, it was predicted that the residents were more likely to interact with other residents through interactions along territorial edges reproductive females would have lower values for social network metrics due to avoidance of conspecifics who may predate offspring.

Those male-female clusters that appeared stable tended to last about three years before dissolving with new cluster formation around a single male. Resident male tigers may interact with the same group of neighbours over time, whereas a floater male (with no established territory) may interact with various new individuals while in search of a territory. Furthermore, interactions among solitary carnivores might be influenced by homophily, defined as the tendency to seek out or be attracted to those who are similar to themselves and to preferentially interact. Through interactions with conspecifics, individuals can regulate conflicts, create affiliative bonds, cooperate, transmit information, and learn. These interactions also help individuals cope with ecological constraints specific to their living environment. For example, resident male tigers may interact more often with females within their territory than with neighbouring resident males. In contrast, dispersing males may more likely encounter other males than females as they challenge resident males for their territory. Competition between adult males can sometimes end in injury or death for the defeated animal. The disappearance of male may have precipitated a breakdown of the network, leading to disjunct individuals in the following year. Indeed, the removal of certain individuals can have disproportionate effects on network structure, depending on the network position of those individuals.

Alternatively, climatic changes, such as cyclonic flooding or fluctuations in prey distributions, may shift tiger activity and associations, thereby restructuring the network through time. Although the networks are highly variable across years, some small clusters maintain their integrity across several years. Males are usually the nodes maintaining cluster integrity, as they more often connect nodes together than females because they move long distances to find and mate with multiple females, while simultaneously excluding other males from mating with the females within their territory. Interactions between conspecifics are all important linkages between individual behaviour and group-level dynamics. While using regular travel routes, tigers constantly look for scent and visual evidence of other tigers. Fights and overt aggressive encounters are generally avoided; though these may occasionally result in serious fights, even culminating in death. For example, scent marking likely helps them discriminate conspecifics as neighbours or strangers or as potential mates or competitors, thereby determining their level of aggression or tolerance toward conspecifics. Insights on conspecific interactions of tigers might also help predict the effects on translocating a tiger to a new population or the outcomes of reintroducing tigers to a new location entirely (Lahiri, 1974).

The Bengal tiger's social networks assume that the co-occurrence at camera trap locations represents an association between conspecifics. These networks were found to be fickle, remaining stable for about three years before dissolving, i.e. these networks were found to be highly variable through time. Most associations were temporary, lasting one season. Tiger clusters in the overall network tended to include about three to four animals. However, the identity of these animals changed often through time indicating high turnover of individuals in the clusters.

It was found that males were more likely than females to form bridges between other tigers, and resident tigers were more central in the networks than non-residents. In addition, interactions between two animals were more frequent if they were of the opposite sex or were both residents.

In general, three types of the tiger's social structures are found in the wild:

1. Solitary tigers: Adult male solitary tigers often live alone and defend their own as well as their breeding partners' territory. Solitary females, on the other hand, are often less territorial and share some of their territories with related females.
2. Transient tigers: They are also known as floaters or dispersing tigers, are typically young males who have left their natal range in search of their own territory. They may move through several different areas before settling down and are more likely to be involved in conflict with other tigers.
3. Communal living: This type of structure is a rare one, but was documented in a few cases. Female tigers and their progeny can establish small family groups, which typically consist of the mother and her cubs. These family units can shield the young from the predators increasing their survival rate.

Tigers have a wide range of social structures from solitary living to short-breeding related relationships between males and females to family groups composed of mothers and cubs. Because the tigers establish different short and long term alliances, defend their territories and look out for their cubs, these social systems are crucial for their survival, day-to-day activities including hunting, protection, ability to procreate and raise the offspring. The strongest and most capable male tiger receives the comparatively better habitat as domain, and also has more mate partners, according to dominance hierarchies. This lessens bitter rivalry and conflict.

Tiger's social behaviour and communication is influenced by a variety of factors, such as habitat, prey availability and social structure. They can more easily adapt to the specific challenges they face in their various habitats. Generally speaking, tigers roam alone when they reach adulthood because they obviously need maternal care while they are cubs. Nevertheless, they are very territorial and maintain ranges of territory characterised by the dense vegetation, availability of water and enough prey. Its territory constitutes an exclusive area of distribution that contains one or several dens in which they rest or in which the females give birth and take care of their young.

The males dominate, and they are larger than females. The females tend to be more tolerant of each other, and they live in smaller territorial ranges than the males. However, the mothers are incredibly fierce when protecting their offspring so they even scare males, which could overpower them, to leave their cubs alone.

Having females within males' territory makes it more convenient for mating because they come into contact with each other. Besides mating, tigers are likely to take separate ways. Tigers can be social with each other though depending on what is going on. Some of the vocal sounds they make indicate that they aren't giving a warm welcome. They may hiss, meow or growl to get others to get out of their territory. Other times they will purr and make growling sounds to indicate interest and to draw others to them, especially for mating purposes.

In 1793, the Sporting Magazine article [Sporting Magazine 2 (July 1793): 199] introduced the letters from the companions of Munro, the deceased victim of tiger depredation in the Sagar Island, by first presenting material describing the ferocious characteristics of the tiger and supplementing this with an account of commonly held beliefs about the supposed effect of fire on the animal. The opening paragraph stressed the tiger's ferocity:

The tiger is allowed to be the most rapacious and destructive of all carnivorous animals. Fierce without provocation, and cruel without necessity, his thirst for blood is insatiable: though glutted with slaughter, he continues his carnage. He fears neither the sight nor the opposition of man, whom he frequently makes his prey, and it is generally supposed that he prefers human flesh to that of any other animal. The tiger is, indeed, one of the few animals whose ferocity can never be subdued.

Pennant writes of tigers in his book History of Quadrupeds (1793; 1: 278-79):

If they are undisturbed they plunge their head into the body of the animal up to their very eyes, as if it were to satiate them with blood, which they exhaust the corpse of before they tear it to pieces. There is a sort of cruelty in their devastations, unknown to the generous lion; as well as poltroonery in their sudden retreat on any disappointment.

Territoriality

Tigers are territorial animals. Territoriality plays a crucial role in the population dynamics of tigers. However, previous models of tiger population dynamics have not adequately incorporated different territoriality of males and females. Female tigers adapt their territories to changes in prey biomass and the presence of adjacent female territories, while males adapt their territories to the number and location of nearby female territories and the presence of adjacent male territories. These territories are vacated or reoccupied every time a tiger dies or a territory is changed or lost.

On the contrary, the relationship between humans and tigers should also be judged in this perspective. One might think that tiger population size would have an inverse relationship with population density, but it appears that tigers may establish habitats regardless of human presence, perhaps because of their dependence on having a sufficient prey base. Even though tigers tend to avoid humans, they are very territorial towards intruders, which mean they may stay put regardless of lingering human presence; this makes them even more vulnerable when humans enter their territory trying to expand our own habitat, because instead of leaving, they will be provoked to attack to defend their territory.

The tigerlands are facing problems like dense populations and rapid human development, which poses a significant threat to any remaining viable tiger habitats. Between habitat loss from destruction of forest land to agriculture, commercial logging, and human settlement, and the tigers' requirement of a large and sufficient prey base, the needs of humans and tigers have clashed more frequently over the years, which then put tiger populations in danger from hunting. As a result of their dietary requirements and unwillingness to leave their territory, tiger attacks on livestock and people have led to an intolerance and fear of tigers by neighbouring human settlements. This then presents a challenge in gathering support for tiger conservation within their native habitats. Tiger attacks often lead to retaliatory killings, even though tigers were simply defending their territory.

The following factors are linked to change in occupation of a territory (Carter *et al.*, 2015):

(i) Mortality: It depends on sex, age, and on whether the tiger is a territory-holder or disperser. When a female with dependent offspring (i.e., cubs and juveniles) will die, then her offspring may die as well. Tigers die when they reach about 15 years old, considered their maximum age in the wild. Mortality events were segregated initially into four categories viz., natural mortality, mortality due to diseases, human-caused mortality and inconclusive causes. These are discussed below (Nigam *et al.*, 2016).

(a) **Natural mortality**: The cases classified into this category included mortalities due to following:

• Intra-specific aggression: Mortalities caused by other tigers due to territorial aggression or infanticide by males.

• Inter-specific aggression: Mortalities caused by other species, either by prey during hunting sessions or by other sympatric carnivores.

• Cub mortality from starvation or separation from mother: Cubs are also killed by the male tigers. Similar incident took place in the Bandhavgarh TR in August, 2023. A seven-month-old female tiger cub's carcass was located next to the pugmarks of another tiger. Another female tiger cub that was 7-8 months old was murdered earlier in April during an attack by a male tiger.

Schaller (1967) calculates that the average number of young raised per adult tigress per year is about one, on the assumption that each has a litter every two years and an average of two cubs from each litter survives to independence. Thus if a tigress has her first litter at four years of age and a life span of 18 years the potential turnover rate of a single tigress would be 14 animals under ideal conditions, but it is pointed out that this rate is never achieved. The major reason for this non-realisation of potential turnover rate in tigers appears to be natal mortality. In Schaller's small sample of eight cubs, three cubs (38%) died as yearlings. Of these four cubs of one tigress survived because of easy availability of domestic cattle and of buffalo kills offered to them. The available information indicates that at best not more than two 1arge cubs survive into second year, which is not always the case. Even accepting this 33% mortality and assuming another10% mortality in the second year, mortality in tiger till maturity at three years may be taken as 40%. Parturition conception interval of a tiger is said to be about two years as the tigress does not come into estrus during the period of dependency of her cubs and the cubs are dependent on the mother for about two years. Thus the average production rate of a tiger is three young in about 28 months or say, in two and half years or thirty months, working out to be 1.2 young per year. This would be limited by a combined natal and juvenile mortality of 40% which would mean a potential recruitment rate of .7 young or 70 young to the breeding population per hundred adult tigresses. But even this potential is further limited considerably by the fact that only a few of the tigresses reproduce and bear cubs. For example, for the eight tigresses identified and one whose tracks were noted by Schaller only three or one third had cubs.

• Senility: Mortalities caused by old age related factors, with no other detectable specific disease processes.

(b) **Mortality due to diseases**: The tigers have become increasingly vulnerable to the impact of diseases. The cases classified here included mortality due to both specific infectious diseases and non-specific causes like emaciation. Perhaps a decade back in Dayapur, news of a tiger trapped in a ditch was reported by villagers. The FD patrol team soon located the tiger and used their tranquilisers to sedate the animal. The tiger was kept in the Sajnekhali Forest Office inside an iron cage for treatment, and nobody was allowed to visit the place except forest staff and doctors. The tiger ultimately died in captivity, as the medical intervention failed to save his life. The doctor disclosed that Sundarban tigers not only get killed by poaching but they also die because of innumerable diseases and changing climatic conditions in the delta and the declining share of food inside the forest. We should not simply correlate the decline in tiger population with growing human settlement in the islands, although they are also part of the cause.

Infectious disease events recorded in the Indian tigers (*Panthera tigris*)(Nigam *et al.*, 2016)

Pathogen	Diseases documented
Bacterial	*Colibacillosis, Hemorrhagic septicemia/ bronchopneumonia, Leptospirosis, Shigellosis, Salmonellosis*, and *Tuberculosis*.
Rickettsial	*Anaplasmosis*, and *Ehrlichiosis*.
Viral	Avian Influenza, Canine Distemper, Feline Calicivirus, Feline Coronavirus, Feline Immunodeficiency Virus, Feline Panleukopenia, and Rabies.
Prion	Feline Spongiform Encephalopathy.
Fungal	*Coccidioidomycosis* and *Microsporum canis*.
Protozoan	*Babesiosis, Cytauxzoonosis, Eimeria (E. harmani, E. novowenyoni), Isosporafelis, Sarcocystis* spp., *Trypanosomiasis*, and *Toxoplasma gondii*.
Cestodes/ metacestodes	*Pseudophyllidea* spp., *Taenia* spp., *Taenia hydatigena* and *Taenia taeniformis*
Trematode	Dicrocoeliidae spp., Hymenolepididae spp., *Paragonimus* spp., *Paragonimus westermani* and *Platynosomum fastosum*
Nematode	Ancylostomatidae spp., *Aelurostrongylus* spp., *Bronchostrongylus subcrenatus, Capillaria* spp., *Capillaria aerophila, Dirofilaria immitis, Galonchus perniciosus, Gnathostoma spinigerum, Molineus* spp., *Mammomonogamus* spp., *Ollulanus tricuspis, Physaloptera* spp., *Spirocerca lupi, Strongyloides* spp., *Toxocara* spp, *Toxocara cati, Toxascaris* spp., *Toxascaris leonina*, and *Trichuris* spp.

Canine distemper virus (CDV)

Among the pathogens reported in wild tigers, only CDV has been associated with mortality in breeding adults coincident with population declines. Although only a small number of cases have been diagnosed, the presence of infection in at least five of the 10 tiger range states suggests that all populations are probably at risk of infection. Canine distemper is caused by an enveloped, single-stranded RNA virus in the genus Morbillivirus (giving rise to its contemporary common name, canine morbillivirus). The viral envelope reduces environmental stability (particularly in warmer temperatures), limiting opportunities for indirect transmission. However, CDV is readily transmitted during direct contact, via inhalation of aerosolized respiratory droplets or possibly during predation of infected individuals. Once infected, the virus replicates in immune cells (leading to immunosuppression) and then in epithelial tissue (causing respiratory and gastrointestinal disease). Tigers that survive these initial stages of infection can progress to develop severe neurological disease as the virus becomes established in the central nervous system. Survivors develop an immune response that should protect them from further infection. While immunologically naïve; tigers of any age can die from infection with CDV, mortality of breeding adults has the greatest impact on population viability.

New pathogens

The emergence of new pathogenic threats has posed fresh challenges. Only one of the 17 pathogens included in the serosurveys is known to cause mortality in free-ranging tigers (CDV), a further five have caused mortality in captive tigers (feline parvovirus [FPV], feline coronavirus [FCoV], feline herpesvirus [FHV], feline calicivirus [FCV] and influenza A virus), and another (FIV) in a captive lion *(P. leo)* and a captive snow leopard *(P. uncia)*. Several of these pathogens (FPV, FCoV, FHV, FCV and FIV) have been associated with reproductive failure or mortality in neonates.

Helminthasis

Helminthasis is common in wild animals but reports on both *Toxocara cati* and *Teania hydatigena* infestation in Bengal Tiger *(Panthera tigris tigris)* from wild nature is scanty. A wild male adult Bengal tiger was observed lying alive on forest floor in the core area near Pirkhali of STR, about 100 feet away from river bank since noon on 9th June, 2012 with the symptoms of severe weakness, unable to stand properly, staggering gait along with dull and depressed appearance. Its reflexes were poor and not interested in taking food. Ultimately the big cat succumbed that night (Bhattacharya et al., 2012). The male tiger was aged about 16 years having body weight 72 kg, length (tip of the nostril to end of the tail) 2 m 52 cm, height 93 cm. The carcass was emaciated, dehydrated, having a rough hair coat, shrunken eyes and there was no external injury except some lacerated superficial wounds on the inner side of thigh and abdomen. Teeth including canines were broken, eroded and having caries in some of them. Stomatitis in the upper palate and gingivitis were also observed along with retracted claws.

Postmortem examination revealed anaemic mucous membrane and liver was slightly enlarged, mottled, congested partly with miliary necrotic foci. Heart was slightly enlarged with thickening of heart muscles. Kidneys were congested, hard with hemorrhagic lesions in cortico-medullary junction. Mesenteric lymph nodes were hyperemic. A plenty of live round worms were present in the stomach and duodenum and few tape worms were found in the ileum. Altogether there were 121 numbers of roundworms and 07 numbers of long tapeworms. Intestinal mucous membrane was inflamed and thickened sporadically with presence of semi-digested faecal materials.

Histopathological examination revealed chronic interstitial hepatitis and fibrosis in liver, exudative pneumonia in lung, oedema and infarction in heart, chronic interstitial nephritis in kidney and focal cholecystitis in gallbladder. Tigers may be infected from wild ruminants and wild boars by feeding on them.

(c) **Human caused mortality**: The cases classified in this category included resultant mortalities from-

• Accidents: Mortalities which are caused by accidents, fire or drowning.

• Poaching: Mortalities which are intentionally caused by poisoning; electrocution; fire arms or by using snares/jaw traps.

• Legal elimination: Tigers are killed as part of conflict mitigation strategy, due to the threat posed by them to human life. As per the standard operating procedure (SOP) prescribed by NTCA a tiger is to be removed from the wild (preferably captured) if it was found responsible for deliberate and consistent human attacks. Most attacks on humans were accidental when humans ventured into tiger habitat in search of NTFPs. However, some tigers did become man-eaters and it was only these individuals that were eliminated.

(d) **Inconclusive:** The cases classified in this category included mortality without any confirmatory diagnosis (excluding senile/emaciated animals) or supporting documents.

Necropsy procedure

Actual necropsy procedure begins with thorough external examination, that can support and lead to some key findings. These are briefly discussed below.

1. Necropsy interval: Estimation of time of death [necropsy interval (NI)] is crucial as it influences the appropriateness of the conduct of necropsy procedures. As many factors operate soon after death, post mortality changes vary to a great extent in each individual.

Evidence for estimating NI may come from "Carcassial evidence" i.e. evidence that is present in the body such as rigour mortis, state of decomposition, presence of insects (forensic entomology) etc. Additional information from environmental and associated evidence (that are present in the vicinity of the body) and anamnestic evidence (in monitored animals, evidence based on the animal's movements, last known sighting and location etc.) may help in assessing the NI.

2. Rigour mortis: Rigour mortis is one of the recognisable signs of post mortem changes, its onset and duration is governed by temperature (lower ambient temperatures accelerates the onset of rigour and prolong its duration, whereas the opposite is found in warmer temperatures), metabolic state of the body (vigorous activity immediately prior to death) and prior disease if any. Even though time of appearance and disappearance of rigour mortis cannot be generalised, it can be still used to

establish an estimate range. Typically, the stages observed over time lapse are as follows:

Stage	Time-lapse since mortality event
Carcass is fresh and rigour mortis is absent (flaccid state)	> 1 hrs
Rigour setting in	1 to 6 hours
Rigour complete	6 - 24 hours
Rigour passing off	12 - 36 hours

3. Changes in the eyes

Hours	Changes in eyes
0.5	Eye lens and fluid fully transparent. Light reflected from within the eye appears brilliant luminous green. Pupil size however depends upon the prevailing light intensity at time of death. Pupils fully dilated shortly after death as the muscles are relaxed.
0.5- 6	Lens and fluids remain transparent. Luminosity and colour may decrease slightly. Pupil fully dilated and is almost round (as rigour progresses the pupil gradually flattens).
6-10	Colour changes toward grey. Luminosity fades. Pupil width may decrease to about one half original diameters.
12-18	Colour fades to dull grey. Luminosity fades away. Pupils may be narrowed to one-third or less of their original diameter.
30- above	Colour and pupil diameter remain the same. Hazy blue colour appears over brown coloured iris after about 48 hours. Normally on the third to sixth day after death, the eyeball partially collapsed.

4. State of decomposition: Decomposition occurs due to enzyme breakdown of tissues (autolysis), and bacterial activity (putrefaction). In

general, carcasses can be assigned to following stages, based on state of decomposition.

Stage of carcass	Changes observed
Fresh	Characterised as the beginning of decomposition immediately following death. There are relatively few changes occurring during the fresh stage and the odour associated with the remains will still be the natural smell of the body. The fresh stage continues until the first signs of bloating begin, which is highly variable depending on the external environmental temperatures and conditions.
Early Decomposition	Early decomposition is marked by the beginning of bloating in the abdomen region and the skin begins to take on a green discoloration. During the bloating period, the body begins to purge decompositional fluids and blood is released from the natural orifices. Bloating can occur rapidly in warm temperatures and last 2-5 days, however, it is extremely temperature dependent.
Advanced decomposition	Advanced decomposition starts with the end of the bloating phase; the tissues will sag and the abdominal cavity will have a sunken appearance. Skin will take on a "wet" appearance where the liquefaction and disintegration of tissue begins. The abdominal cavity will remain moist while other areas of the body such as the extremities will exhibit mummification or partial skeletonisation depending upon external environmental conditions. The odour of decay during this phase is strong and putrid and can be detected over long distances.
Skeletonisation	Early phases of this stage are identified when the majority of soft tissue has decomposed or when mummified tissue begins to break down to reveal bone. Odour is minimal and takes on a musty or mouldy smell. It can take place as quickly as two weeks in hot and humid environments but takes much longer to reach in areas characterised by cold and dry climates.

Stage of carcass	Changes observed
Added information	The condition of the heart may also be used as an indicator of the time of death and degree of postmortem change. The presence of unclotted blood in the left ventricle usually means that death was recent and rigour mortis has not yet set in. A disintegrating clot and dark haemolysed blood in the left ventricle is seen when an animal has been dead for 24 hours or more. This indicates that rigour mortis has occurred and that extensive post-mortem changes can be expected.

(ii) Aging: Change in age (cub, juvenile, sub-adult, adult) leads to change in territoriality. While the minors are attached to their mothers, the over-aged tigers are displaced into the marginal and sub-optimal habitats by the stronger opponent tigers due to territorial fight.

(iii) Breeding female: Upon reaching the breeding stage (3-4 years), females move to and select a location where they establish the origin point of their territory with the highest mean prey that has no other female territory within 2 km and have no other transient female present. In the wild pairs were seldom found together for more than two days, although prolonged associations may be related to inexperience. Courtship in tigers is risky and familiarity appears to be an important prelude to mating. Once the female is ready to mate, copulations occur frequently (e.g. 17-52times/day). Females are thought to be induced ovulators; that is, they require a certain number of copulations within a limited time period to stimulate ovulation. Conception rates are low, about 20 to 40%, and if conception does not occur the female will come into heat again in about a month.

(iv) Breeding male: Upon reaching breeding stage (4-5 years), males select a location to begin looking for available females that do not belong to any male. The male cannot move to that location if another dispersing male has already moved to it. This ensures that young males from the same cohort do not all clump on the same female. The natal range is defined as the centroid of the dispersing male's mother's territory at birth. Breeding territories of males are established by direct takeovers and sometimes expanded after fights with neighbouring males. Males who do not hold territories do not breed. Turnover rates of territorial males are high. The average reproductive life is normally extended up to eight years.

(v) Extended territory: Females try to extend their territory until the total amount of prey available reaches a certain threshold. They select new territory based on their prey availability and presence and rank, which are correlated to age, of other females. Within a time frame, females can try up to 48 times to add a new one. In other words, females can potentially add up to 3 km² to their territory in a timestep, which is approximately the area added per month observed in the field (Sunquist, 1981). These territories function to ensure that females have exclusive access to food, cover, and other resources needed for survival and the successful rearing of young. Females who do not hold territories do not breed. Site fidelity is strong; females spend their entire reproductive lives in these territories, although some females shifted so that part of their range was acquired by their daughters. The tendency for daughters to try to settle next to their mothers resulted in clusters of females that were on average as closely related as lionesses in a pride.

(vi) Female-starvation: Females die if the total prey production within their territory is below 76 kg/month (derived from Miller *et al.*, 2014) and the food within their territory has not increased.

(vii) Parenting (female only): When a female gives birth, this determines when the female becomes fertile again and her cubs have to leave her.

The reported home ranges may be as small as 20 km² as in Chitwan (Nepal) where prey is abundant to as large as 400 km² as in the Russian Far East that has low prey density (Sunquist and Sunquist 2002; Goodrich *et al.* 2015). Similarly, tiger densities were reported to range between 17- 19 individuals per 100 km² (India's Kaziranga and Corbett NP) to as low as 0.13-0.45 individuals per 100 km² (Russia's Sikhote-Alin Mountains) (Jhala *et al.* 2011, Soutyrina *et al.* 2013). Adult female home ranges seldom overlap, whereas male home ranges typically overlap with 1-3 females, a common felid pattern of social organisation. Male territory size is dependent on the assertiveness of the resident and adjacent territorial males, whereas the female's territory size is dependent on available food resources to meet demands of offsprings.

Singh *et al.* (2020) examined how the home ranges of deceased Bengal tigers would be filled in the Ranthambore NP, the westernmost habitat of the subspecies. Nine home ranges of tigers were vacated (two males and seven females) because of death, translocation, emigration and dispersal. Vacated female home ranges were filled by neighbouring tigers; 57% of neighbouring females were related to each other and after females vacated their ranges, their daughters acquired these home ranges. Mother tigers shared their home ranges with their daughters to increase the reproductive success of the latter. The home range of an adult male tiger (TM-02) was estimated to be 73 km² using camera traps. Vacated male home ranges were filled by four transient individuals that were not related to each other.

Territories are not imposed but emerge from the tigers' perception of habitat quality and from their interactions with each other. All of them set their own territories and boundaries - based on several factors including the availability of prey and mates. They are also known to travel to great lengths in search of mates or other biological needs. The occupant of a territory is subjected to change when the occupants are ousted by a stronger rival or by death. Male occupants of territories change more often, and females tend to hold a territory for longer times. It is also well accepted that the territory of a male overlaps with those of one or more females. A tiger needs everything to be in place and thus the presence of a tiger tells us the state of good health of the forest. However, for the tiger, being on top of the food chain also means "life is tough". A quarter of the males dies fighting each other, for mates and territory. While a tigress in an average lifespan of 15 years can give birth to 12 cubs, survival of whom is uncertain. Still they are seen around in the wild, thanks to some of the wonderful conservation efforts from the government, individuals and civil society. But as tiger numbers increase, and dispersing tigers seek new territories, there will be more incidences of some of them being displaced outside PAs.

Five categories of the scent-marking are so far recorded- urine spraying; scraping with deposits of urine faeces, and anal gland secretions; clawing; cheek rubbing; and vegetation flattening. The territorial marking fluid of the male Bengal tiger, *Panthera tigris*, consists of a mixture of urine and a small quantity of lipid material that may act as a controlled-release carrier for the volatile constituents of the fluid (Burger *et al.* 2008). Male tigers mark spaces with pheromones and violently guard their territory from rivals.

Each of these exclusive territories in the mangrove forests forms a "biological" barrier against the intruders from outside.

In 1954, Locke first pointed out that tigers expel upwards and backwards a strong-smelling secretion from beneath the tail. The tail is then raised vertically with a surprising force. Traces of this fluid can be detected from the surrounding vegetation. He noticed this in the male only. Later Schaller (1967) observed males and females spraying the fluid. He described this fact more elaborately and termed it as anal gland secretion which he thought might have some informative connotations. On the basis of close observation of a pet tigress Khairi of Simlipal forest of Odisha, Chowdhury (1979) and Brahmachary and Dutta (1981) proposed that spraying takes place through the urinary channel and there is little chance of the spray to mix up with anal gland secretions. Tigers and tigresses very frequently spray an odorous lipid-rich fluid through the urinary channel upwards and backwards (as opposed to ordinary urination which drips downwards).

Smith *et al.* (1989) summarised the scent marking activities in free-ranging Bengal tigers:

Urine spraying and scraping were the predominant forms of marking in this population. Tigers marked more heavily at territorial boundaries than in the interior of territories. Furthermore, in border areas marks were highly clumped at contact zones where major routes of travel approached territorial boundaries. This pattern appears to be a result of the density of vegetation which channels travel. The intensity of marking in these zones represented a higher frequency of marking rather than an increase in time spent in these areas. Marking was most intensive when tigers were establishing territories, and animals on adjacent territories appeared to mark in response to each other. Females marked intensively just prior to oestrus; this behaviour was reduced during oestrus. Males marked more frequently when females were in estrus than during other stages of the females' cycle.

Poddar (1993) explained that there is a considerable difference in the posture or stance; while spraying the fluid, tiger (or tigress) is standing and the tail is raised as a prelude and while urinating, the animal is in a equating posture, more so in the female, by putting the hind feet slightly apart. This fluid is mentioned as Harking Fluid (MF). This MF consists of a set of volatile molecules including a very characteristic aroma and nonvolatile lipids which serve to fix the volatile molecules and make it long-lasting in Nature (Brahmachary and Dutta, 1987).

But territoriality in the mangrove Sundarbans is quite different. The Sundarbans have diurnal tidal regimes where the mangrove forested islands are washed by tidal waters at least twice a day. This makes marking of territories difficult. Mallick (2021) wrote:

Unlike tigers on the mainland, the Sundarbans tigers comfortably migrate across borders, swimming from one island to another. In this process, their scent marks, usually indicating territory, are swept away by tidal waves. This means that the Sundarbans tigers have far more fluid territories, or none.

A strong smell, sometimes perceptible to human beings, is detected in the tranquilised tigers of Sundarbans. An old-time hunter Dunbar Brander (1923) mentions the strong body odour of freshly killed tigers and the wildlife officers of Sundarbans detected a strong smell in the tranquilised marsh tigers (personal communication), when body odour samples were collected from three adult tigers- one male in 2008, one female and one male in 2010 (Gopal Tanti and Subrata Pal Choudhury)- who rubbed the upper hind portion of the body. Poddar-Sarkar *et al.* (2013) investigated the skin secretions of three such tranquilised tigers and detected saturated, unbranched fatty acid methyl esters and a few benzenoid compounds. Tigers usually rub their head, face and body on tree trunks and roll on grasses and herbs. Body smell may be a pheromone for a mating pair and a kairomone for the prey species. Pheromone, which connotes a sense of "carrying" (pherein) excitement, is concerned with a breeding tiger's transmission of "excitement" (here, a chemical signal) to another individual.

Tigers use the flehmen response when investigating different scents left by other tigers. Although the sense of vision is extremely well developed among felids, it does not seem to play a crucial role during courtship. Several days before oestrus, females spray urine more often than usual and secretions are enriched with sexual hormones. Males are very responsive to the scent left by a female: they collect droplets of scent on the tongue and smell it via Jacobson's organ. The Jacobson organ is a pouch-like structure located directly behind the front incisors. It has two small openings that direct scent particles from the air as the tiger inhales to nerves located within the structure. The nerves then transmit the message to the olfactory region in the brain that identifies the scent.

Tigers exhibit a behaviour called flehmen, consisting of opening the mouth with the head raised and sticking its tongue out briefly while the lips are curled and drawn back. They pick up a scent on their upper lip and curl it upwards towards their nose to detect scents. This behaviour makes the tiger appear to be snarling but without any sound. This behaviour is accompanied by turning up the muzzle and teeth baring. To conclude the course of the first stage of courtship, the sense of smell allows for non-visual contacts between potential sexual partners and announces that a female has a desire to mate (Jaroš, 2012). It is performed by a male to check whether a female tiger is in the area, or for rival male tigers encroaching in their territory.

A male tiger can actually tell if a female is ready to mate by smelling her urine. The molecule 2 acetyl-1-pyrroline (2AP) is present in tiger urine (marking fluid or MF)(Poddar-Sarkar 2013). The marsh tigers of the Sundarbans spray MF in the tidal estuarine habitat. The marking is subject to twice-daily tidal inundation, and is thus likely to be washed away. Perhaps together with the fixative lipids of MF, the waxy surface of certain mangroves further help slow down the release of the odour molecules. Of three mangroves *Excoecaria* sp., *Sonneratia* sp., and *Heritiera* sp., *Heritiera* sp. best retains the smell. Some hold the view that marking does not last at all, while others are of the opinion that tigers spray MF only on trees above the high-water mark (Montgomery 1995). The lipidic composition of the MF might be correlated with the wax coating of the leaf for sustenance of the aroma. Deer are not particularly scared of the MF smell on the shrubs or tree trunks, they only show curiosity, but on detecting the scent of the body smell of the tiger deer become alert or panic and flee.

There is another stage, when a male and a female come into visual and physical contact. The second stage may last several days. Generally said, it is now the female's turn to make her mind and accept or drive off a male. There is little known about the determinants of a female's choice. There may be a possibility that females follow their individualistic choice when accepting a mate (i.e. preference does not follow a strict general rule).

Although the tigers may breed in any time of the year, the tigresses come in oestrus every three to nine weeks, and her receptivity lasts three to six days while the cubs are born during all the months, and the cubs are also born during all the months, there are peak periods of matings and births depending on the climatic conditions. According to Mallick (2013), the mating season in the Sundarbans starts late in the monsoon season because then the forest is not disturbed by human intrusions and continues up to March-April. During this period, the males often fight with each other, but there has never been any report on fatal fights in the Sundarbans. The tigress in oestrus becomes restless and moves about frequently and sometimes does not feed. The tigress in oestrus squirts scent, sniffs, moans, and roars in low tone. The minimum gap between two consecutive oestrus periods recorded in the Alipore Zoological Garden was 26 days and the last oestrus phase was observed up to the age of 16 years (Das, 1980). This suggests that menopause starts in a tigress after 16 years. Variations in the peak oestrus, mating and birth period may be due to climatic and other conditions.

The gestation period is 95 to 110 days and the litter size of one or two is very common and rarely three or more cubs have been sighted. Usually, health conditions, availability of adequate nutritious food, stress and strain in the mangroves may have some role in litter size of a tigress. Usually the cubs stay with their mother up to two-three years, but in the Sundarbans it is seen that they are separated by the time they are approximately two years old. Inter-cub interval of the tigress is about three years, but not much observation has been made in this regard. The females attain maturity at the age of three-four years and the males at four-five years or even earlier. There are many individual variations in the age at which the tigresses become sexually mature.

When a pair of tiger mate, the explosive sounds of their courtship would continue intermittently, usually for three days and nights. In the beginning, the female approaches the male very cautiously, sniffs the male with a puffing sound. If the male is in a mood for mating he responds with a purring noise. If the male is not in a mood for mating a fight may occur.

However, the male may accept the female the next day. When mutual confidence is established, the female approaches the male, purrs, rolls on her back on the ground in front of the male and pats at him playfully. The male sniffs the genitalia of the female, purrs, and squirts scent frequently. He follows the female when she walks with a lashing tail and starts playing with her to reduce the normal antagonism between them. The tigress sometimes comes very close to the male and rubs her head, body and mouth to the head, body and mouth of the male. She sometimes lies on her belly stretching her forelimbs fully on the ground and her hindlimbs remaining half folded. The male approaches from behind the female and arches his back, bringing his penis in contact with the genital region of the female. At this time the female begins to tread by pushing against the ground with her hind legs and when vaginal contact is made she bends her tail sharply to one side exposing her genital passage and at the same time she turns her hind part in the direction of the stimulus. Then the male holding the scruff of the neck of the female by his teeth begins a series of vigorous pelvic thrusts at the female's urinogenital sinus. Intromission of penis is signalled by a loud copulatory cry from the female and a tremendous roar is produced by the male as he presses his genital region tightly against that of the female. Ejaculation occurs during this interval. The actual time of sexual intercourse varies from seven to ten seconds. The female then pulls forward, turns abruptly on the male, hisses and paws him and throws him off her back. Sometimes fighting takes place between them, inflicting scratch wounds. There may be unsuccessful attempts at mounting prior to successful mating.

After copulation the male moves away from the female and walks about or usually lies on the ground. Sometimes he passes stool and urine and in some cases drinks water. The tigress after copulation also rolls on the ground and sometimes goes to the water for bathing. It was observed that the female goes to the water body three to five times a day during the heat period. After a short (average 7.1 minutes) interval, the female again approaches the male and the whole process of courtship and copulation is repeated. The female's urge is much more during this period. The same process described in the foregoing paragraph is also repeated and the entire process of copulation lasts for one to three minutes. The average heat period is five to nine days. The range of mating is two to fifty two times per day and the average is 22.2 times per day. The highest number of matings observed in a single head period was maximum 235 times and minimum 20 times.

However, the tigress does not become pregnant after matings in each heat period. The chances of pregnancy is estimated to be ±50%. Usually two to three cubs are produced per litter. Sex ratio of tiger cubs at birth varies widely. However, the female's ratio is higher than that of the males. Schaller (1967) records 4:1 in favour of females. Data from shooting records suggesting 2:1 may be biassed as the males are more daring, conspicuous and first to approach kills. For the present we may assume a middle ratio of 3:1 in favour of females and based on this a tiger population of 10 individuals would have 7.5 tigresses (75%) of which one third will breed at the rate of .7 young per year giving a recruitment rate of 1.75 or 17% young. Judging from the relative infrequency of natural tiger deaths reported in the past and also bearing in mind that a dominant predator like tiger is fairly at the end of the food web in nature, natural mortality in adult tigers may be taken as not exceeding 5%. Thus the overall net increment of tiger populations is obtained as about 12% annually under adequate food supply.

Changes in fertility and recruitment

Population density is primarily geared to productivity of the habitat in terms of usable food. And the mechanism for maintaining the equilibrium between population density and the food productivity of the habitat seems to be operating through changes in fertility and recruitment rate. There is a strong case in favour of the view that the reproductive rate has evolved in relation to the availability of food for the breeding adult. The fact that the litter size in tigers may occasionally reach up to seven, while three or four cubs are quite common, might indicate that the reproductive rate in tigers has evolved in conditions of abundant populations of ungulate prey. Decline of the latter is a relatively recent occurrence. A review of the reproductive processes, recruitment rate and final turnover of tiger populations already dealt with would show that population homeostasis in tiger presently operates through reduced breeding and pre- and postnatal mortality.

Breeding potentiality

As regards whether or not the non-breeders are physiologically capable of breeding, once the dominant breeders are eliminated, indirect evidence points to the possibility that the non-breeders are capable of doing so. Although evidence shows that one-third of the females do not breed in the wild, this feature is not reflected in captivity where most of the females reproduce. This shows that it is mostly lack of enough male contact that leaves a sizable proportion of females infertile. It is known that the sex ratio in tigers is heavily weighted in favour of females. Coupled with this there are also the factors that the female is in receptive oestrus only for about five to seven days and that ovulation is perhaps not spontaneous, but rather induced by copulation. There is also the factor that many of the tigers are territory holders with fixed home range, where there might be a number of tigresses claiming his attention. It is probably due to this that estrus tigress wanders widely in quest of more males and mating opportunities. But due to the contribution of all these circumstances, particularly that or adverse sex ratio, the end result is that a sizable portion of tigresses do not breed, although most of them are physiologically capable of breeding.

Thus, it appears that social behaviour is an important limiting factor in the tiger population. This seems to be operating through the spacing mechanism determined through dominance, particularly in the male. Although a number of tigresses could live in the territory of a resident male, another male is tolerated there only as a transient. This perforce leaves a small or large part of the male population, depending upon the total habitat area and conditions, as transients or floating. Territory holders are clearly dominant over the non-territory holders and the former thereby comes to have access to a number of tigresses, much more than he ever needs. Given this territorial behaviour of male in a population where there is already a great disparity in sex ratio, this further lumping and batching of females drastically reduce the reproductive potential to such an extent that it is not relieved by the behavioural antidote of wandering by the estrus female and roaming transient males.

Tiger as predator

Tiger continues to be a predator least understood! Tigers are solitary beasts rarely with social organisations except during matings or at sharing kills. Tigresses with cubs behave more socially than the male counterparts for the initial period of 19-20 months. There have been observations of adults of the first litter coming close to the mother tigress, signifying social behaviour which is more prominent in the initial two to three years. Hence, the only long lasting bondage in a tiger's life cycle is the relationship between a mother and its offspring. However, there have been numerous instances of a resident male (which has sired the litter) sharing a kill with the mother and cubs.

In Ranthambore, George Schaller's observation in the 1960s puts on record a kill shared by tigress and cubs with a male- an example of social organisation. He observed that tigers appear to socialise more at kills than on any other occasion. It is a well conceived thought that adult tigers readily join for brief periods, particularly at a plentiful food supply, but their association rarely persists longer. Schaller rendered some interesting observations in 1964 of a kill shared by a tigress with cubs, an adult male: "I tied a buffalo to a stump at 16:30 hours and waited in a blind 80 feet away. At 19:40 hours, a tigress attacked the animal which died eight minutes later. Five minutes after the cubs (three) arrive at the kill, the tigress appears and right behind her is the male tiger. The male rises at 22:50 hrs and walks to the kill. Two cubs nuzzle his face and neck...".

This gave the first probable, the cubs, for 19-20 months has always overshadowed the "fatherhood status of male tigers". Being a surreptitious species, such behaviour in male tigers, being hardly noticed or very less documented, the 'parental protection' provided by the males deserve special place in tiger behaviour. In Ranthambore, T19 female with three cubs were in the bigger home range of their presumed father T28. The territory of T28 has increased or varies with the movement of T19 and her three cubs, signifying reach of parental protection by the males. On 18 March 2012, the T19 female with 2 cubs was sighted. One of the cubs stood up and moved close to the male T28 and sat beside him to get affection.

The T26 tigress with three cubs stay in close vicinity of T20, an old male. T31 with two cubs is frequently visited by T23, a male. T11 with three cubs is protected by the male T33. T30 with a litter of three cubs are being protected by T3. T9 with two cubs is protected by T33 male. These associations signify social behaviour in tigers, especially considering the fact that 5 out of the 7 mother tigresses are with cubs 15-18 months old. Such a long association of the male tiger in each of the "families" shows affectionate behaviour in the tiger, demanding fatherhood recognition to the male tigers. Another possible explanation could be the apprehension of infanticide by unrelated males. The father of the cubs are providing parental protection to prevent infanticide of their siblings and gradually extending their supremacy with related ones.

Ranthambhore surpassed all these incidents in the tigerlands across the world with an incidence in 2011. A male Tiger T25 was seen to rear two orphan cubs in the wild, opening a new chapter of parental care and protection by male tigers. On 29 January 2011, T5, a female tigress who was being tracked for the past few days, was sighted with two cubs. The mother tigress died of a physiological problem within 10 days on 9 February, close to the Kachida chowki, leaving the 3-4 months old cubs unfortunately christened "orphans". The cubs of 3-4 months were traced by the video camera specially installed in the home range of the mother T5. A camera trap picture in May revealed that T25 male closely following one of the cubs, 5-8 km away from the home of the cubs. On one occasion, T25 has been sighted coming in direct confrontation with T17 female, to protect the two cubs. Such amazing behaviour of the male tiger in tiger ecology marks the unexplored area of behavioural ecology in tigers.

With the observation of the male T25 marking the climax of male protection to its cubs, it has become amply clear that male tigers do display affectionate behaviour, resorting to parental care or protection to establish their strong genes.

Male tigers usually let the females and their cubs finish a meal before they get their turn. This is because the females are responsible for raising the cubs and need the food to provide for them. The males, on the other hand, are solitary and can hunt for food on their own. In STR, Dr. Kalyan Chakraborty described a family lunch in the Bagmara forest block, where a mother with two cubs was observed to feed on the carcass of a wild boar, when a male tiger came there and sat for long three hours watching their eating. The male waited till the female and her two cubs had finished their lunch and left. Only then he began feeding on the remaining carcass. Similar incident took place at Netidhopani, where a tiger waited for four hours when the mother and cubs were busy eating a chital's carcass. Such behavioural features of the male tiger indicates that the tigers usually live alone but that does not mean that they are unsocial.

Some research indicates that tigers can recognise each other. They may be willing to share prey they have killed with other males that are related, females that have had their cubs, and even those that have been in the area before and they recognise their scent. This behaviour contrasts with most types of cats, especially lions, that the males eat first then what is left can be shared by those remaining as they go down the social hierarchy.

Chapter 11
Interspecific Interactions and Relationships

"When a man wants to murder a tiger he calls it sport; when a tiger wants to murder him he calls it ferocity." – George Bernard Shaw

Predator-prey dynamics

Predator and prey populations cycle through time, as predators decrease numbers of prey. Lack of food resources in turn decreases predator abundance, and the lack of predation pressure allows prey populations to rebound.

Tiger's diet by species

Each type of tiger has a different diet based on the geographical location where they reside. I have tabulated each known tiger species and what do they eat in the wild below:

Siberian or Amur tiger

In the cold terrain, Siberian tigers eat red deer, wild boar, Manchurian elk, and sika deer. Also, tigers attack and eat grizzly bears and black bears.

South China or Amoy tiger

They eat serow, tufted deer, sambar deer, muntjac deer, and wild boars.

Indochinese tiger

Their main diet includes sambar deer, antelope, wild pigs, buffalos, serow, young gaur, baby rhinos, turtles, fishes and baby elephants. When food is scarce, they even go after porcupines, hog badgers (*Arctonyx collaris*), macaques and muntjac deer.

Malayan tiger

Malayan tigers' diet includes deer species (sambar and barking), wild boar, bearded pigs, and serow. Malayan tigers also prey on sun bears (*Helarctos malayanus*), young elephants, and rhino calves. Due to human encroachment, livestock such as domestic cattle become prey to these tigers.

Sumatran tiger

They eat fishes, monkeys, crocodiles, fowl (bird), wild pigs, and deer.

Bengal tiger

They prefer hunting large mammals such as deer (chital, sambar, barasingha), water buffalo, nilgai/blue bull, serow (goat-like), and takin (gnu goat). On the medium-sized species list, they frequently kill a wild boar and rarely hog deer, muntjac deer, and grey langur (black-faced monkeys). When Bengal tigers encounter small prey species such as porcupines, hares, or peafowl, they will kill them for snacks. Due to humans invading their habitats, Bengal tigers also prey on domestic cattle.

Conventional predator

Mangrove tiger *(Panthera tigris tigris)*

Tiger sign and sighting encounter rates were higher on the Indian side, especially in the core zone of NP as compared to other areas. Tiger numbers, estimated by camera trapping using SECR within the minimum bounding camera trapped polygons and the multiple regression model for the remaining part, was 182 (145 to 226) in the total tiger occupied area of 6,724 km²; with 106 (83 to 130) tigers on the Bangladesh side and 76 (62 to 96) tigers on the Indian side in 2018.

Extinct prey species

The mangrove landscape of ancient Sundarbans has been changing rapidly since the mediaeval period due to large-scale reclamation and shrinkage of habitat (estimated at 5,363 km²; base year 1770)(Mallick, 2015). Moreover, the gradual decrease of freshwater flow and increase in salinity has led to near complete destruction of the grasslands and death of the less tolerant tree species. As a sequel to this ecological disaster and unrestricted hunting, a wide range of grassland-dependent herbivore prey species, extant even in 1875, late 19th or early 20th century, got extinct (Mallick, 2011).

Rhinoceros

It appeared that the great swampy forest of Sundarban provided ideal shelter to both the greater and lesser rhinoceroses.

(i) **The Indian rhinoceros** - *Rhinoceros unicornis* Linnaeus, 1758- went extinct much earlier from the Indian Sundarbans as Ghosh *et al.* (1992) reported subfossil remains of this species for the first time from southern Bengal, which indicated that the greater one-horned rhinoceros thrived in the riverine grass-jungles close to the present Sundarban mangrove swamps about 3,000-2,500 years ago. Pirazzoli (1996) also reported soil stability of the Sundarbans by that time.

In May, 1990 a fisherman, while venturing to deepen a puddle in Ramchandrapur village (c.22°25′30″N, 88°24′42″E) under Bon-Hugli Panchayat, PS Sonarpur, South 24-Parganas district, some skeletal remains (broken mandible with body, having eruptive secondary I_2 on each side, P_4, M_1, M_2 and embedded M_3 on the right ramus, sternal bone, three fragmentary ribs; 7th thoracic vertebra; distal end of right radius; cuneiform of right manus; right astragalus) of one ponderous animal were unearthed from under the depth of 4 m of a silty pit, which, on closer examination and comparison, appeared to belong to some massive sub-adult rhinoceros, akin to the great one-horned rhinoceros.

(ii) **The Javan rhinoceros** (*Rhinoceros sondaicus inermis* Lesson, 1838)

Javan rhinoceros was the prime prey of the tiger in the Sundarbans. Baker (1887), returning to a dead rhinoceros shot the day before in the Sundarbans, described: "The ground round this carcass was thickly marked with the footprints of tigers, several having visited it since noon of the day before; but beyond attacking the softer parts they had not yet made much impression."While there are a few early records, almost all available information pertains to the 19th century, and it is believed that the rhino must have become extinct in the area during the first decades of the 20th century. There are three earlier reports of rhinos in the Sundarbans. Around 1630, Sebastien Manrique passed the island of Xavaspur in the estuary of the River Meghna, and came across many rhinos. On 16 January 1664, the Dutchman Wouter Schoutens (1676) passed (the River JiIlisar, where the shores of the Ganges arc covered with bushes, inhabited by rhinos and other animals. Pennant (1914:153-154) in his accounts of the Sundarbans islands confirmed that during the 19th century one-horned (Javan) rhinoceros was very common in those islands and swampy places, and is a frequent concomitant of the tiger. But an adult rhino was not scared of the tiger. The big cat usually preyed on the rhino calf, aged only four months and weighing about 137 kg, opportunistically (Mallick, 2023a). Baker (1887) shot numerous rhinoceros dead in the Sundarbans and increase in predation of calves by tigers also contributed to their decline in numbers. During the 1850s Simpson also hunted a number of rhinoceros in the grassland of Piyali Island. When de Poncins visited the Eastern Sundarbans in January, 1892 he found rhinoceroses in five islands (Lot nos. 165, 169-172) but he did not hunt any one of them. The last footprint was reportedly observed in 1900; but, when E.O. Shebbeare started working in the Sundarbans, he did not find any rhinoceros and, thereafter, the rhinoceros was reportedly extinct.

On 19th April, 2000 some bones of unknown animals were recovered from Mollakhali village under Gosaba Police Station of Sundarbans while excavating a pond in the village, from a soil depth of 2.7 m. After getting the information from a villager the staff of Sajnekhali WLS Range went to that spot and took over the bones from the owners of the pond. The bone pieces were skull, ribs, legs etc. The bones were sent to the Zoological Survey of India for proper identification. Scientists of ZSI, Kolkata reported that the bones were of a Javan Rhino.

Again on 14th May, 2000 few bones were recovered from a pond in Tentulia village near Pathankhali within the Gosaba Police Station of Sundarbans, where people were excavating a pond of depth about 7 m. The bones were sent to ZSI, Kolkata and were identified as the bones of *Rhinoceros sondaicus*.

Bovids

(i) **The gaur** (*Bos gaurus* Smith, 1827), body mass of adult male 440-1,000 kg and adult females 588-1,500 kg: Recorded to be extinct from the Bangladesh Sundarbans without any sighting reference (Aziz and Paul, 2015).

(ii) **Wild water buffaloes** (*Bubalus arnee* Kerr, 1792) 800-1,200 kg

The wild buffalo roamed about in the Sundarbans till 1885 and died out by the end of the 19th century. The last reports of evidence proving the presence of Wild Buffalo in the Sundarbans mangroves dates back to 1890.

On 3rd March, 2001, some bones were recovered from Netidhopani (Compartment 1) within STR, which was badly eroded by recent storms. The bones were sent to ZSI for identification and were identified as the bones of wild buffalo. One skull of an animal was found opposite Bidya village, which is also the headquarter of NP West Range in Pirkhali forest block in an eroded area on 4th March, 2002. The skull was sent to ZSI for identification. 'The skull was identified as that of a Wild Buffalo. Groups of 8-10 buffalos were sighted in tall grass of the riverbanks in Sarankhola Range, Bangladesh Sundarban, until 1925-1930 (Khan 2004).

Deer

(a) **Barasingha** (*Rucervus duvaucelii* G. Cuvier, 1823) 170-180 kg: Probably got extinct in the Indian Sundarbans by 1930. Barasingha was extinct in Bangladesh Sundarbans in 1950s.

(b) **Barking deer** (*Cervus muntjak* Zimmerman, 1780) 20-30 kg

Extinct in the Indian Sundarbans during late 1970s, last seen in Haliday Island and Bulcherry area; and

(c) **Indian hog deer** (*Axis porcinus* Zimmermann, 1780) 20-30 kg

It was reported from Sundarbans till 1945.

Extant prey species

Only a few primary prey species like chital, wild boar, rhesus monkey, etc. now survive by virtue of their physical toleration and behavioural adaptation. Due to low prey species diversity and reducing populations have forced the Sundarbans tigers to become an omnivore or opportunistic general feeder by choice, lacking only carnivore specialisation, and sustains itself from a number of mixed food sources. Unlike the stenophagous (feeding on a limited variety of food) Bengal tigers inhabiting the mainlands, the mangrove tiger is an euryphagous (feeding on a large variety of active prey) carnivore. Here its natural terrestrial prey species are chital, wild boar, rhesus macaque, otters, lesser cats, birds (mainly red jungle fowl) and others. These species are discussed separately below.

Non-conventional predators of the big cat

Sharks

Although bull sharks are found swimming in the mangrove swamps along with six other shark species, Bengal tigers and bull sharks are not known to interact with each other in the Sundarbans mangrove forest, but the more common interaction of two prime predators often takes place between tigers and crocodiles.

Crocodiles

The Sundarbans is home to three crocodile species in the family Gavialidae-

(i) **Gharial** (*Gavialis gangeticus* Gmelin, 1789) or fish-eating crocodile

It occurs mainly in aquatic habitats with open canopies, including freshwater marshes, the margins of large rivers and lakes, tidal marshes, and mangrove forests. No prey-predator relationship between this species and tiger in the mangroves is known.

(ii) **Mugger** (*Crocodylus palustris* Lesson, 1831)

The mugger crocodile is a smaller species of crocodile found in the Sundarbans. It is recorded in the Sundarbans on 16th January, 2016 and on 19th July, 2018. They are typically around 3-4 m in length and weigh around 150-200 kg. Mugger crocodiles are mainly found in freshwater habitats and marshes. They eat fish, other reptiles and small mammals, such as monkeys. The prey-predator relationship between this species and tiger in the mangroves is poorly known. However, Somerville (1924) observed regarding tiger shooting in the Sundarbans: "It is well-known that the tiger (…) on making its kill seldom eats the flesh of its victim the very day. He (…) drags the carcass into some secluded spot, to which he returns a few days later to feast on the partly decomposed flesh." Knowing also the tiger's weakness for the flesh of the crocodile [called mugger by him], Somerville proposed: "First shoot a decent-sized mugger, next drag the body well up on to the bank, but not too far inland, as this will excite suspicion. Select a spot where fresh pug marks leave no doubt in your mind as to the probability of a tiger finding it, and then either sit up or dig yourself in and await his arrival."

(iii) **Estuarine crocodile** (*Crocodylus porosus* Schneider, 1801)

In the Sundarbans, the tiger is the top mammalian predator, but the largest estuarine crocodile of the world is also found in the Sundarbans and it is the top aquatic predator in the Sundarbans. Estuarine crocodile is one of the rarest species to be photo-captured, however, one should keep in mind that camera traps are not the most ideal equipment to evaluate this semi-aquatic reptile's distribution. The Estuarine crocodiles seemed to have a patchy distribution with more signs and sightings recorded on the Indian side.

Males can grow up to a length of 6 m, rarely exceeding 6.3 m, and a weight of 1,000-1,500 kg. Females are much smaller and rarely surpass 3 m. Rare leucistic saltwater crocodiles are occasionally spotted here.

In water, it can even overpower a swimming tiger. It is so powerful that it is known to have killed even tigers in its aquatic kingdom (Pandit, 2012). Most prey animals are killed by the great jaw pressure of the crocodile, having the highest bite force with a value of 16,414 newtons, although some animals may be incidentally drowned. The tiger may sometimes kill the crocodile. There are instances of fights between crocodile and tiger in Ranthambore TR over prey, but in the end the tiger won the battle. Juvenile crocodiles may fall prey to tigers in certain parts of their range, although encounters between these predators are rare and cats are likely to usually avoid areas with estuarine crocodiles. While crossing a wide river, a tiger had been chased by a crocodile. It thrashed its powerful tail on the nose region of the tiger to cause bleeding. Still the clever tiger avoided confronting its enemy under water but was swimming on and on to land the riverbank ahead. Reaching his own territory, he turned to face the attacker, came up under the crocodile's belly, flipped and ripped it open.

In May 2008, a male tiger later named 'Raja' got into what is described as a 'territorial fight' with a crocodile when he was crossing the River Matla in STR. The lower part of his left hind leg was bitten off. A tiger with this kind of injury would be very unlikely to survive in the wild as they would not be able to hunt for food. Raja was already 11 years old and life expectancy for this species in the wild is usually not more than 15 years. It was bleeding profusely and sent to Alipur zoo, Kolkata for treatment. After three months, he was translocated to Khayerbari Rescue Centre in North Bengal. Here, he developed a love of bathing which he did four or five times a day.

Crocodiles hunt by stealthily stalking their prey from water. Once it has caught its prey, a crocodile will then drag it into the water and drown it. For the first time in the Sundarbans, a tigress was found killed by an estuarine crocodile on 9th August 2011. The carcass was found at Pirkhali-5 compartment in STR on the mudflats of Dobanki Khal (Creek) near Dobanki Camp at about 8.45 am. Although its vital reproductive organ was eaten up by the crocodile, its sex was confirmed by seeing four slightly large teats.

Dobanki is a perennial habitat of the estuarine crocodile and a few large-sized crocodiles were often seen during patrolling by the FD staff. The Dobanki area was also the territory of three tigers, which used to swim across the creek frequently and migrate from one island to another like other tigers of the Sundarbans. A few months earlier, there was an incident of a spotted deer killed by a crocodile at almost the same location where the tiger was killed. It was also crossing just 60-70 m away. Its carcass was lying in a lateral recumbent position on the small mudflat of Dobanki creek and a very large estuarine crocodile approximately 14 feet in length was moving around about 150 m away from the spot, repeatedly trying to reach the victim, but forest staff prevented it from doing so. Signs of movement of the crocodile on the site of the carcass of the victim were also noticed. Decomposition of the carcass was not observed by the forest staff, which clearly indicated that the incident took place only a few hours before detection.

During the post mortem examination of the carcass conducted at Dobanki Camp, it transpired that the carcass was of a tigress, which was approximately 5-6 years old. Externally the rigour mortis stage of the carcass was over so time of death was estimated to be 24-28 hours before post mortem, i.e., in the afternoon of 8 August 2011. It was observed that all four canines were intact; no abnormal discharges from the nasal opening or ear orifice were detected. Length of the body from nose tip to lumbar vertebra was 108 cm and height was 84 cm. Tail and hind portion had been eaten by the crocodile so full body length could not be measured.

Several conical-shaped wounds were found on the body. Besides, twenty-two scattered, elliptical, deep piercing wounds were observed on the right upper abdomen (nine numbers), left upper abdomen (two), right dorsal foreleg (seven), dorsal thoracic region (three) and dorsal posterior neck region (one). Jaws of the crocodile were so powerful that it penetrated through the rib of the dead tigress. The body beyond the second lumbar vertebra was totally torn off and absent except for the dislocated femur up to the paw of the left hind leg, which was attached to the body by a flap of skin. However, the kidneys and lungs were found to be intact and there was a large torn vent in the diaphragm. All visceral organs like heart, trachea and oesophagus were not damaged, but the pleural cavity was completely damaged through the big diaphragmatic vent, where blood-tinged fluid was noticed. The morphology of the liver was normal except only a few patchy whitish marks. Gallbladder and ducts were identified in the carcass. Lymph grand was not identified due to massive loss of texture of visceral organs. Pericardial sac was full of blood tinged fluid. Heart muscle was shrunk but not discomposed, numerous haemorrhages evidence was noticed in the pericardium of both auricles. All the three chambers of the heart were empty except the left ventricle which was full with clotted blood. In the oesophagus a skin or tail portion of the water monitor lizard was observed. Small and large intestines were not found during post-mortem as it was eaten by the crocodile. Reproductive organs and ordinary bladder were also not found. From the second vertebra to posterior all the skeleton is absent except the left hind leg which was detached from the tabular cavity through the portion of skin.

Therefore, it was proved that the death of this tigress was due to a crocodile attack. It appeared that while the tigress was swimming across, the said crocodile attacked her hind portion and killed her by repeated jerking and drowning. There were signs that the tigress tried to fight back but could not save itself because she could not land on the solid dry ground.

Sympatric or co-predators

The sympatric or co-predator refers to those carnivore species that primarily obtain their food by hunting tiger prey species or prey of relatively smaller body size.

(a) **Large co-predator leopard** (*Panthera pardus fusca* Meyer, 1794)

The Bengal tigers are mostly sympatric with leopards across their distribution range, overlap substantially in their diet, and are both nocturnal. The subdominant leopard is believed to coexist with tigers via several mechanisms like spatial segregation, temporal avoidance, and differential prey selection. Top predators like tigers often shape their communities through intraguild predation (IGP), that is, feeding interaction between two consumers that share the same resource species, as is commonly observed in natural food webs. However, the Sundarbans is the only TR in India without leopards- the last leopard sighting in the Bangladesh Sundarbans was in 1931. Mallick (2023) recorded traditional hunting system in the village of Maheshwar pasha, Daulatpur, Khulna, by the local professionals using nets made of jute, which were eight to nine feet in height, propped up by bamboo pieces, the base being stuck to the early by bamboo hooks closely set. A stretch of jungle was encircled with this net and the leopard was driven into it and hunted. Such practice indicates the existence of the leopard in the northern belt of present Sundarbans. Paucity of empirical data on the Sundarbans leopard has caused controversy over its existence.

In fact, whenever the mainland tiger populations rise in certain areas, the leopard populations decrease to the point where the leopards are rare in habitats where the tiger density is high except when the prey is abundant or when the spotted cats hunt smaller games than their striped cousins- only then the tigers and the leopards coexist peacefully. In the case of the Sundarbans, the optimal mangrove habitat is always dominated by the Bengal tiger and the leopards never colonise the mangrove tigerland; rather they used to live in the periphery of the mangrove forests. Mostly the scattered forest intermixed with grassland has been defined as the potential habitat for the leopards in the historical Sundarbans. Hence, they were allopatric in the mangrove tigerland.

Pennant (1798:153) in his zoological account of the Sundarbans islands recorded the presence of a melanistic leopard or panther. He wrote:

"...of a deep black colour, with the spots of a more intense black, was taken in these forests, and added to the menagerie in the tower of *London* by Mr. Hastings".

The Bengal Delta was originally occupied by vast stretches of grassland filled with saline marshes and tropical wetlands and rich forests- the Bengalian Rainforest. Originally, the Sundarban had covered a much wider area and the northern fringes of the forest were not typical mangroves, but Freshwater Swamp Forests surrounding the Sundarbans Mangroves in India and Bangladesh, in the zone where the freshwater from the rivers push back the saline waters of the intruding ocean. Some experts believed that the leopard was once found along the edge of the Sundarban (Curtis, 1933). Earlier, the district gazetteers of Bakerganj (Jack, 1918) and Jessore (O'Malley, 1912) recorded occurrences of the leopard everywhere, but it was very rare in the northern part of 24-Parganas. An occasional animal is heard of from time to time as frequenting some jungle or waste lands, but as soon as it takes to killing goats and young cattle, its doom is assured (O'Malley, 1914). In the Khulna Sundarbans, the leopards were numerous and in newly reclaimed land, where they take up their quarters in thickets near human habitations and carry off cattle and other animals (O'Malley, 1908). The checklist of mammals in the Bangladesh Sundarbans and adjacent areas also includes the leopard (Seidensticker and Hai, 1983). In fact, leopards fear tigers, especially when they enter core areas where tigers lurk. Hence, they are mostly found in fringe areas in the mainland forests.

(b) **Smaller co-predators**

According to Eric Wikramanayake, a Sri Lankan landscape ecologist and conservation biologist, the larger common leopard was another holdout in the Sundarbans and several smaller predators are there like the Jungle Cat, Fishing Cat, and Leopard Cat. Illegal hunting, large-scale clearing and habitat loss have caused local extinction of the key predator species in the Sundarbans Freshwater Swamp Forests.

There are three small co-predators in the Sundarbans mangrove forests and edges.

(i) **Fishing Cat** (*Prionailurus viverrinus viverrinus* Bennett, 1833)

The fishing cats dwell amidst the dense vegetation associated with the marshes and mangroves (Mallick, 2010c). The encounter rate of the fishing cat was 0.03/km². The medium sized fishing cats thrive in the mangrove delta because of ample supply of fish, the main prey of the animals. The adult feline weighs around 5-16 kg. The males are heavier than the females.

Their length measures between 57 and 78 cm. The unique feature of the fishing cat is its partially webbed feet which help swim in water and move on wet soils. The claws are retractable so that they can protrude as they are incompletely sheathed. The fishing cat resembles more of a leopard cat. They are differentiated from the leopard cat by the retractable claws. It is a semi-aquatic animal, which covers an area of hardly 2-3 km². These nocturnal creatures have an interesting technique for catching fish. They sit at the edge of water bodies, skim the water surface with their paw, mimicking the movement of an insect. Any hungry fish coming to the surface to investigate is caught. Scat analysis has revealed that one fell prey to the tiger in Sundarbans.

The trap cameras installed in December 2021 and retrieved in January 2022 revealed that a total of 450 images were captured which included some doubtful images, while some others were captured twice. A healthy population of fishing cats was captured throughout the Indian Sundarbans. However, there were fewer photo-captures in NP East range in the core area. In 2022, it has been reported that STR has at least a population of 385 fishing cats. Sajnekhali WLS Range has 97, NP East 60 and NP West 98. The number of fishing cats detected in the Chilika region in June, 2022 was 176, which was less than 50% of the number detected in the Sundarbans.

(ii) **Leopard Cat** (*Prionailurus bengalensis bengalensis* Kerr, 1792)

Type locality: The coast of Bengal; restricted to South Bengal, India.

Holotype: From Pennant (1781; 164) a male that "swam on board a ship at anchor off the coast of Bengal and produced young afterwards with female cats in England". The specimen's remains were seen at Hammersmith. It has a short, thin coat, and a tail slender in winter months. Ground colour ranges from ochreous buff to buffish white on flanks, but darker on head and back. Spots large and well spaced, sometimes solid and may run in chains.

Leopard cat was captured throughout the landscape and its abundance is indicated by the numerous capture hotspots both in STR as well as SBR. There is no record of being hunted by the tiger in the Indian Sundarbans, but Khan (2008) recorded this species as a prey of the tiger.

(iii) **Jungle or Swamp Cat** (*Felis chaus kutas* Pearson, 1832)

Jungle cat was captured only at the fringe areas of the mangrove forests, with its highest photo captures in the buffer zone of Basirhat range. There is no record of being hunted by the tiger.

The smaller predators generally avoid the larger one, though the final outcome often depends on prey abundance.

(iv) **Golden Jackal** (*Canis aureus* Linnaeus, 1758)

Golden jackal was captured for the first time in the Indian Sundarbans and seems to be a recent coloniser restricted to the mangrove forest edges of Sajnekhali WLS and islands of SBR adjacent to the human habited islands. There is no record of a jackal being hunted by the tiger.

(v) **Bengal Fox**

The Bengal fox (*Vulpes bengalensis* Shaw, 1800) is recorded from SBR, but it is very rare and not known to be hunted by the tiger.

(vi) **Indian Grey Wolf**

The Indian grey wolf (*Canis lupus pallipes* Sykes, 1831) is recorded recently from the fringe areas of the Indian Sundarbans in 2017 and Bangladesh Sundarbans in 2019 (Akash *et al.,* 2021). After 70 years, a solitary male of was killed in retaliation in June 2019 as livestock predation events erupted and lasted for a month after a severe cyclone had swept coastal Bangladesh. The specimen was about 119 cm from nose to tail tip with a skull length of 26.23 cm. Two molecular markers, mt D-loop control region and 16S rRNA, and 54 cranial parameters consolidated the identity. Bayesian inference and maximum-likelihood analyses indicated its intraspecies position. The locality of conflict, 450 km eastward of the easternmost population of *C.l. pallipes*, is adjacent to the Sundarbans in the Ganges estuary that presents formidable tidal rivers as dispersal barriers. In 2017, another wolf was sighted from the Indian Sundarbans vicinity. It is very rare and not known to be hunted by the tiger.

(vii) **Otters**

Sundarban is home to three species of otter, viz. smooth coated otter (*Lutrogale perspicillata*), Eurasian otter (*Lutra lutra*) and oriental small clawed otter (*Aonyx cinereus*). They are known as 'tigers of rivers'.

However, it was not possible to identify the individual species from the camera trap photographs. Hence, the status depicts the varying intensity of photo-captures of all three species combined together. The capture hotspots were primarily in the core zone of NP beside wide channels. Otter in general had a higher presence in the Bangladesh Sundarban as compared to the Indian side with ranges of Chandpai-Sarankhola and Khulna recording highest encounters. Aziz (2018) surveyed approximately 351 km (97 km in Satkhira block, 78 km in West WLS, 101 km in East WLS, and 75 km in Chandpai block) of water courses including 30 river segments, with 50 m to 800 m width in the Bangladesh Sundarbans, when 53 individuals of small clawed otters were recorded in 13 groups, with a mean group size of 4.08± SE 1.13. Mean encounter rate of combined sighting, footprint, and spraint was 0.06/km of rivers surveyed, with higher abundance along the eastern regions of the Sundarbans. Otters were found predominantly feeding on mudskippers (*Periophthalmus* sp.) on the exposed river mudflats, particularly during ebb tide.

Otters were rarely sighted (encounter rate 0.009/km²)(Mallick, 2011). However, Manjrekar and Prabu (2014) conducted a river transect survey in STR over an area of 237.8 km (117.3 km in the Sajnekhali range and 120.5 km in the NP West range) in which four sightings of smooth coated otter groups were recorded and otter signs at seven different locations (tracks at six locations and a spraint at one location) were recorded on the mud banks of the forested islands. Of the four sightings of smooth coated otter groups three were sighted in Sajnekhali range and one was sighted in NP West range. Among the otter signs recorded, four were found in Sajnekhali range and three in NP West range. Recorded mean group size of the smooth coated otter groups was 2.75 (S.E. = 0.85, Range = 1-5) and the encounter rate of the otter signs was 0.03/km (S.E. = 0.01). The fluctuation in water level due to tides obliterates the signs deposited on the shoreline of the islands which in turn affect the detection probability of the signs. However, they could not identify the species of otter based on the pugmarks on the mud banks. Otters are recorded as being hunted by the tiger.

(viii) **Civets**

The Asian palm civet (*Paradoxurus hermaphroditus* Pallas, 1777) was captured only at the edges of the buffer zone of Basirhat range, adjacent to the human habited islands. Small Indian Civets (*Viverricula indica* Geoffroy Saint-Hilaire, 1803) are very rare in the mangrove forests. There is no record of a civet being hunted by the tiger.

Predator-prey relationship

The predator-prey relationship consists of the interactions between two species and their consequent effects on each other. In the predator-prey relationship, one species is feeding on the other species. The prey species is the animal being fed on, and the predator is the animal being fed. The predator-prey relationship develops over time as many generations of each species interact. In doing so, they affect the success and survival of each other's species. The process of evolution selects for adaptations which increase the fitness of each population. Predator-prey relationships greatly affect the populations of each species, and that because of the predator-prey relationship, these population fluctuations are linked.

Tigers in general possess only about 500 taste buds compared to a human's 9,000. Therefore, taste buds are speculated to have a minimal role in their survivability. They seem to be able to taste salt, bitter and acidic flavours and to a lesser degree sweetness. The honey gatherers claim that the tigers have developed a taste for honey. A tiger first rolls in mud and lets the mud dry on its body, creating a mud shield against bee stings. Then it leaps into the air to knock down a part of the hive. As most of the bees swarm back to the remaining part of the hive, the tiger rushes to the river with the broken hive and drops it in the water. This drives the remaining bees out, leaving the tiger to enjoy a rare drink of golden liquid in this saline marshland.

Wild pig's presence was recorded throughout the landscape with higher encounter rates on the Indian side. Prevalence of more capture hotspots in the SBR concurs with higher sightings of wild pig during boat transects in this area.

Rhesus macaque was captured more or less throughout the Indian Sundarbans landscape with highest concentration of photo-captures in Sajnekhali beat of Sajnekhali WLS. Its sign and sighting encounter rate was higher in the East range of the Sundarbans NP as well as in the buffer zone of the Basirhat range. The capture hotspot in this beat coincides with the location of the beat and range offices where there is maximum instantaneous human presence inside the TR on any given day as these offices issue tourist and fishing permits.

Conventional extant prey species

(i) **Chital or Spotted Deer** (*Axis axis axis*)

Chital's faecal pellets are cylindrical with the bottom end rounded and the apical region tapering to a point. Adult chital pellets range in size from 1.2 to 1.8 cm. It is the principal prey species of tiger in both Indian and Bangladesh Sundarbans. This species, like the tiger, lives on the edge of its natural range in the Sundarbans. They are most abundant in those habitats where extensive grassland and scattered forests of *Sonneratia* sp. occur. Open grassland and *Keora* appears to be favoured habitat and they have preference for *Keora* leaves and fruits. This habitat is available in all the PAs in the tigerland.

In the Bangladesh Sundarbans, spotted deer sign and sighting encounter rate was more or less uniform across the landscape with pockets of higher encounter rates in the Sarankhola and Khulna ranges. Hard size of the spotted deer varies considerably with the season and availability of food and water. Reza (2000) recorded a total of 889 groups in Katka-Kachikhali, with varying group size 2-137 (mean 7.36), group density 9.56 groups/km², species density 70.4 individuals/km² and biomass and metabolic biomass were 3,870 kg/km² and 2,903.2 kg/km². The density increased subsequently by 7.4 kg/km² and biomass by 411.75 kg/km² and the reasons were attributed to reduced poaching due to departmental protection.

Dey (2007) recorded about 83,000 deer in the Bangladesh Sundarbans with a varying density in different habitats- 3-4 in *Heritiera fomes*; 6-7 in *Excoecaria agallocha-H. fomes*; 10-13 in *H. fomes-E. agallocha*; 12-14 in *H. fomes-Xylocarpus mekongensis-Bruguiera gymnorrhiza* and *E. agallocha* forest; 15-16 in *E. agallocha-Ceriops decandra-X. mekongensis*; 14-18 in *X. mekongensis-B. gymnorrhiza-Avicennia officinalis*; 43-55 in *C. decandra-E. agallocha-Sonneratia apetala*; and 112-195 in *Sonneratia apetala-E. agallocha* open grassland associations. The population density decreases with the increase in canopy closure because such closure is affected by tree heights and canopy widths and takes into account light interception and other factors that influence microhabitat. The deer density was lowest (2-4/km²) in pure *H. fomes* where canopy closure was 70-80% and highest in *S. apetala-E. agallocha* open grassland associations with canopy closure 40-50%.

The radio-tracking based home range of both males and females varied from 140 to 200 in April-September, which is their peak breeding season, although the chital breed throughout the year, but population decreased in October-January and population remained more or less stable during March-July with average male:female:fawn ratio 15:60:25 as against the peak season's ratio of 15:50:35. The home range of the males increased during the peak breeding season (295 to 410 ha), whereas the female's home range increased to 1.5 times higher than the rest of the year. A male shared home-ranges of 4-5 females during the rutting period.

Chital was captured throughout the Indian Sundarbans landscape with highest concentration of photo-captures in Sajnekhali WLS. Two of the capture hotspots in this sanctuary coincide with the location (Dobanki camp) where chital had been released in the late 2000s.

Category of age-sex classes for identification

Type	Description
Adult male	Prominent antler and bigger in size as well as in body weight.
Adult female	Absence of antler and more than one year but bigger than the juvenile.
Juvenile	More than four months but less than one year and smaller than the female.
Infant	Less than three-four months and attached to the mother.

The empirical fact of a deer's curiosity is based on a controversial theoretical supposition that there are alluring properties of the tiger's coat. On the other hand, it is asserted that the deer's curiosity is an exception from a general rule, which is to avoid a predator whenever possible. Both have justification in the exploration of pre-predator encounters in the wild. A high percentage of deer killed by the tiger are the direct result of the prey's curiosity. The tiger's striking colour pattern may serve to attract deer's attention. A tiger displaying its half concealed striped body would induce deer to investigate the object: one or two inquiring steps toward a tiger can be lethal. Schaller (1967) describes several occasions when deer displayed curious and incautious behaviour. On one occasion a lead chital doe (*Axis axis*), followed by six other animals, investigated a bush after she smelled the scent of a tiger. Similar behaviour was recorded for the barasingha (*Rucervus duvaucelii*), which was found in the Sundarbans during the early twentieth century. Typical behaviour of the deer consists in an attempt to circumvent situations when a tiger could be encountered. Sometimes it was seen in a different situation, when a stag, coming slowly to a trail, was rapidly alerted by the scent of a tiger and immediately trotted off.

Deer have different gaits and can jog, trot or walk at any speed up to their top range. Their front and back legs are coordinated, work in opposing rhythms to continuously launch them off the ground and move forward. The main reason deer are so fast is related to their muscles. Generally speaking, their muscle fibres are divided into two types:

Type I: These are called slow twitch muscle fibres. They can produce small amounts of power for long periods of time without getting tired. In other words, they're for endurance.

Type II: These are called fast twitch muscle fibres. They can produce short bursts of maximum power but get tired quickly.

Scientists have discovered that deer have a much greater proportion of type II muscle fibres than humans and many other animals. As a result, they're capable of great bursts of speed, which is what allows them to run so fast. Tracking a deer over long distances is usually a better strategy than chasing after it.

A relatively low rate of successful hunts in the case of tigers inhabiting the Sundarbans mangroves is supposedly caused by the fact that prey detect the predator by their senses of smell and hearing. Deer have a large number of visual and auditory channels for communication of danger (raising a tail, thumping) so if an individual becomes aware of a predator the whole herd is alerted (this applies even for mixed interspecific assemblages). Even in an abundant year, tigers were successful only once from a number of stalks accomplished. Chitals often have trouble distinguishing motionless forms.

On the other hand, when the outline of a feline body is easily recognisable and is in contrast to the background, deer exhibit alertness with a tendency to flee. In addition to this, they are very sensitive to slight movements and pay close attention to the source of disturbance. It is crucial for a tiger during a stalk to avoid making rush movements when it faces the sight of prey. If the prey looks towards the predator's location, the predator stops moving and the image of its parts partially exhibited through vegetation comes under the static view of crypsis. At the moment the cat launches an attack, the camouflage does not account for the success of the hunt. Either the prey is disturbed and tries to escape immediately after sensing the predator, or it is not aware of the attack and looks in a different direction. Visual factors are crucial for the hunt before a predator's run is started, after that only speed and swiftness decide whether it would stay alive. A very scared prey will run faster than one that is not as afraid.

Deer have several adaptations that allow them to run without tiring that much. One of the main adaptations is their cardiovascular system, which is highly efficient at delivering oxygen to their muscles. They also have large lungs, which allow them to take in more oxygen with each breath. Additionally, deer have strong leg muscles and tendons, which help them run quickly and efficiently. They also have a unique skeletal structure that allows them to bound through the forest, conserving energy. Overall, it is a combination of these adaptations that allows deer to run without tiring that much. For example, the Axis deer can run at speeds of up to 60-65 km per hour and can jump over obstacles up to six feet high. But in the Sundarbans their speed may be reduced due to difficult terrain. However, the deer are strong swimmers and can swim for long distances if they need to. They usually swim with their heads above water, but they can also dive underwater to escape predators. The fastest recorded swimming speed for a deer is 13 km per hour.

Deer anatomy drives their speed. Their hooves correspond to our fingers or toes, and you could argue that this is where their speed begins. Those small, hard points have a relatively small surface area, which reduces friction in contact with the ground. Besides increasing speed, this also decreases noise when they're walking through crunchy leaves. For added traction, deer have two vestigial toes, called "dew claws" that occur farther up their leg and usually contact the ground only in mud. A deer's hooves attach to their lower legs with a special ligament. As they plant a hoof on the ground, the ligament stretches out, with the deer's own weight supplying the force. This stores energy like a stretched rubber band. When something startles the deer into a leap, that ligament snaps back. The stored energy adds power to their high jump. In a photo captured mid leap, you can see that the hooves are curled back due to the contraction of that ligament. Moreover, a deer's rear "knees" bend backwards. Deer lack a collarbone at the top of their front legs. This means that their left and right shoulders can pivot independently. With rear legs providing thrust and front legs steering, deer hit speeds of up to a very high speed and turn on a dime. Deer don't always use their speed, though. Sometimes they just stand still looking confused. Eyesight is not a deer's most important sense. Deer ears are incredibly powerful, and like their front legs, can move independently of each other. By turning their ears this way and that, deer can hear sounds from behind them, and also pinpoint where sounds are coming from. But hearing alone is often not enough for deer to identify danger. That's where the smell comes in. Once they catch a whiff of a predator on the breeze, those ligaments start springing. They have developed all these amazing adaptations. It is their role as a prey animal that sharpened their skills.

Tigers can go up to the highest speed of around 65 km/hr, but they are not able to keep such speed for long. They have to bring down the prey with the run of about 100 m or their speed would start coming down. I saw a tigress in hunting mode. First it would stand in a fixed position as long as it takes for prey to get distracted so that it could move forward to attack. A tiger would concentrate only on the prey, nothing else, for achieving the goal with a minimum number of attempts. In the Sundarbans, the big cats primarily prey on deer, which are diurnal with activity peaking during morning hours. Elsewhere Bengal tigers are active mostly at night, an adaptation in synchrony with its primary prey, the nocturnal sambar deer.

(ii) Barking deer (*Cervus muntjak* Zimmerman, 1780)

The faecal pellets of barking deer are elongated and crinkled often with a twisted look and dented surface. The pellets taper towards both ends giving a spindle shape, the tips vary from being long and hook-like to blunt. In a clump it is often common to encounter a large variation in size and shape between individual pellets from the same animal. They range in size between 1-1.8 cm and about 0.4 to 0.7 cm in diameter.

Barking deer is a small (20+kg) and solitary species appearing to be confined to the north and north-east in the Bangladesh Sundarbans. Although observed around Dhangmari, Karamjal, Jhongra stations in Chandpai Range, they are not seen in southern Sundarbans. In contrast with the spotted deer, this species appears to avoid grassland and is thus better suited to the denser woodland in habitats to the north. Dey (2007) recorded a population of about 2,150 with variable density- 1-2 km^2 in *H. fomes*; 2-3 in *H. fomes-E. agallocha* and *X. mekongensis-B. gymnorrhiza-A. officinalis* forest; 3-5 km^2 in *E. agallocha-X. mekongensis-B. gymnorrhiza;* and 4-5 km^2 in *E. agallocha-H. fomes* forest. Both males and females have tusks-like upper canine teeth, which are longer, well developed and distinctly visible from a distance in males. The territorial males use them along with their short antlers for fighting against each other.

The home-range also varied from 45 to 90 ha during November-February but during the rutting period, a male's home-range increased from 80 to 170 ha. During the breeding season a female's home-range increased to 1.25 times larger than that of the non-breeding season. A male shared his home-range with those of 2-3 females during the rutting period. Whereas the spotted deer are active mostly during day time, the barking deer are more active during the night.

Barking deer is also a major ungulate prey of the tiger in the Bangladesh Sundarbans (Khan, 2008).

(iii) **Rhesus monkey** (*Macaca mulatta*)

Rhesus macaque is one of the principal prey species of the tiger in both India and Bangladesh. Foraging on the banks of smaller or narrower channels close to the thickets was not usually observed because they are prone to sudden tiger predation there.

Rhesus macaque was captured more or less throughout Sundarban landscape with highest concentration of photo captures in Sajnekhali beat of Sajnekhali WLS. Their sign and sighting encounter rate was higher in the NP East range as well as in the buffer zone of the Basirhat range on the Indian side. Reza (2000) recorded a total of 37 groups in Katka-Kachikhali, the largest group size was 41 and the smallest one only three, mean group size 11.92 individuals/group, and biomass of the animals was 126.7 kg/km² and 2,903.2 kg/km². Dey (2007) recorded 51,000 rhesus monkeys in the Bangladesh Sundarbans. Hasan *et al.* (2013) recorded a total population of 966 individuals distributed in 41 groups (mean size 23.6 ± 5.2; range 14-31; adult male-female ratio 1:2.84 and adult-immature ratio 1:1.64 including sub-adult male, sub-adult female, juvenile and infant) in the Sundarbans (southwestern).

Mallick (2011) recorded the rhesus density in STR as 1.2/ km². The Rhesus population densities (extrapolated) in six selected sites in the Bangladesh Sundarbans (Mallick, 2019b) are given below.

Location	Relative density	Absolute density
Katka-Kochikhali	2.42 (± 0.77)	6.5
Hironpoint	2.36 (± 0.81)	6.3
Mandarbaria	2.40 (± 0.69)	6.4
Harintana	1.18 (± 0.59)	3.2
Burigoalini	0.85 (± 0.47)	2.3
Chandpai	0.83 (± 0.45)	2.2

Mallick (2019b) also recorded the highest concentration (based on personal observations) in the western Sundarbans, particularly in the PAs and surrounding RFs of Sajnekhali WLS (362.40 km²) between the rivers Peechkhali and Gomdi, Haliday (5.95 km²) on the River Matla and Lothian Islands (38 km²) at the confluence of the River Saptamukhi and the Bay of Bengal. Troops were sighted at the popular tourist spots-cum-supplementary feeding sites with well maintained sweetwater ponds, e.g. Sajnekhali, Sudhanyakhali, Choragaji, Dobanki and Netidhopani (STR). A small troop was camera-trapped in South 24-Parganas FD (Das *et al.*, 2012). The monkeys also inhabit the northern anthropogenic (civil) areas (e.g. Gosaba). Mallick (2019) also referred to the earlier records of *M. mulatta* in the eastern Sundarbans by Hendrichs (1975) (29 January-21 April 1971),

Green (1978) (July-November 1976), Gittins and Akonda (1982) (early 1980), Khan (1985) (1980 and 1982) and Feeroz *et al.* (1995) (February 1990-June 1993). Sightings were more common in the East WLS (312.26 km²) than the West (715.02 km²) and South (369.70 km²) WLSs as well as ten individuals camera-trapped in the East WLS during 6 September-4 December 2006 (Khan 2012).

The estimated rhesus population in STR was about 38,000 in the 1990s (Mallick 2011). No such estimate for the South 24-Parganas FD is available, but the population appears to be comparatively low (about 10,000) on the basis of counting the population at Bakkhali in January 2001 and then extrapolating the figure for the entire division (Mallick in Bahuguna & Mallick 2010). The Bangladesh population was between 40,000 and 68,200 (recorded in Mallick, 2019b) against the higher estimate of 88,000 to 126,220, but the current estimate is 40,000 to 50,000. Due to shrinkage of habitat, the population has declined over the last two decades.

Density of the Rhesus population in various mangrove vegetation types in the Sundarbans was studied during 2011-2012 (Mallick, 2019b). The results are shown below.

Major pure and mixed genera (Number of species)	Forest zone	Area (in km²)	Population density (%)
Sonneratia-Excoecaria-Oryza (5)	River flat	82.86	60
Ceriops-Excoecaria-Sonneratia (6)		648.07	15
Excoecaria-Heritiera (2)	River flat-Ridge forest	1,816.76	6
Excoecaria-Ceriops-Xylocarpus (5)		346.04	5
Xylocarpus-Bruguiera-Avicennia (9)	Ridge forest-River flat slope-River flat	40.30	5

Major pure and mixed genera (Number of species)	Forest zone	Area (in km²)	Population density (%)
Pure *Excoecaria* (1)	River flat	215.20	4
Heritiera-Xylocarpus-Bruguiera (7)	Ridge forest-River flat slope	95.56	4
Pure *Heritiera* (1)	Ridge forest	749.92	1
Total		3,994.71	100

The monkey distribution in the substratum between the high and low tides is determined by the extent and depth of flooding. Six inundation classes in different tidal zones, leading to constant remodelling of the landscape, were recorded (Mallick, 2019b). These are stated below.

Zone	Inundation conditions	Habitat	Periodicity	Mangrove elements
I	All the high tides up to 7.5 m	River flats/slopes	30 days/month	Major and minor elements (32 + 69 species) of mangrove and mangrove associates
II	Medium high tides	River banks/ridge forests	20 days/month	
III	Normal high tides	Ridge forests (flat land and dense vegetation)	15 days/month	Minor elements, mangrove associates and back mangrove
IV	Spring tides	Ridge forests (flat lands with dense/ sparse vegetation and salty patches)	10 days/month	

Zone	Inundation conditions	Habitat	Periodicity	Mangrove elements
V	Abnormal/ equinoctial tides (monsoon/sum mer)	Ridge forests, sparse vegetation and reclaimed/ naked areas	5 days/ month	Back mangal
VI	Above the tidal reaches	Deltaic region	-	Non- littoral plants

Table: *Inundation classes of monkey habitats in the tidal zones*

A rhythmic response of the macaques to the different levels of tidal fluctuations was observed. For example, they take the opportunity to feed on the river banks and in low water at the time of low ebb and prefer arboreal feeding and shelter, when the high tide floods this feeding ground. The short flood-tide (2-3 h) and long ebb-tide (8-9 h) during the dry season, when the higher land does not get flooded, is advantageous to the foraging monkeys.

Like the tigers, the rhesus macaques feed on hermit crabs to supplement their diet and for this purpose, they have even to dexterously manoeuvre their movements to avoid landing their feet on needle like piercing pneumatophores, which speaks of their obligatory dependency on this diet.

The preferred biotope or permanent abode of the monkeys is the well-drained or less predation- prone arboreal habitat, but the treeless reed swamp does not support these animals. Increase in the numbers of monkeys per km² was observed in the areas above general tide level compared to those frequently inundated and below tide levels. Ecologically, the scattered small patches of Keora *S. apetala* (2% of the mangrove forests) are most important in the mangrove food chain . The monkey troops were primarily found on this quick-growing, tall emergent (10-20 m) and spacious tree with heavy foliage relative to other natural mangrove genera of lower height and foliage. Keora is common along the border of islands near estuaries, intertidal zones of several creeks and channels towards upstream swamps and invariably in places affected with fresh and brackish water mixture in association with *Avicennia, Ceriops, Excoecaria* and *Xylocarpus*. When the floodwater is drained, a rich foraging and hunting ground is exposed on the intertidal mudflat, sparsely vegetated or with no vegetation and exploited by the monkey troops. The third biotope, i.e. the estuary or creek with characteristically low-saline brackish water system, is foraged during the low tide for fish, crab, floating tender leaves or ripe fruits. The brackishwater is also taken in when sweetwater in the forest is scarce in the vicinity (out of 41 sweetwater ponds dug in STR for rainwater-harvesting, only 12 are well maintained). During the dry season, the monkeys were also seen to negotiate the narrow channels in quest for new feeding sites.

The biology and behaviour of Rhesus monkeys has been modified due to ecological stresses (Mukherjee, 2006). They adapt efficiently to various elements of the complex habitat, such as anomalous tract, swampy/muddy terrain, seasonal, lunar, dial or daily phenomenon, thermal dynamics, strong winds and rains, increasing salinity in vertical and horizontal planes, fluctuating tidal current (opposite, unidirectional and oscillating) and flooding (Mallick, 2019b).

These islanders appeared to be smaller than the mainlanders (Head and Body up to 590 mm and Tail 280 mm). An average weight of 4 kg recorded in the Sundarban mangroves is also markedly lower than in non-mangroves (7-8 kg). The advantages of this insular dwarfism are assumed to be lower food requirement, ability to move quickly through the muddy terrain while foraging or negotiating waterways with minimum loss of energy and quick escape from sudden predatory attacks (Mallick 2011).

Primates have trichromatic colour vision and thus are able to distinguish green and yellow colour. The deer react (by snorting, foot-stamping) most strongly to the tiger's presence in the vicinity because they are considerably alert when it catches sight of the tiger.

When the tidal water receded back exposing the clay bed, they were observed to come down the shoreline in small troops for ground and semi-aquatic foraging, but always remained alert against sudden attacks by tigers. In response to such danger, the macaques flee into the trees. While foraging at Sudhanyakhali, the alpha male, sensing a tiger lurking around, became tense and emitted a shrill bark (alarm call), thumping his feet a few times on the ground to warn other monkeys of the approaching predator, running a short distance and then climbing the nearest available tree at lightning speed. All the female monkeys, along with the sub-adults and juveniles, followed him.

The deer were also observed to respond to the monkeys' repetitive alarm calls ('kech-kech') and begin a persistent barking of their own to escape predators (Bahuguna and Mallick, 2010).

Unlike in other habitats, the tigers in Sundarban hunt scarce prey, including the rhesus macaque. Scat analysis has revealed that the monkey is the third highest component [6.89% (n = 10) to 7.19% (n = 13.57)] in the tiger's diet. During predation by the tigers, alarm responses and apparent mobbing of a tiger by *M. mulatta* have also been reported.

(iv) **Wild boar** *(Sus scrofa cristatus)*

Wild pigs were distributed throughout the Sundarban landscape. Prevalence of more capture hotspots in SBR concurs with higher sightings of wild pig during boat transects in this area. The habitat of the wild boar is the tangled mass of Garan, whose extensive breaks harbour the sounders. Population status of the wild boar was studied in the Bangladesh Sundarbans. Reza (2000) recorded a total of 133 wild boar groups in Katka-Kachikhali, with varying group size 1-15 (mean 2.21), group density 3.58 groups/km^2, species density 7.9 individuals/km^2 and biomass 300 kg/km^2. The mean biomass and the metabolic biomass were reported as 330.22 kg/km^2 and 247.67 kg/km^2 respectively. Dey (2007) recorded 28,000 wild boars in the Bangladesh Sundarbans.

Wild pigs have an excellent sense of smell but poor eyesight. Still they are not easy to be hunted by the tigers, as they can run up to 40 kmph and can attain their top speed in just a few strides. They do not usually run in a straight line. They often zigzag in order to escape tigers. They are sprinters and can also jump at a height of 140-150 cm. The wild boars' primary defence is speed, but when cornered, they can behave quite fiercely. An adult tiger is often killed in a fight with a wild boar in the tigerland. Weighing up to 100 kg, Boars use their skull as a battering ram and come armed with sharp tusks. They are very aggressive and fiercely defend themselves, especially if there's a piglet to be saved. Incidents of wild boars chasing tigers have also been reported.

Wild boars are no pushovers. So, deciding to target a heavy tusker prey is a big deal for a tiger. It uses up quite a lot of energy, especially if the preyed-on animal decides to make a run for it as this wild boar does. Sometimes things get a little muddy in the swamp. It cannot always match the speed and technique of the tiger. The Sundarbans tigers won't hesitate for a moment to swim in the water for prey. The strength of a gigantic tiger may be observed as he expertly dispatches the wild boar while knee-deep in mud and water. The tiger quickly catches it up and wrestles it to the ground. The big cat uses a suffocating throat bite just below the junction of the jaw and neck. The tiger manages to spin around and find the right angle to bite into the boar now that it's in its paws. This effectively crushes the animal's trachea so that it cannot breathe. Even when the prey is pinned to the floor by the tiger's powerful front paws, the wild boar continues to struggle and makes a heart-breaking squealing sound of terror. Gradually, amid the huffing and puffing of the tiger, the boar's shrieks start to falter. The boar's resistance continues to weaken, and within moments, its movements stop entirely. Within a few seconds, it is all over and the tiger emerges victorious, his prize secured, even if he bears the marks of a muddy battle. Then he makes a start on its meal.

(v) **Monitor lizard** [*Varanus* spp. (Reptilia: Varanidae)]

Sundarbans is considered as the most potential habitat for the Asian water monitor (*Varanus salvator*), length 250 cm, weight 19.5 kg. It is a rather common lizard, although not very easily spotted in the mangrove swamps. Due to its large size and high abundance this species may form an important component of the tiger's diet in the Sundarbans. The tanks at Sudhanyakhali, Netidhopani and Haldi are the favourite haunts of this animal. They were also seen swimming in those tanks.

Prey species of tiger and the tiger both prefer areas proximate to perennial water sources. The capture hotspots of the monitor lizards in STR were in the Sajnekhali WLS, the buffer zone of Basirhat range and the core area of the NP West range with very few photo-captures in the South 24-Parganas FD of SBR. Monitor lizard sign and sighting encounter rate was highest in the Chandpai-Sarankhola range of Bangladesh, while its presence was not recorded in many areas.

A total of 14 species of ungulates, carnivores, omnivores and Galliformes were photo-captured in the India Sundarbans. Wild pig and chital were apparently the most common species where it took only five and 10 trap nights respectively to capture one photo while estuarine crocodile and common palm civet were the rarest photo-captured species. These data are shown below.

Species	No. of photos per 100 trap night	No. of trap nights to get one photo
Chital	9.90	10
Common palm civet	0.01	12298
Estuarine crocodile	0.01	12298
Fishing cat	4.77	21
Golden jackal	0.03	3514
Jungle cat	0.06	1757
Leopard cat	3.57	28
Monitor lizard	2.26	44
Otter	0.11	911
Red jungle fowl	6.41	16
Rhesus macaque	6.01	17
Rodent	0.09	1171
Tiger	3.34	30
Wild pig	18.78	5

The camera trap photo index (RAI) for various species should be viewed in the context of using lures for camera trapping in the Sundarban landscape (lures are not used at other sites). The lures enhance chances of attracting species of carnivores and wild pigs to camera traps. Therefore, these RAI can be used for future population trend comparisons within these species in Sundarbans but not for interspecies relative abundance or comparisons between other sites. However, species like chital and red jungle fowl are unlikely to be affected by lures and their RAI are comparable across sites (Jhala *et al.*, 2018).

Bangladesh Sundarbans

Prey species density in different habitat types of Katka-Kachikhali (Reza *et al.*, 2004)

Habitat type	Area surveyed (km²)	Spotted deer	Wild boar	Monitor lizard
Forest	20.1	48.8	2.2	9.7
Meadow	37.9	79.9	2.3	0.8
Sea beach	19.7	100.4	0.6	0.2
Forest edge	8.1	41.0	15.8	21.1
River bank	7.2	31.5	3.3	6.1

Non-conventional prey species

Snakes

(a) **Non-venomous**

Normally the tigers are not a snake-eater. However, a tiger once completely ate up a large python *Python molurus* (length up to eight metre and weight two quintals), which earlier had swallowed a chital stag entirely with antlers measuring 16″ and 18″ (Gupta, 1966). The python was shot by Mr. A.K. Gupta. He left the site to bring his men but when he returned to the spot with men to carry it, by that time the tiger finished the double lunch.

(b) **Venomous**

Mallick (2011, 2013, 2015) gave a vivid description of the first record of an exceptional incident of consumption of two venomous snakes (Squamata: Elapidae) by a tigress in the mangrove Sundarbans after the super cyclone 'Aila'. The hostile ecological conditions make the Sundarban predators more hardy and agile when compared to their counterparts in other PAs. As the predator was aged, it might be possible that it could not hunt the fast moving prey animals. The officials ruled out that the big cat's death was due to snake venom or poaching as there was no injury or gunshot mark on its body.

The cyclone 'Aila' struck both the Indian and Bangladesh Sundarbans on 27th May, 2009. Thereafter on July 17, the forest officials posted in Netidhopani of STR came across a dead tiger near their field office. On 18 July 2009, an autopsy of the carcass of an adult female tiger, about 14 years old, with no external injury revealed that it ate two poisonous snakes. The undigested parts of three uncommon prey species were found in the stomach contents of an adult dead tigress at Netidhopani in STR. As this tigress was old (12-14 years) and the habitat became hostile and inundated that it could not hunt the ungulates by chasing. Some opined that when the king cobra consumed the cobra, the tiger came on hunting. But if that was so, normally there should have a biting sign of the king cobra on the head of the cobra, but such a sign was, in fact, could not be seen. However, hunting of the live venomous snakes successfully in the wild proves the exceptional skill of the Bengal tiger.

Two uncommon prey species of the Sundarbans tiger, i.e. reptiles, are-

(a) A king cobra (*Ophiophagus hannah*)- sex unknown, probable length two m+; although the biggest one is recorded to be about three m.

The species was first reported in 1836 by Danish naturalist Theodore Edward Cantor who described four such specimens, three captured in the Sundarbans and one in the vicinity of Kolkata. It is the world's longest venomous snake with maximum length of 5.85 m and an apex predator and dominant over all other snakes, poisonous or non-poisonous, except large pythons); it is found mostly in deep forests. It is fierce, agile and can produce large amounts of highly potent venom in a single bite. After the supercyclone, its habitat was damaged and it might have come out to the forest edge in search of food.

(b) A Monocled Cobra (*Naja kaouthia*) - sex unknown, probable length 1 m+ whereas the length is 3.18-4 m.

Swamps, and mangroves are their preferred habitats and they have become semi-aquatic in this type of habitat, the neonates feeding mostly on amphibians whereas adults prey on small mammals, snakes and fish.

The post-mortem of the tigress revealed that it could have died due to a bacterial infection in the liver, as its necrotic foci in the heart, lungs and other body parts had enlarged, which might aggravate the problem after consumption of the poisonous snakes. Cobra hoods were found to be intact while there were pieces of body parts of snakes in the big cat's viscera. Difficulties faced in hunting may be the cause of such alternative food habits.

Decapod crustacean

The crabs are the major macrofauna and the ecosystem engineers of mangroves occupying bi-ways of burrows amid mud-roots of mangroves. They use the relative safety of the mangrove to hide from the tiger, but often expose themselves to certain death by traversing the mud banks. A half-digested crab (unidentified species of Brachyura) was also found in the stomach of tigress detailed above (Mallick, 2013). The Sundarbans tigers are also reported to consume the mud crab (*Scylla serrata*) (Mallick, 2015). The mud crab is also a recorded prey species of the tiger in the Bangladesh Sundarbans (Khan, 2008).

Dolphin

Dolphin sign and sighting are in general low with higher presence recorded in the Bangladesh side especially within the declared PAs for the species.

(a) Ganges river dolphin (*Platanista gangetica gangetica* Lebeck, 1801)

According to Mallick (2011), the Gangetic dolphin is frequently found in the eastern side, particularly in the river Raimangal, the upper part of which is fed by some freshwater flow. It descends during the monsoon season when salinity is low but appears to decrease in number during the summer. This migration also seems to be associated with the migration and dispersal of fishes, which are their main prey.

Mallick (2015) reported an incident when a tiger had eaten the carcass of a Ganges river dolphin (*Platanista gangetica* Lebeck, 1801), weighing ±100 kg, caught in a fishing net, thrown away and lying on the bank.

(b) Irrawaddy dolphin (*Orcaella brevirostris* Owen in Gray, 1866)

The Irrawaddy dolphins are not found in the more saline Matla-Hooghly estuary. Seventeen individuals were seen in the pre-monsoon and twenty two in the post-monsoon months. The Irrawaddy dolphins were recorded mostly in the lower reaches of the river Raimangal and a few were also observed at the Jhilla and Amlamethi. There was no sighting record of this species in the river Matla, which is a habitat of the Ganges dolphin. The black finless porpoise were rarely found in the rivers near the estuary.

Khan (2008) recorded a similar incident of tiger depredation in the case of an Irrawady dolphin (*Orcaella brevirostris* Owen in Gray, 1866) in the Bangladesh Sundarbans.

(iv) **Whale**

Mansur Fahrni's in his personal communication (2005) to Neumann-Denzau, observed numerous tiger tracks going back and forth to the carcass of a whale (possibly a Bryde's whale) in December 2004, a week after it was stranded at the shore of the Bay of Bengal in the southeastern Sundarbans of Bangladesh. Dark and oily tiger faeces in the surroundings indicated further that the big cats had eaten from the whale.

(v) **Porcupine**

(a) The Indian crested porcupine (*Hystrix indica* Kerr, 1792), weighing 11–18 kg; and

(b) Asiatic brush-tailed porcupine (*Atherurus macrourus* Linnaeus, 1758), weighing 1.5-4 kg; both are recorded to be eaten by the tiger in the Bangladesh Sundarbans (Khan, 2008). There is no field record of the porcupine as a prey species in the Indian Sundarbans.

(vi) **Birds**

(a) The Sundarbans tigers often hunt the Indian red jungle fowl (*Gallus gallus murghi*), weighing 1-1.5 kg (Mallick, 2015). It is also a recorded prey species of the tiger in the Bangladesh Sundarbans (Khan, 2008).

(b) The lesser adjutant (*Leptoptilos javanicus* Horsfield, 1821), weighing 4 to 5.71 kg, are higher in the Indian counterpart. It is reported to be one of the main prey species in the Sundarbans West WLS (Khan, 2008).

(c) Even feathers of the Green-billed Malkoha (*Phaenicophaeus tristis* Lesson, 1830), usually about 50-60 cm long and weighing 100-128 g, were found in their scats.

(d) Greater coucal *Centropus sinensis* Stephens, 1815 (47–56 cm; male 208–270 g, female 275–380 g) and

(e) Jungle myna *Acridotheres fuscus* Wagler, 1827 (weight 24cm; 72-98 g).

(vii) **Turtle** (species unidentified)

It is a very rare prey of the tiger. Tigers are found in areas of dense vegetation and are usually near water where their prey is located. However, a tiger is opportunistic and will not hesitate to eat other animals available to them. It was seen attacking and capturing an adult Ganges Softshell Turtle (*Nilssonia gangetica*) in the morning of April 2, 2021 at Ranthambore NP, Rajasthan. This turtle can weigh up to 50 kg. Tigers, mostly cubs, were observed "playing" with turtle shells often, but rarely with living turtles. And they lose interest as soon as the turtle recedes into its shell. On April 2, a sub-adult tigress called Riddhi sat at the edge of Rajbagh Lake. She had recently separated from her mother's guardianship but was still residing within her mother's home-range. She would avoid any confrontation with her more powerful mother by staying well away from her. A few minutes later, alarm calls from a distance indicated that another tiger (most probably her mother – Arrowhead) was approaching. Riddhi instantly got up and started moving away from the direction of the calls. On the shore of the lake Riddhi got that turtle. Some moments later, the turtle tried to run away, but proved to be a huge mistake. The tigress got up to take one step forward and grabbed the turtle by the right hind leg that was protruding out of the shell. She then picked up the helpless turtle and carried it to an opening in the grass, a few feet away, and started chewing on the leg that she was holding. Within the next minute, she had opened up the shell around the right hind leg to eat up the entire turtle. That's when the alarm calls of Spotted Deer started up once again, and Riddhi left the turtle and walked away to get away from whatever was approaching her. However, the forest guards informed that the tigress had come back to finish her meal.

(viii) **Fish**

Eating fish is definitely not the tiger's first choice. However, fish are one of their backup options if their regular food sources are a bit sparse because it is hard for them to catch fish. They will look for other options before they resort to eating fish. Ghosh (2020) recorded the tiger's preference for live fish (usually large carps, also called as Gangetic carps) like Catla *Cirrhinus cirrhosus, Labeo rohita* etc, which comes to the aquatic mangrove forest for breeding purposes. These fishes are found in the river water during early monsoon season when the tigers too have their feast. Using their claws and fangs to catch a moving fish under the water can be challenging for them at times. When fish are swimming around in the rivers and lakes, they don't really have a strong smell to be detected by the tigers. The tiger often simply jumps into the river and catches the fully grown fish with the help of their mouth. Sometimes the tiger gets in the water and locates the fish. Then it startles the fish with a blow of its paw, although the fish never feels the same power blown by the tiger that a deer or a boar would feel. This is because the water blunts the power of the blow. Still, the fish underwater feels a sudden movement in the water. The fish is then startled and this is when the tiger picks up the fish with its powerful jaw. The tiger struggles to hold the fish inside its mouth. This is because they tend to go for their prey's neck or throat. It is instinct to them. However, they don't have that advantage when fishing as fishes don't have necks. After getting picked up, the fish tries its heart and soul to be free again. It starts to jump inside the mouth of the tiger. However, its efforts go in vain as the tiger gives its head continuous enormous shakes. The fish dies soon after being out of the water and getting wounded by the tiger's sharp teeth. Finally, the tiger enjoys its meal. However, fishes are small prey and don't provide enough protein. Moreover, catching a fish is more difficult for them.

Gupta (1966) wrote: "Driven by hunger, the Sundarbans tiger is known to catch and eat fish in creeks".

(ix) Frogs

The tigers can also eat smaller animals such as frogs if they come across them. However, they only do so when they are very desperate and large sources of meat are not available because the tigers cannot follow prey for long distances.

(x) Unidentified others

There are a small percentage of unidentified prey species of the tiger.

(xi) *Phoenix paludosa*

It is recorded in the scats of the tiger 14 times.

(xii) Fruits

Tigers occasionally eat fruits and parts of plants for dietary fibre. Definite proof of this food habit in the Sundarbans is not available.

Mukherjee and Sen Sarkar (2013) recorded frequency of occurrence, estimated biomass consumed, and estimated numbers of individuals consumed by tigers in STR, based on the contents of scats collected from 1999 to 2001 (n = 113). An examination of the 214 scats [113 (phase 1) + 101 (phase 2)] led to the detection of 305 prey items in terms of occurrence. Spotted deer made up to 53.79% (phase 1) and 52.51% (phase 2) of all the prey items detected, but occurred in 69.02%-72.27% of all scat samples. Wild boar made up to 19.31% (phase 1) and 30.93% (phase 2), rhesus monkey 6.89% (phase 1) and 7.19% (phase 2), and water monitor 9.65% (phase 1) and 4.31% (phase 2) of all the items detected in the collected scats. Fish, crab, and birds together contributed to only 9.63% (phase 1), 5.02% (phase 2) of the detected items. Fishing cats and turtles were detected only once in the scat of the tiger, proving that they are not a common dietary preference of the Sundarbans tigers. The leaves of *Phoenix paludosa*- a common mangrove palm of the Sundarbans was found to occur commonly in the scats that are known to be part of the tiger diet to help it regulate bowel movements.

Aziz *et al.* (2020) collected a total of 512 scat samples during a survey of 1984 km² of forest across four sample blocks [Satkhira Block (SB), West

WLS, East WLS and Chandpai Block (CB)] in the 6,017 km² of the Bangladesh Sundarbans. Analysis of scat composition and prey remains reliably identified five major prey species (spotted deer *Axis axis*, wild pig *Sus scrofa*, rhesus macaque *Macaca mulatta*, barking deer *Muntiacus muntjak* and water monitor *Varanus salvator*), of which spotted deer and wild pig contributed a cumulative biomass of 89% to tiger diet. Tiger preference for prey species was highly skewed towards spotted deer and wild pig, but the relative contribution of these two species differed significantly across the four study areas, which spanned the Sundarbans, demonstrating important spatial patterns of tiger prey preference across the Sundarbans landscape. On a spatial scale, spotted deer were the predominant prey species at EWLS and CB, while wild pigs predominated in the SB and WWLS. Of the remaining three prey species, rhesus macaques were found across four sites, with higher relative counts in SB and CB. Water monitors were relatively rare across all sites, and barking deer were entirely absent in SB and WWLS. The mean ± SE group size of spotted deer was 4.63 ± 1.74 (n = 38) whereas it was 2.88 ± 0.55 (n = 26) for wild pig and 12.10 ± 1.25 (n = 37) for rhesus macaque. Barking deer (n = 4) and water monitor (n = 13) were encountered only as single individuals. Overall, spotted deer were tigers' preferred prey, with wild pigs as a secondary alternative in their diet. However, sample block-wise results showed that preference for wild pigs was higher in SB and WS, whereas the preference for spotted deer was highest in ES and CB. The remaining three species were least preferred by tigers across all sample blocks

Recently Variar *et al.* (2023) reported prey selection by the Indian tiger (*Panthera tigris tigris*) outside PAs in less protected RFs of Begur Range, a wildlife corridor in the Malenad–Mysore Tiger Landscape (MMTL) in the Western Ghats. The study recorded seven tiger prey species from the study area. The percentage composition of different prey species revealed that chital (*Axis axis*) constitutes 63.15% of the overall diet, whereas sambar (*Rusa unicolor*) and gaur (*Bos gaurus*) constituted 26.33% each. Contrary to their prediction, the relative biomass of different prey species showed that chital, sambar, and gaur constituted 95% of the overall diet. The presence of small prey species such as the Indian crested porcupine (*Hystrix indica*) and small Indian civet (*Viverricula indica*) indicates opportunistic hunting behaviour and diverse prey selection by tigers.

Tiger-human conflict

Humans are the most notorious predator of the tiger. The encounter rate of human sign and sighting was higher near the edges of the forest and overall higher in Bangladesh Sundarban which is further exacerbated by the usage of river channels for transportation of commercial vehicles.

An estimated 80,000 Bengal tigers were killed during the period 1875 and 1925 and probably more numbers were hunted before 1875 and after 1925 which would cross one lakh victims of hunting and poaching in India and Bangladesh (Mallick, 2023).

Hunting and poaching is the most important threat to the tigers spurred by the illegal trade, traditional beliefs, rituals and increasing population pressure. The tiger has historically been a popular big game animal and has been hunted for sports and securing a prestigious trophy. Dating back to the early 16th century thousands of tigers were killed in elaborate hunts by the Indian and British nobility before hunting was outlawed by the Indian government only in 1971.

At the close of the 19th century, when Rudyard Kipling penned the Jungle Book, between 50,000 and 100,000 tigers were thought to roam the Indian subcontinent. According to historian Mahesh Rangarajan, "over 80,000 tigers…were slaughtered in 50 years from 1875 to 1925. It is possible that this was only a fraction of the numbers actually slain." Not all were trophy-hunted in the Sundarbans mangroves; the big cats were considered vermin during the early British Raj and systematically exterminated with incentive from government bounties.

The Sundarbans tigers have acquired a reputation for being aggressive and attacking people unlike those elsewhere in the subcontinent. The terai is possibly an exception. Since the inception of STR the annual human toll by tigers had varied between 25-35 people. During the end of the 20th century it has lowered significantly due to rigid control over people's access to the PA, effective monitoring and quick response to a developing situation. During a period often years from 1963 to 1972, a total of 360 people had lost their lives to tiger attacks, with the Chamta block registering the highest figure of 80 deaths. Considering the vocations, honey collectors accounted for the highest toll of 58% followed by the Golpata and Hental collectors 23%, coupe workers 13% and fishermen 6%.

Human deaths and tiger's profile (Barlow *et al.*, 2013)

Between 1984 and 2006, 490 human deaths from tiger attacks were recorded in Bangladesh Sundarbans; these were used to create the tiger profiles. The estimated mean number of tigers killing humans, based on the number of compartments recording human deaths was 8 tigers/ year (SD = 5, range = 0-16). Using t = 12 months, an estimated total of 110 tigers killed humans in the SRF over 23 years, with a mean five victims/tiger (SD = 7.1 range 1-39). Approximately 50% of tigers only killed one person and tigers that killed more than one person accounted for 81% of total human fatalities. The mean time a human-killing tiger was operational was 8.2 months (SD = 13.4, range 1-68). The number of tigers killing humans was not distributed evenly across the area; most tigers that killed humans were clustered in the west, but there were also patches of tigers killing humans in the south and north. People killed while working illegally in the forest or those that later die due to injuries, tend either not to be reported by their companions, or not catalogued by the authorities. The ecological correlates of the human-tiger conflict (e.g. intensity of area use by humans, incidence of tiger killings by humans, variation of prey abundance, or conditions otherwise suboptimal for tigers) are considered in order to better understand the importance of landscape and human factors influencing the spatial pattern of the conflict. However, no inferences can be made about the health, age, or sex of the tigers involved, or any ecological factor that may have contributed to the frequency and distribution of tiger attacks. However, human-killing still remains a problem in other areas in Nepal, Sumatra and the Russian Far East.

Retaliatory killing of straying tigers

A key factor driving the loss of tigers in India is so-called "retaliation killings" by livestock owners frustrated by the threats — perceived or real — that these big cats pose to their livestock. During the several initial years of STR the local resident community had an extremely adversarial relationship with the reserve management. It has taken a great amount of sustained effort in painstakingly earning the trust of the people. The management strategies addressing protection, habitats, ecological requirements and the human interface via innovative eco-development programmes have demonstrated their effectiveness. The overall effectiveness has resulted in people having stopped taking retaliatory measures.

In Bangladesh Sundarbans Inskip *et al.* (2014) found that killings are not purely retaliatory in nature (i.e. driven by a desire for retribution following livestock depredation or attacks on humans by tigers), and that previous negative experience of tigers is not the sole determinant of villagers' acceptance of killing behaviour. Inter-related socio-psychological factors (risk perceptions, beliefs about tigers and the people that kill tigers, general attitude towards tigers), perceived failings on the part of local authorities whom villagers believe should resolve village tiger incidents, perceived personal rewards (financial rewards, enhanced social status, medicinal or protective value of tiger body parts), and contextual factors (the severity and location of tiger incidents) motivate people to kill tigers when they enter villages and foster the widespread acceptance of this behaviour.

Reasons for the man-tiger conflict

The identified reasons for the man-tiger conflict in the Sundarbans are discussed below.

Scarcity of Prey Animals

It has further been noted that the abundance or scarcity of prey animals are not directly related to the human casualty figures of tigers. Cattle in the reclaimed areas of Sundarbans serve as a supplement to the tiger's natural preys, which actually contribute significantly to the human-tiger conflict in the fringe villages, but such depredation figure is the lowest in the Sundarbans than other conflict areas in the country (Chauhan, 2011). The main prey animals of tigers in the forests are cheetal and rhesus monkeys (social preys), wild boar (solitary prey), and occasionally birds like lesser adjutant, red jungle fowl, etc and monitor lizards (*Varanus bengalensis, V. flavescence, V. salvator*) or even fish, crabs and turtles. Although no detailed study on the prey-base of the Sundarban tiger was conducted, preliminary studies indicate that their prey-base is stable (Deuti and Roy Choudhury, 1999).

According to Jackson (1990), there is hardly any such food shortage in the Sundarbans. Taking into account the age-old (even 100 years) history of tiger-straying in the Sundarbans, they have a natural tendency to migrate as are known from other tigerlands in the country. However, Dr. Y.V. Jhala of WII commented that the cheetal density in Indian Sundarbans is 12-14 per km^2, which also supports the observation that there cannot be a large population of tigers (WWF-India 2011: 11). According to him, the present estimated population of 70 (tigers) can barely be maintained by this low level of prey abundance of the cheetal and wild boar there. But the tiger population in SBR has increased to minimum 100 in 2022 and their prey requirement has also correspondingly increased, although the poaching of prey species has now decreased effectively.

Difficulty in Hunting

The tiger depends on its stealth to stalk and ambush upon a prey. But the Sundarban terrain provides an unfavourable condition for this purpose due to sticky, deep mud all around. Stepping a foot forward cannot be done without making a distinct sound, depriving the predator of its stealth. Hence, the hunting success rate in the Sundarbans is far low (only about 20%) as compared to other areas. As recorded at various observation sites in Sajnekhali, Sudhanyakhali, Dobanki, Chamta, Haldi and Jhingakhali, the cheetal is very cautious while approaching the sweet water ponds and, interestingly, they are not seen at any sweet water pond on a day on which a tiger has visited the pond (Mukherjee, 2006). On 14 March 2009 an adult tiger (13-14 years) was trapped at Samsernagar (Hingalganj), which had a deep wound (septic) on the hind leg, who could not even walk properly, obviously failed to hunt and came to the village for easy prey.

Moreover, unlike other tigerlands in the country, the village cattle are not grazed inside the Sundarban, depriving the tigers of their easy prey or additional biomass.

Proximity of Reclaimed Human Settlement to the Tiger Habitat

In some areas of Sundarbans human habitation and tiger habitat are divided by very narrow creeks. For example, in the Arbesi block of STR, the probable reason is the blurred boundaries as the river dividing the forest and villages is silted up and during the low tide period at places there is no water. Besides, the cattle sheds are situated almost at the fringe of the villages and on the banks of the narrow creeks. After crossing the creek, it spots these sheds and after successfully hunting the easy prey, it develops the habit of becoming a regular cattle-lifter. Generally catching prey in the Sundarbans is quite difficult for the tigers due to the geomorphologic conditions. Sometimes, the tigers stray inside the villages mostly at night in search of easy prey like the cattle or dog and get back to forest within a few hours of this hunting operation. This possibility is applicable to the places like Kalitala and Kumirmari villages where villages and tiger habitat are divided by narrow canals like 'Shakuni Khal' and 'Bagna Khal'. Cattle in these areas serve as a supplement to natural prey (Chauhan, 2011). In villages like Samsernagar and Kalitala, a very small river is the only boundary between the forest areas of Arbesi-1 and the villages and during low tide the river gets entirely dried up. So the tiger easily walks across the river to catch the easy prey of cow and goat.

Name of villages affected by tiger straying in STR

Mostly affected	Moderately affected	Least affected
Samsemagar, Kalitala, Rajat Jubilee, Jamespur, Kumirmari, Chargheri, Lahiripur, Dayapur, Jharkhali, Hemnagar, Luxbagan	Pakhiralaya, Patharpara, Pargumti	Parasmoni, Sardarpara, Sandhupur, Sonagaon, Choto Mollakhali, Kalidaspur, Santigachi, Enpur

24-Parganas (South) FD

In 24-Parganas (South) FD, Deulbari, Dongajora, Bhuvaneswari are the areas of human-tiger conflicts due to closeness with the nearest RF area and the adjoining rivers of two villages- Melmel and Gomor- are highly vulnerable due to its proximity to RF (Chatterjee, 2022).

Tigers do not Stray in the Village to Kill Easy Prey like Humans

Although the Sundarbans bear the bad name of 'the land of man-eating tigers', yet the view that tigers stray in the villages of the Sundarbans to kill human beings is totally wrong. Among the 94 cases recorded till 1995, there were only four cases where the tiger had killed a human being. From 1997 to 2005, there was only one such case. All these incidents took place between January and February 1990 at Samsernagar. It indicates that it was the same tiger that was responsible for all these human killings. Man-eating takes place in the forest and the victims are fishermen, honey collectors and wood cutters, where the tiger continues to view man as easy prey and it cannot be considered aberrant behaviour. However, Chaudhuri and Chakrabarti (1980) observed that the man-eater did not cross into human habitation for predation, which is a striking feature of the Sundarban tiger.

Embankment Protection Mangrove Strips of the Villages are Confused by the Tigers as their Own Habitat

In the fringes of some villages, embankment protection mangrove strips have been developed. Tigers sometimes confuse those mangrove strips with their own habitat (NEWS, 1996). Some villages have small patches of mangrove forests, like in Jharkhali village which is adjacent to Pirkhali Block, so the tiger gets into these forests by losing direction. Thus, tiger straying may take place due to this reason.

Littering Female Strays in the Paddy Field to Protect her Cubs

It is also argued that the female tigers stray into the paddy fields around the villages during the littering season apparently to protect the new-born cubs from the aberrant behaviour of the males. But this is a baseless assumption because the villages do not provide any safe or secluded site for reproduction.

Paddy Fields Confused with *Porteresia coarctata*

In the Sundarbans, the tigers enter several kilometres inside the dry high paddy fields (De 1990). During the late monsoon or post-monsoon when paddy in the field around these villages ripe, the migratory tiger naturally gets confused as to whether there is a forest on the other side of the creek or not. To add to its confusion, the ripped paddy looks somewhat similar to *Porteresia coarctata*, commonly known as "shali ghas".

Generally Old Tigers Stray for Easy Prey

It is popularly believed that the aged or injured tigers, who cannot capture any prey in the forest by chasing, stray out into the villages to capture easy prey. The tigers with paw injury cannot hunt any agile prey and have to depend on the easy kills like domestic animals and sometimes human beings. Mallick (2007) recorded that some of the strayed tigers suffered from cataract or had broken canines. But this is not wholly true because it is hardly found in less than 10% cases among the rescued animals. Besides, most of the strayed tigers were young and healthy, showing any visible signs of injury.

Straying Due to Washing Out of Pheromone by Tidal Waves

The researchers have different ideas in this respect. Every day, the Sundarbans witness high tide and low tide twice. It may be that the tigers' territory marking and tracking instincts are frustrated by the muddy mangrove forest, leading to aberrant behaviour. Some authors (Sanyal, 1999) suggest that these tigers are hyper-territorial. Pheromone sprayed by tigers on the tree trunks to mark its territories are washed out by the tidal waves everyday and tigers get confused and stray inside the human habitation. Kalyan Chakrabarti observed that the Sundarban tigers are territorial because they carefully scent-mark only those areas which are not inundated by the high tide. Recent data from the radio-collared tigers reveal that the animals are using specific areas, indicating territoriality.

The Male Tiger Losing Domain to the Aggressive Male Tiger may Stray

The possibility of territorial conflicts among the Sundarban tigers is higher as twice every day there are high and ebb tides causing a change of the habitat by the tigers within the intertidal forest, when a male may intrude upon another's territory resulting in conflict. Occasionally, an old male tiger is driven out by the dominant male and the former takes shelter in the adjacent mangrove plantations near the village. In August 1974, a young male tiger moved into a populated area from STR and killed one woman and a number of livestock. It was captured using immobilising drugs and released in a different area, but less than a week later it was found dead from wounds evidently inflicted by another intolerant tiger (Seidensticker et al., 1976). However, as the female tigers also often stray out to villages, this hypothesis cannot be a major one.

Fog Factor

Samik Gupta has observed that most of the straying, approx 40-45% occurs during the winters, particularly in the month of November to January. During winter, thick fog covers making it difficult to differentiate the land and waters. The tigers have *Tapetum lucidum* in their eyes which help them have a better and clear vision at night than in the morning, but it seems that it does not help much in dense fog. Most probably the tigers cannot detect the right route (forest or village island) and hence enter the villages by mistake.

Adventure

Many young tigers were also seen to stray out. It is argued that some of them may do so for adventure. According to the villagers, it is very difficult to ascertain the age of the straying tigers as in most cases straying takes place during night. Hence, it cannot be confirmed that the sub-adult males straying inside the villages losing domain before aggressive adult males.

Impact of Environmental Change

Sea level and associated changes in the Sundarbans have been identified as the most important threats to the ecosystem (Hazra *et al.*, 2002; Hazra, 2010). The 2005 Tsunami has led to shifting of the Indian plate causing altitudinal variation and salinity in the western Sundarbans. Moreover, the rivers and freshwater sources are shifting towards the west. Hence, the riverine grasslands are also growing more in South 24-Parganas FD than STR. Lack of suitable habitat for the herbivores is decreasing the prey base for the tigers in Sundarbans. In the southern part of Sundarbans, about 20% forest area has been reduced during the last 30 years, leading to shrinkage of the tiger habitat. Between 1969 and 2009, it has lost 210.247 km, of which 65.062 km has been lost in the current decade. As a result, the tigers, living in the southernmost areas, shifted northwards following the prey species. The statistical records show that during the 1980s the tigers could be sighted from the Haldibari watch tower on the southern tip of the tigerland. During the 1990s, the tigers were frequently seen from the Netidhopani tower in the central part of the tigerland. During the 21st century, they are being sighted mostly from the Sudhanyakhali tower in the northern part of tigerland. For increasing density of the tigers in the northern habitat there might be a neck-to-neck competition for the prey within a limited area. Under these circumstances, the tiger often stray out to the nearest reclaimed areas in search of suitable food. This might be a valid reason for the increasing trend of tiger straying in the north-western part.

Use of chained pet dogs to save human beings from man-eating tigers (Khan, 2009)

In 2016, approximately 120 miles east of New Delhi, in Barbatpur village, a dog died a heroic death after saving its owner from a ferocious tiger attack. Gurdev Singh, a farmer, had been sleeping in the open outside his home, when Jacky, the dog, picked up the scent of the approaching predator. Although Jacky managed to alert his owner, the owner took a little while to gather his wits, but the tiger had made its move. However, before the tiger could strike his target, Jacky intervened immediately, which gave Gurdev Singh time to get a stick and raise an alarm. As soon as Singh's family came to his aid, the tiger escaped, dragging Jacky away with him. The four-year-old dog's body was found a while later some distance away and he was given a formal burial by the village. In 2019, in a testament to dogs' unconditional love for their owners, a pet dog in Madhya Pradesh's Seoni rescued his owner from a tiger. The incident took place in Piparwani village where 22-year-old Pancham Gajba and his brother had gone to the nearby forest to relieve themselves. A big cat pounced on Gajba, his pet dog kept barking at the tiger until the wild animal retreated. Even though the canine managed to rescue his owner, Gajba sustained serious injuries on his hand.

Hence, keeping one pet dog (chained) with each group of people (and one big stick with each person) was found very effective in saving humans from man-eating tigers in Bangladesh Sundarbans, because the dogs warned people about the presence of the tiger around them. The dogs were particularly useful for honey gatherers, because when they smoke the honeycomb their visibilities become very poor, and they become very vulnerable to tiger-attack.

The dogs that were used were of local breeds, available in the villages along the edge of the Bangladesh Sundarbans. 40 dogs were selected on the basis of tests. A total of 67 dogs went through the test. Each of these dogs was taken to the Sundarbans in an exploratory visit when the performance of the dog in detecting wild animals was observed. Based on the performance, 40 dogs were finally selected. After the selection, informal training was given to these 40 dogs by forcing them to smell the tiger trails and drawing their attention to any wild animal in the forest. Dry food was given immediately after any successful performance.

During the 18-month study period (August 2005 to January 2007) a total of about 30 days were spent observing each of the 40 dogs in the field. All the 40 dogs could not be sent to the Sundarbans and observed at the same time due to limited manpower, but the observation took place in different seasons in different parts of the Sundarbans (except the three sanctuaries, because people are not normally allowed to work there).

In the field the signals of the dogs were recorded. The signals were classified into two groups:

(i) signal for any wild animal, including the tiger, characterised by sudden excitement, and

(ii) signal (apparently) for the tiger, characterised by fear and low noise.

The signals were verified either immediately, by observing the animals or their signs around, or in the next day (in order to avoid the risk of meeting the tiger face-to-face), by observing the signs (pugmarks in case of tigers).

The responses of the 40 dogs were recorded and verified in the field and was found that the dogs could successfully detect the presence of any sizable wild animal around them (success rate: 92.4 ± 4.8 %), but they could not always distinguish tiger from wild boar *Sus scrofa* or spotted deer *Axis axis* (success rate: 61.6 ± 17.5 %).

ID no. of dog	Success in specifically signalling presence of tiger		
	No. of dog's specific response of fear (for tiger)	No. of times presence of tiger confirmed	Level of success (%)
1	13	9	69
2	19	11	58

ID no. of dog	Success in specifically signalling presence of tiger		
	No. of dog's specific response of fear (for tiger)	No. of times presence of tiger confirmed	Level of success (%)
3	5	4	80
4	10	6	60
5	8	6	75
6	18	12	67
7	14	10	71
8	9	6	67
9	12	8	67
10	17	12	71
11	16	12	75
12	8	5	63
13	15	11	73
14	14	10	71
15	5	2	40
16	18	13	72
17	24	14	58
18	11	4	36
19	27	18	67
20	8	5	63
21	11	9	81
22	2	0	0

ID no. of dog	Success in specifically signalling presence of tiger		
	No. of dog's specific response of fear (for tiger)	No. of times presence of tiger confirmed	Level of success (%)
23	15	7	47
24	12	8	67
25	16	11	69
26	18	13	72
27	6	3	50
28	9	6	67
29	21	14	67
30	5	0	0
31	13	9	69
32	8	5	63
33	14	9	64
34	15	8	53
35	13	8	62
36	11	6	55
37	10	7	70
38	6	3	50
39	5	4	80
40	16	12	75
Mean ± SD	12.4 ± 5.5	8.0 ± 4.0	61.6 ± 17.5

Table: *Effectiveness of pet dogs in signalling people about the presence of tigers and other wild animals in the Bangladesh Sundarbans Community tolerance for tigers*

Inskip *et al.* (2016) indicated that beliefs about tigers and about the perceived current tiger population trend are predictors of tolerance for tigers. Positive beliefs about tigers and a belief that the tiger population is not currently increasing are both associated with greater stated tolerance for the species. Contrary to commonly-held notions, negative experiences with tigers do not directly affect tolerance levels; instead, their effect is mediated by villagers' beliefs about tigers and risk perceptions concerning human-tiger conflict incidents. This finding is consistent with the results of a study in Chitwan NP, Nepal.

Chapter 12
Conservation Status: Global, National and Local Levels

"The Tiger in many countries, particularly India, represents the apex of the animal pyramid and the protection of their habitat should be a worldwide priority." – Dr. Karan Singh

Importance of Tiger Conservation

Tiger conservation within and beyond the PAs is based on the individual components of biodiversity- genes, species, and ecosystems on the one hand and on the other, immediate and durable aesthetic values; but the landscape is vulnerable to natural disasters (for example round one third of the Indian Sundarbans was destroyed due to super-cyclones Amphan in May 2020 and Yaas in May, 2021), absence of alternative livelihoods forcing the people to go to the forests to collect honey, catch fish and crabs and, above all, anthropogenic activities particularly poaching and illegal trade (an active hub for tiger poaching and illegal wildlife trade because of their proximity to China and Myanmar due to the demands of tiger parts in Chinese medicine), as well as development works and pollution violating the existing rules. These are major reasons behind the slow increase in tiger population in this region in comparison to other Indian states like Madhya Pradesh, Karnataka or Uttarakhand. The situation got worse with the current Covid-19 pandemic and its associated economic hardships. But without the tigers, people would go to the forests more often, depleting its resources. It is essential to realise that if there are no tigers, there would be no Sundarbans.

Economic valuation of Sundarbans

The ecosystem values are divided into three types-

1. Direct Use Value: Employment generation, agriculture, fishing, fuel wood, grazing / fodder, timber, non-wood forest produce;
2. Indirect Use Value: Carbon sequestration, water provisioning, water purification, sediment regulation/retention, nutrient cycling/retention, biological control, moderation of extreme events, pollination, nursery function, habitat/refugia, cultural heritage, recreation, spiritual tourism, research, education and nature interpretation, gas regulation, waste assimilation; and
3. Option Value: Gene-pool protection.

Conservation of tiger, an umbrella species, ensures the protection of habitats of several other associated species while ensuring continuity of ecological and evolutionary processes in the wild. The Sundarbans also offers a wide range of ecosystem services having social, cultural and economic benefits to the society. However, these ecosystem service benefits at various scales like local, regional, national and global were not quantified before.

India (STR)

Considering the importance of contribution of TRs not just for biodiversity conservation but also in generating ecosystem services, a pilot study was commissioned by NTCA for economic valuation of select TRs including STR (Verma *et al.,* 2015). The information on forest cover of STR including area under different forest density classes is as shown below.

Land Cover/Vegetation Class	Total Area (km²)
Forest (Mangrove)	1,538
Very Dense Forest	730
Moderately Dense Forest	663
Open Forest	145
Non-Forest (Waterbody)	1,047
Total	2,585

The estimated value of the ecosystem services of STR is worth Rs.12,800 million annually providing flow benefits worth Rs.0.50 lakh per hectare annually. The monetary estimates (in Millions INR/Million USD/Year) for the 15 ecosystem services are specified below.

1. Employment Generation - Through management and Community-based Ecotourism: 36.22/0.06
2. Fishing: 1,600.00/25.10
3. Standing Stock 6,28,700.00/347.86
4. Non-Wood Forest Produce (NWFP) 5.50/0.09
5. Gene–Pool Protection 2,870.00/45.02
6. Carbon Storage 24,100.00/292.63
7. Carbon Sequestration 462.08/7.25
8. Biological Control 101.5/1.59
9. Moderation of cyclonic storms 274.83/4.31
10. Pollination 276.84/4.34
11. Nursery function for various fish, crabs and shrimps 5,170.00/81.10
12. Provision of habitat and refugia for wildlife 359.89/5.65
13. Recreation 37.00/0.58
14. Gas Regulation 110.74/1.74
15. Waste Assimilation 1,500.00/23.53

Source: Centre for Ecological Services Management (CESM), IIFM, Bhopal (2016).

Distribution of Value: Approximately 16% of flow benefits accrue at the local level, 39% at the national level and 44% at the global level.

Local level

The mangrove forests act as a natural shelter belt and protect the hinterland from storms, cyclones, tidal surges, sea water seepage and intrusion.

The mangroves serve as nurseries to shellfish and fin-fishes and sustain the coastal fisheries of the entire eastern coast.

There are cosmic and mundane aspects of marginalised seafaring communities who depend on the delta's backwaters and the mangrove forest for livelihood. Annu Jalais in her book 'Forest of Tigers: People Politics and Environment in the Sundarbans' (2011) puts forth a political narrative that differentiates between the "local" tiger as opposed to the global cosmopolitan tiger. In the local mythology, the reigning deity of the forest is *Banbibi*, sent by *Allah* to restore peace between humans and animals. The sage-turned-tiger, *Dakhhin Rai*, seeks her protection as does *Dukhe*, a village boy abandoned by his family inside the jungle, symbolic of islanders depending on the forests for their survival. In Bonbibi's eyes, both the tiger and the islanders are equal: they need the forests. The mythology dictates that the fishermen, honey and crab collectors must enter the forests with a pure heart and take no more than what is necessary for survival.

The islanders believe they share a "parallel history" with the tiger. They believe in the story that tigers, after being pushed out of Indonesian islands of Java and Bali, migrated to the Sundarbans (although the present DNA analysis has established the close connection between the Central Indian and Sundarbans tigers). This mirrors the reality of islanders and their "history of rejection" by the state. Some of them were landless farmers and adivasis brought by the British to start cultivation and generate revenue. A majority migrated from then East Pakistan following India's independence and partition.

In 1979, the erstwhile communist government of Bengal enacted a massacre in the Marichjhapi islands. After partition, to manage the influx of refugees from East Bengal, the government settled them in the semi-arid Dandakaranya (parts of Odisha, Chhattisgarh, Maharashtra and Andhra Pradesh), a region far removed from where they belonged. After the communists came to power in Bengal, the refugees had made their way back and settled in the Morichjhapi islands of the Sundarbans. When they were forcibly removed by the then Left Front government of West Bengal, an estimated 4,128 families perished due to cholera, starvation, their boats drowned by the police or simply shot to death. "The ease and brutality with which the government wiped out all signs of bustling life of the past 18 months were proof for the islanders that they were considered completely irrelevant to the more influential urban Bengali community, especially when weighed against the tigers," Jalais writes.

In the islanders' understanding, Morichjhapi marks the beginning of tigers becoming "arrogant". The tigers, they would explain, were disturbed by the violence and started attacking people. Many contended that corpses of killed refugees floating in the water gave them a taste for human flesh. Notwithstanding the veracity of such belief, in local perception, Morichjhapi marks the entry of the cosmopolitan tiger armed with state protection. Jalais brings out similarities between seemingly opposite ways of the colonial ruler and the present conservation exercises. In the colonial literature, the Sundarbans tiger was reasoned a man-eater because the superstitious residents were too afraid to stand up to it. The British killed them at the rate of 1,200 per year.

While in the colonial times the British were more concerned about revenue and killed the tiger with impunity, a century later, "the prestige of hunting" was replaced by prestige of "save the tiger". What was always ignored, argues Jalais, was the local perception of the tiger. According to Jalais, neither the colonial nor the conservationists' representation of the tiger permits an engagement with the local ways of understanding tigers. For the islanders, the new cosmopolitan tiger is a high status animal enjoying state protection through investment in tourism and sanctuaries. "It is this tiger, one that is constructed by the state, conservation NGOs and urbanites, that the islanders argued they were trying to fight against, not their old tigers," Jalais writes.

Regional level

The mangrove forests trap debris and silt and stabilise the near shore environment. Certain mangrove species also act as bio-filters as they have been found to bioaccumulate heavy metals. They filter ground-water and storm-water runoff which often contains harmful pesticides. They recharge the ground-water by collecting rain-water and slowly releasing it to the underground reservoir.

National Level

It constitutes over 60% of the total mangrove forest area in the entire country and has 90% of the total Indian mangrove species.

The plant species comprise of true mangroves or major elements, minor elements of mangroves or mangrove associates, back mangrove trees and shrubs, non-halophytic non-mangrove associates in the area, halophytic herbs, shrubs, and weeds and epiphytic and parasitic plants.

It is known as a kingfisher's paradise as out of the 12 species of kingfishers found in the world nine species are found here.

Inaccessibility and absence of human habitation provides a pristine habitat for the biodiversity within the mangrove forests.

Global Level

The Sundarbans has been classified as a Tiger Conservation Landscape of global priority, as it is the only mangrove habitat in India and Bangladesh, which supports a significant tiger population.

The mangrove forests is home to a large number of endangered and globally threatened species like the fishing cat (*Prionailurus viverrinus*), Gangetic (*Platanista gangetica*) and Irrawady Dolphin (*Oracella brevirostris*), King cobra (*Ophiophagus hannah*), water monitor lizard (*Varanus salvator*) etc.

It harbours the population of the Northern river terrapin (*Batagur baska*), in a captive breeding facility in STR, which was once believed to be extinct.

It is the nesting ground for marine turtles like Olive Ridley (*Lepidochelys olivacea*), Green sea turtle (*Chelonia mydas*) and Hawksbill turtle (*Eretmochelys imbricata*).

A number of heronries are formed here during monsoon, which harbour large bird populations, which come and breed here. Also, during the winters it is home for Trans-Himalayan migratory birds. Goliath heron (*Ardea goliath*) is another important bird found in the area.

Two species of horseshoe crabs (which are considered as living fossils as they are thought to be more than 400 million years old), i.e. *Tachypleus gigas* and *Carcinoscorpius rotundicauda* out of the four species found in the world are found here.

Type of Value: Indirect Use 65%, Option value 22% and direct value 13%.

Type of ecosystem services: Regulating 64% and Provisioning 35%.

Advantages

The advantages of the habitat are-

Within the core or buffer there is no (i) human habitation; (ii) livestock grazing pressure; (iii) crop cultivation; and (iv) encroachment.

Disadvantages

(i) Although seasonal collection of NTFP on permission is confined to only part of the buffer area, some illegal fishing and honey collection in STR are recorded.

(ii) Staff in position is partly satisfactory.

Bangladesh (Million USD/Year)
Provisioning Services
Fisheries (fish, crab) 0.2-356
Timber 0.4
Fuelwood 0.06-29
Honey 0.011-19.25
Fodder 9.51
Wax 0.004
Construction materials 0.06
Regulating services
Storm protection 4.59
Erosion 0.75
Supporting services
Nursery and habitat function 3.27
Cultural services
Ecotourism 0.042-5.

Unique habitat

The uniqueness of the mangrove habitat, its biodiversity and a plethora of services (both tangible and intangible) associated with the Sundarbans at local, regional and global levels, have made their protection and management a conservation priority, where the tigers require large and interconnected territories for shelter, prey abundance and breeding. The Sundarbans tiger prefers baen (*Avicennia*) and keora (*Sonneratia*) plants over hental (*Phoenix*) and goran (*Ceriops*) as habitat. The tigers possibly prefer baen and keora habitat in low lands because of higher prey availability and ease of hunting in these habitats. Hental and goran habitats on relatively higher grounds give them shelter during high tides and often act as cub-rearing refuges.

While the tigers represent a keystone species necessary for their ecosystem to survive, as top predator, they balance populations of herbivorous prey species, which then balance plant populations, allowing the mangrove trees to become the dominant vegetation in this flourishing delta. The tigers are also considered a landscape species since they are widely distributed, highly mobile or dispersed in different habitat types and best recovered by managing natural and anthropogenic threats associated with habitat loss or degradation at a landscape scale.

National and Global Heritage

The Bengal tiger is the national animal of both India (declared in April, 1973 and notified on 30th May, 2011 vide no. 25-1/2008-WL-I) and Bangladesh (1972). Tigers are a symbol of marvellous, numinous, predominant and integral part of the mangrove ecosystem. Sundarban is a UNSECO's World Heritage Site in both India (NP: 1,330.10 km^2; 1987) and Bangladesh (1,395 km^2; 1997). Sundarban wetland is also a Ramsar site in both the countries (Bangladesh No.560, 6,017 km^2, 1992; India No.2370; 4,230 km^2, 2019).

Last stronghold vis-à-vis ecological isolation

The critically endangered marshy mangrove ecosystem in the Sundarbans forest is one of the largest natural strongholds of the Bengal Tiger. About 6,000 years ago (in the mid-Holocene), the shoreline of the North-eastern Indian Peninsula was situated to the west of the present shoreline of Northern Bay of Bengal, which was comparatively closer to the Himalayan foothills. Therefore, the present Sundarban area did not exist for terrestrial life until the shoreline had moved eastward, and tigers could arrive there from Peninsular India. The Sundarban tigers are, however, genetically more close to the Central Indian big cat population than the North Indian tigers. Three haplotypes have been detected within the Sundarbans tigers, of which one is unique to this population and the remaining two are shared with tiger populations inhabiting the Central Indian landscape.

During the 16th century, the main flow of the Ganges shifted almost totally eastwards within the river Padma, as a sequel to which two tidal rivers-

Matla and Bidyadhari- got completely cut off from the sweetwater source and since then are being fed only by the seawater, witnessing a sharp increase in salinity. This has resulted in loss of the riverine grassland and extinction of four large herbivorous prey species of the Bengal tiger from the Sundarbans expedited by "random" hunting during last two centuries:

(i) Wild water buffalo (*Bubalus arnee*)(1885)[800-1,200 kg],

(ii) Javan Rhinoceros (*Rhinoceros sondaicus inermis*)(1888)[900 kg],

(iii) One-horned Rhinoceros (*R. unicornis*)[1,600-2,200 kg](recent fossil, estimated to be about 3,000 years old, discovered at Ramchandrapur of South 24-Parganas district), and

(iv) Swamp deer (*Rucervus duvaucelli duvaucelli*)(1930)[170-180 kg].

600 to 800 years ago, anthropogenic activities started to impact the tiger habitat. The known ecological history of northern Sundarbans dates back to the late 17th century, when Kolkata (erstwhile Calcutta) was the northern boundary of Sundarbans. It was reported that Warren Hastings had killed a Royal Bengal tiger on the spot where the St. Paul Cathedral stands today (Cotton, 1907).

Now the Sundarban tiger inhabits a unique mangrove habitat type in isolation from the neighbouring tiger populations [e.g. Similipal (21°50'N 86°20'E)] by hundreds of kilometres of agricultural and urban land. All these factors probably caused the complete isolation of Sundarban tigers. Under the circumstances, the tiger population in Sundarban mangroves is an isolated, genetically and ecologically differentiated group of the Bengal Tiger. Therefore, the genetically unique Sundarban tigers need comprehensive conservation in order to preserve their morphological and behavioural distinction in the long run.

Whereas the entire landscape is connected by a meshwork of innumerable water bodies and tigers can easily cross between India and Bangladesh, ecologically the tiger population is not a separate one; they live in a trans-national ecosystem. An approach with closer cooperation on sharing the forest resources, tiger habitat and upkeep of freshwater supply in the area is important. The population could still survive in the region if conservation of tigers and the local ecosystem is made a priority by (a) designating larger PAs for tiger conservation, (b) making more corridors for tigers to move around safely, and (c) remaining vigilant about poaching and hunting in the area to avoid any further decline in their population.

Eco-Tourism

Ecotourism is a tool in order to sensitise and make the common mass realise the importance and significance of the landscape and the need to conserve its biodiversity including the tiger. Tigers are a major tourist attraction in the Sundarbans and help generate revenue through ecotourism. This revenue supports the local communities and contributes to the economy. A very good infrastructure exists in order to promote ecotourism in SBR- at Sajnekhali, Sudhanyakhali, Dobanki, Netidhopani, Burirdabri, Bonnie Camp and Kalas. But tourism has proven detrimental to the natural environment due to habitat destruction for hotel construction, air and water pollution caused by garbage disposal, oil spill, poor sanitation, and noise caused by mechanised boats.

Tigers in Culture

Tiger is deeply embedded in the rich cultural heritage of the Sundarbans, a source of local myths, legends, worshipping among the communities living next to the Sundarbans.

Conservation challenges

The Bengal tiger is a conservation-dependent species that needs good quality habitat with sufficient prey-base and undisturbed breeding grounds. Tigers today face incredible challenges as their numbers shrink in the wild due to poaching, encroaching human population, and loss of habitat as well as traditional prey. Saving the Bengal tigers is critical for the Sundarbans and without tigers, the entire ecosystem would collapse. Therefore, habitat management using scientific-based ecological data and protection of tigers and their prey from poaching are the prime prerequisites of tiger conservation in the Sundarbans.

Prioritisation of Tiger Conservation Landscape: Decadal Promotion from Class III (Long-term Priority) to Class I (Global Priority) during 21st century

Tigers are a treasure of the Earth and important players in the ecosystems where they are found. The recent tiger conservation strategy emphasises the need to expand conservation efforts to include more than one meta-population, leading to identification of "Tiger Conservation Landscapes" (TCL). These include a number of PAs interconnected by corridors that could potentially support viable populations. A TCL is a block or cluster of blocks of habitat meeting a minimum size threshold specific to habitat-type, where the tigers-

(a) Have been confirmed to occur in the last 10 years;

(b) Have a minimum core area in the largest block of habitat that is specific to the habitat-type in which it is found; and

(c) Need not be restricted to PAs, but have corridors to disperse and become established in the entire landscape.

The tiger corridors are categorised into five classes: very low, low, medium, high and very high, indicating the importance of the pathways of the corridor. The very high values indicate good connectivity whereas 'very low' values indicate low connectivity.

TCLs are identified according to their Class and Priority. Class (I-IV) refers to a TCL's current conservation status and probability for achieving conservation success in the next decade. The scientifically estimated populations in different classes of TCL are -

   i. Class I: ±100 with evidence of breeding;

  ii. Class II: ±50;

  iii. Class III: Some tigers (indeterminate) and

  iv. Class IV: Insufficient information on three or more conditions.

Priorities represent relative measures of quality and incorporate explicit representation of the major habitat types. The Sundarban mangrove forest is considered a TCL of Global Priority because it is the only representation of a mangrove tiger habitat. It is also the single largest continuous habitat of the tiger anywhere in Asia, but spotting the tiger here is harder than in any other forest because of tough and difficult terrain as well as absence of adequate infrastructure. Sundarban TCL fulfils all the above three (a-c) criteria and belongs to the highest category for the breeding population.

While setting priorities for the global tiger conservation (2005-2015), the Sundarban mangrove landscape was provisionally considered a Class III (low priority) TCL with insufficient habitat, high threat level, low tiger population, too diminished prey bases to recover within a decade, and lack of commitment to tiger conservation by local people and government in the TCL. The Class I TCL is determined on the basis of (a) a viable breeding population, (b) a sufficient prey base, (c) sufficient area (scaled by habitat type), (d) some successful conservation activities, and (e) minimum-moderate threats. Though higher status is considered important, that would likely take a sustained effort of more than 10 years for rebuilding habitat and connectivity to the required state of improvement. After a decade of successful conservation efforts, the Sundarban mangrove landscape was 'promoted' to the Class I (highest) TCL of global priority to represent mangrove habitat for long-term conservation in the region.

On an absolute scale of the current quality of tiger conservation, the Sundarban mangrove is a successful landscape to ensure representation of 'mangrove habitat'. Even though Sundarban is one of the main habitats for the Bengal tigers, their density on both the Indian and Bangladesh side is relatively low compared to most of the mainland landscapes. Both India and Bangladesh face similar problems of poaching of tiger and its prey species. Apart from this, there are also high frequencies of human-tiger conflicts in the Sundarbans primarily due to the fact that thousands of humans regularly venture into the tiger's domains for reasons such as extraction of various forest resources. Additionally, in recent years, rise in sea levels due to global warming has also been identified as a major threat to this landscape and the tigers living therein.

Conservation actions

The Bengal tiger is most important for the sustainability of the biodiversity in the mangrove forest. This mangrove landscape was first brought under scientific management as early as 1892 and the first working plan came into force in 1893-1894 for commercial exploitation. Although the tiger is a conservation-dependent species, no species-specific action has been taken during the colonial period. On the contrary, during the 19th century, tigers were treated as a pest and their hunting was felt necessary to reclaim the Sundarban for revenue generation by the Britishers. Special measures were taken up in 1852 for destruction of tigers in 24-Parganas district and the reward was raised from Rs. 5, which afforded very little profit, to Rs. 10. Government sanctioned the increase in 1853 and doubled the reward in 1860 for every tiger killed near the Matla in order to aid the work of reclamation there, which was again increased to Rs. 200 in 1909. This led to a large-scale slaughter of tigers. More than 2,400 adult tigers were killed between 1881 and 1912. The reclamation of mangrove forests continued. By 1947, the forests in the Sundarbans were only 50% of their pre-colonial size.

The Bengal tiger has been listed in CITES Appendix I and in Bangladesh it is critically endangered and also enlisted in the Wildlife (Conservation and security) Act, 2012 in Schedule I; so also in India listed as Schedule I species in the Wildlife (Protection) Act, 1972.

To preserve the Bengal tiger population in the region, the establishment of STR stands out as a crucial initiative in 1973 followed by creation of six tiger-bearing PAs in the region. Thereafter, a series of preventive measures have been taken against the man-tiger conflict in the Indian Sundarbans. These are enlisted below.

(a) Digging of freshwater ponds began in 1975.

(b) Issuance of permit for collection of *Phoenix* and *Nypa* was discontinued in 1980.

(c) Electrified dummies and human face masks made of rubber were introduced in 1983 and 1987 respectively (Rishi, 1988). When stalking its intended victim, the tiger takes care to choose the moment of its attack when its quarry is off-guard. Case studies of tiger attacks on human beings in Sundarbans indicated that almost all the attacks were made from behind, the unguarded side of the victims. An alert backward stare may, therefore, reduce the chances of attack by denying the tiger an opportunity to catch its human quarry off-guard. In nature, many organisms use false eye-spots for protection against their enemies. Human face-masks worn on the backside of the head create an illusion of watchfulness on the part of the wearer. It was observed that when stalking its intended victim, the tiger takes care to choose the moment of its attack when its quarry is off-guard. 2500 human face masks made of lightweight rubberised plastic material were distributed among the people permitted to work in the buffer zone of STR during a period of one year from November 1986 to October 1987. The yearlong experiment was divided into three phases, designed to cover one complete annual cycle of human activity and exposure to conditions under which tiger attacks take place. Phase I covered the activities of fishing and wood cutting in timber coupes between November 1986 and March 1987; Phase II covered mostly the activity of honey collection during the hot months of April and May 1987; and phase III covered the activity of fishing between the months of June 1987 and October 1987. Whereas Phase I was the introductory phase of the experiment, Phase II subjected it to the severest possible test for its efficacy as a protective device. Phase III was designed to find out the impact of the experiment and its acceptability to the local people. During the first phase masks were distributed only to the people who came forward for voluntary participation in the experiment, while in Phase II the use of masks was made compulsory for the honey collectors. The demand for masks was treated as an indicator of the interest of the people in the mask as a useful device in protection of human lives and therefore an indicator of the impact of the experiment on the local people. During Phase III masks were distributed only on demand.

During Phase I, out of a total of 4,943 people permitted to work in the buffer zone of TR, 877 workers- 410 coupe workers and 467 fishermen- volunteered to use the masks after the purpose and the method of use of the masks were explained to them. Among the volunteers, one coupe worker and two fishermen lost their lives due to tiger attacks. None of the persons attacked were wearing masks at the time of attack. The coupe worker and the fishermen had been using the masks when working in the coupe or catching fish respectively, i.e. pursuing the activities for which permits had been taken. The coupe worker was attacked when during a lunch break he had taken off his mask and gone to catch fish for lunch 2 km away in a nearby creek. The fishermen were attacked when they had taken a break from fishing and, leaving their masks in their boats, stepped into the nearby bank to collect firewood for cooking their meals. None of the mask wearers were attacked by the tigers, and the results obtained during Phase I encouraged the continuation of the experiment into the next phase.

During phase II 548 honey collectors and 373 fishermen worked in the buffer zone of STR. In all 882 masks were issued to the honey collectors who often needed replacements for masks due to their strings getting entangled in the undergrowth as they crawled through the mangrove vegetation in search of honey combs. The fishermen took 137 masks. Seven honey collectors and four fishermen lost their lives due to tiger attacks. None of the victims were wearing masks at the time of attacks. During the first week of April 1987 two honey collectors lost their lives. Stricter enforcement of the use of masks by the honey collectors while working in the forest resulted in total stoppage of tiger attacks during the rest of the honey collection season; during the entire month of May 1987 no honey collector was attacked by the tigers. All the fishermen who lost their lives during Phase II were attacked when collecting firewood in the forest; none of them wore masks at the time of entering the forest.

During Phase III, 604 fishermen received masks on demand; all of them had also used masks on earlier occasions. There were no casualties among this section of fishermen, whereas 11 fishermen from among those who had not taken masks died between June 1987 and October 1987 due to tiger attacks.

An unusual phenomenon was observed during Phase II of the experiment. The permits for honey collection in the TR are issued from Bagna and Sajnekhali stations of Project Tiger only. Unless there is an accident, the permit holders do not surrender their permits before the honey collection season is officially closed down because their earnings depend on the quantum of honey collected by them. The permit holders from both the stations work in the same areas. The permit holders from villages around Bagna are more orthodox and were found to be reluctant to use the masks for the fear of incurring the wrath of their traditional protective forest deities. On the other hand the permit holders from Sajnekhali, who often get influenced by the radical ideas of visitors to Sajnekhali Tourism Complex and forest stations, needed little persuasion to use the masks in the forest. Consequently, not a single life was lost among the permit holders from Sajnekhali. The permit holders from Bagna suffered all subsequent tiger attacks on honey collectors during this phase. The permit holders from

Sajnekhali complained of frequent sightings of tigers and reported being followed by tigers for periods ranging from half an hour to eight hours. Although they were not attacked by the tigers, the fear created by the sudden appearance of the tigers presently compelled them to surrender their permits when the honey collection season was at its peak. By 11th May, 1987 all but one of the permits were surrendered by the honey collectors from Sajnekhali; the last one was surrendered on 15 May. In spite of the heavy casualties suffered by the honey collectors from Bagna, no one complained of similar encounters with tigers, nor did they surrender their permits till the end of the season.

During informal discussions with the honey collectors, it was revealed that the honey collectors from Sajnekhali were invariably alerted by the tiger itself as it disturbed the undergrowth behind them while following them. None of the Bagna people accompanying the ill-fated tiger victims knew of the presence of the tiger till their companions were attacked. The honey collectors from Sajnekhali also stated that after following them for some time, the tigers invariably came out in the open , sometimes compelling them to return to their boats, and went away after casting a baleful glance at them. No such incident was recalled by the Bagna people. None of the Bagna people held any faith in the utility of mask as a protective device, while a section of people from Sajnekhali felt that masks may be useful in the forest, but the majority of them expressed that the Sundarban Tiger is too clever to be deceived by the masks for long. However, the preliminary results obtained during the yearlong experiment indicate the possibility of developing the mask as a protective device against tiger attacks on human beings in Sundarbans. Phase II of the experiment has given important indications that call for replications of the experiment till the hurdles created in pursuit of the experiment because of the traditional beliefs are overcome, and the uninhibited participation by the local people yields concrete results. Ultimately, this practice was discontinued in 1990.

(d) Following the killing of an armed forest guard by a tiger in the Chamta block of STR in 1981, a special head-and-nape guard of fibre-glass has been designed and introduced with reintroduction of dummies and face masks in 1994 to protect people moving in the forests. Tiger guards were given to the field staff. Interestingly, the head and nape are the targets of the tiger's attack. As the outfit is uncomfortable in summer it is yet to become acceptable to everyone. This and other measures brought down the annual toll of human lives in the TR to 29 in 1982 against the average of 45 during the period 1975 to 1982.

(e) In 1983 a new experiment was launched. In the experiment a life size clay-model of a honey-collector, or fisherman or wood-cutter is placed in a boat near the river-bank or in the forests in an area where a man-eater has been active. These models were set up at Netidhopani, Pirkhali, Panchamukhani and Jhilla forest blocks, which were wrapped with energisers, charged to 230 volts by a 12 volt battery source. Electricity is passed through a galvanised wire round the neck of the model from an energiser connected to a 12-volt car battery. A fuse recorder is put in the circuit. The problem lies in deceiving the tiger into attacking the model. Clothes recently worn by a honey collector, fisherman or wood-cutter are used on the model as an olfactory lure. In June 1983, for the first time, a man-eater suffered an electric shock when it jumped on a fisherman's model in a country boat at Sudhanyakhali. The next attack was on a wood cutter's model at Netidhopani in February, 1984 by a man-eater who had accounted for 24 deaths. When a team of forest officials visited the site the following morning, the man-eater called out in pain and scampered into the dense forest. It is claimed that the electric shock does not seriously hurt the 'shocked' man-eater. The case of the shocked 'Kali Char man-eater' is fascinating. The man-eater residing at Kali Char at the confluence of rivers Pichakhali and Dattar Chhera had killed over 60 men. Every year the first tiger-kill was reported from Kali Char. In early January, 1985 the man-eater attacked again. Fortunately, he picked a clay-model, not a living man. As the model was not energised, it was shattered to bits. It is strange that the man-killing at Kali Char stopped after the incident. Perhaps the tiger didn't like the taste of 'human flesh' and turned to normal prey thereafter. During that period nine man-eaters have been 'shocked' in STR.

Compensation

An *ex gratia* compensation of ₹1 to ₹2 lakhs is paid by the government for every family, a member of which entered the forest in the buffer zone with a valid permit and attacked by a tiger; but for death in the prohibited core area no compensation is paid. In 2016, it appeared that out of 100 fishermen killed due to tiger attacks in the past six years, only five families applied for compensation, of which only three actually received it. The COVID-19 pandemic and disruptions of other livelihood options led the desperate fringe people to enter the core area, often illegally. As other avenues of income dwindled between 2020 and 2022, over 30 people were reportedly attacked by tigers and most of them were located in the core areas.

Now, the authorities have implemented a range of measures, including anti-poaching patrols, Tiger Response Teams, satellite installation and monitoring of tigers to know their roaming areas, resolve the human-tiger conflict by constructing nylon fencing, habitat management, and community engagement, aimed at safeguarding the tiger population. Some of the important actions are detailed below.

Land-based and floating protection camps

Protection is the prime focus as far as proper forest management is concerned. It is not easy to patrol the forests within these islands as regular high and low tides keep the islands slushy. This makes walking through the islands difficult. Furthermore, there is always a threat from tigers hiding in the bushes. In STR, there are 21 land-based anti-poaching and five floating camps, where all members of the camp stay in a boat and use another boat for patrolling through narrow creeks within the islands. With a limited number of staff and patrolling boats it is not possible to cover all the areas at regular intervals. In Bangladesh Sundarbans too, there are a total of 17 stations and 61 running patrolling camps across the four ranges-Sarankhola, Chandpai, Khulna and Satkhira. As the patrolling camps are run by two to three staff each, it is becoming tougher for the FD to guard the entire area regularly and protect forest resources. For acute manpower shortage, the rackets of smugglers and poachers are easily carrying out their crimes. As the forest guards play the key role to protect the Sundarbans' resources, the manpower must be increased to make it sufficient.

Use of specially designed application for patrolling

To overcome the limitations of old practices of using pen and pencil for prolonged and time-bound paper-works, different field camps are involved in electronic data collection related to the following on a day-to-day basis:

(i) Tiger sighting and direct and indirect evidence.
(ii) Other wildlife sightings and indirect evidence.
(iii) Protection related data.
(iv) In 2015, E-patrol/Smart patrolling was introduced in STR and SRF. In this new system, every camp has been given a cell phone having an android operating system with a compatible mobile application installed in it for monitoring and patrolling purposes. With the help of this application, the frontline staff are recording every possible activity like patrolling, monitoring the condition of fences, night patrolling, offence detections, and wildlife sightings.

Use of Unmanned Aerial Vehicle (UAV)

UAV or drone outfitted with a camera has been introduced as a part of the Smart Patrolling initiative in the Sundarbans. With the help of drones or unmanned aerial vehicle (UAV) outfitted with GPS device and hi-resolution cameras, sitting at one point, scanning a forest area of 15-20 km^2 is done along with tracking the movements of animals or entry of poachers, timber mafias or illegal fishers with limited staff. These have proven a remarkably useful tool in patrolling those areas in STR which are otherwise inaccessible. This tool has also proven useful in case of locating a stray animal in a locality, especially a tiger. UAV are also being used to monitor a post-released animal in the wild up to considerable distance inside the forest.

Radio Telemetry

The biology of the Sundarban tiger, its ranging or distribution patterns, presence or absence of territoriality have for long intrigued the field managers and wildlife biologists alike. The dense vegetation-cover and limited access to the forest area on foot or otherwise necessitates the simultaneous tracking using a Very High Frequency (VHF) antenna as well. A study of Radio- Collaring is carried out with scientific and technical inputs by WII, Dehradun. Their research personnel and staff are used for monitoring the radio-collared animals. These have led to valuable insights into the animal behaviour, especially the movement patterns.

Tigers are radio-collared in STR to assess human-tiger interactions. The radio-collaring exercise was carried out by a team of biologists, veterinary experts, FD officers, and field staff. The wild female or male tiger, fitted with a satellite collar, is closely monitored. The radio-collared tigers provide crucial information on spatial ecology, tiger behaviour, their habitat use within the SBR and the extent of negative human-tiger interaction around human settlements. The details of radio-collaring are given below.

SN	Date	sex	Trapped at	Released at	Collar	Terri-tory	Remarks
1	05.12.07	♀	Pancha- mukhani- 3	Pancha Mukhani- 3	GPS	35-42 km²	Functional four months from April, 08
2	24.02.10	♀	Pirkhali- 2 to Sonagaon	Netidhopani- 2, 65 km away	Satellite	80 km linear distance	Functional 1.5 months from April, 09
3	28.02.10	♀	Pirkhali- 5	Pirkhali- 7	Satellite	5 km²	Functional 11 days 1-11.3.10
4	20.03.10	♂	Netidhopani-1	Pirkhali- 7	Satellite	30 km²	Functional 21.3.10-6.4.10
5	22.05.10	♂	Jhilla- 3 to Kalidaspur	Khatuajhuri-1	Satellite	70 km linear distance	Functional 2.5 months at Talpatti, Bangladesh
6	22.05.10	♂	Netidhopani-1	Netidhopani-1	Satellite	30 km²	1. 4 months till 2.10.10
7	30.09.10	♀	Netidhopani-1	Netidhopani-1	GPS	Same tiger at 6 above	2 months till 15.12.10

SN	Date	sex	Trapped at	Released at	Collar	Terri-tory	Remarks
8	15.08.14	♀	Pirkhali-1	Netidhopani-1	GPS+ Satellite	100 km linear distance	-
9	29.01.16	♀	Bali Khal; Tridibnagar	Chora Mayadwip Khal, Gosaba-3	GPS+ Satellite	-	-
10	25.01.17	♀	Kishorimohan-pur	Ajmalmari near Bonniecamp	GPS-Satellite	-	-

The tigers often prowl on either side of the Raimangal River. A male tiger, radio-collared in end-December 2020, travelled from STR all the way to the Bangladesh part of the mangroves, a journey of about 100 km in over four months. In the course of its long journey, the big cat even crossed a few rivers, some of them wider than a kilometre. It was captured from Haribhanga forest, just opposite the Harikhali camp under Basirhat range, and later released with the satellite collar on December 27. After initial movements for a few days on the Indian side, it started venturing into the Talpatti island in Bangladesh Sundarbans and crossed rivers such as Choto Harikhali, Boro Harikhali and even the Raimangal. In over four months - from December 27 to May 11, when the radio collar stopped giving signals - the tiger moved across three islands: Haribhanga and Khatuajhuri in the Indian Sundarbans, and Talpatti island in the Bangladesh Sundarbans, and mostly stayed in the Bangladesh Sundarbans, and did not even come close to human habitats. The last recorded location of the tiger on May 11 was at Talpatti island in Bangladesh.

In January 2017, a tigress was radio-collared and released in the South 24-Parganas FD, the buffer area of the forest. This tiger, too, travelled a linear distance of over 100 km in four months to reach the tip of the Bay of Bengal before finally settling in its territory. Before that, five other tigers - one of which had also ventured into Bangladesh's Talpatti island - were also radio-collared in the Indian Sundarbans. The gadget also had a mortality sensor, which gives signals in case of the tiger's death. But neither that signal nor any static signal from the collar was received, which points that the tiger is safe. In all probability, the collar has slipped off its neck. In the Sundarbans, salinity in the water can also damage radio collars. This tiger was not clicked by camera traps. So, it's possible that it had come from the Bangladesh Sundarbans when it was captured in the Indian Sundarbans for collaring.

Tigers from Bangladesh usually show up on the Indian side during the mating season, which falls between November and January. These are usually lonely males. It is anticipated that they are wandering to find female mates. Other reasons for the straying of tigers from Bangladesh might be the fights taken place among male tigers over a female they fancy. Generally, the male tigers that lose the battle of possession have to leave the area after the fight is over. A tiger that has been weakened by a fight is not capable of hunting in the dense forest, which requires physical strength, so it turns to the villages to find easy prey. In such efforts, the stray big cats also tend to enter the fringe villages. In one such instance, a male tiger entered Mathurakhand village near the forest compartments of Pirkhali II and III on the fringes of STR area late on a January night and killed a couple of cows and buffalos. Since January 2022, the STR authorities have recorded four such instances of straying.

A total of six tigers were captured and radio-collared. One female was radio collared in 2007 and the other five in 2010. Of these, four tigers were residents and captured inside the forest whereas two tigers were the strayed tigers and captured from the villages of Sonagaon near Bidya Range office on 22 February, 2010 and Kalidaspur of Basirhat Range. Out of these radio-collared tigers three were male and three female. Among the two strayed tigers radio-collared one was female and the other was male. In general, the resident tigers captured and released at the same place where their movements were normal and within a definite area in comparison to the strayed tigers. The radio-collared Sonagaon tigress showed a very extensive movement territory before it lost the radio-collar after about two months. The male strayed tiger, which was captured in Kalidaspur village and released at Khatuajhuri forests, moved to Talpatti forest of Bangladesh after a few days. It moved considerably but remained within this island indicating signs of territoriality. It also confirmed the trans-border movement.

Mallick (2010) described the outcome of first radio collaring efforts in STR. If a tiger remained at a particular location within a periphery of only 100 m for more than12 hours during the daytime, it was considered that it must have hunted. Similarly, if it is localised for more than 8 hours at night, the same conclusion is arrived at. On 4 December 2007, a healthy tigress was trapped in Panchamukhani with a goat-bait. It was then tranquilised and radio collared on 5 December and released the next day at Choragajikhali and was being tracked, when she was found resting, moving intermittently and, till the evening, covering a distance of about 2.5 km from the place of release. It seemed that she saved energy by lazing for over 15 to 16 hours a day. During 7-18 December, some signs of her local movement were detected during periodical monitoring, but she was mostly resting. Continuous monitoring began on 20 December afternoon when she was located at Pirkhali forest near the Deulbharani khal, opposite to the River Gosaba. She stayed at almost the same location for the last few days and it was anticipated that she might have hunted her prey. The physiology (internal chemistry) of the tiger is adapted to a regimen of enormous meals, at intervals of four to five days or even longer. The following day (21 December), she moved towards Panchamukhani by crossing the said *khal*. But ultimately she came back to her earlier location at Deulbharani-Gosaba junction. On 22 December, till afternoon, she was resting. At 17:00 h she moved again towards Deulbharani khal. Next day, she did not change her position. Only on 24 December (19:00 h) the signal indicated her quick movement. At about 23:00 h she went on hunting because repeated alarm calls of the deer were heard and the sound of splashing water indicated that the predator was chasing the prey. It was already high tide then. The prey could, however, escape. After ten minutes she again resumed chasing the target in a zigzag way. Suddenly a loud shriek followed by a thud confirmed seizure of the prey by the tigress. She dragged the kill through water towards the dry land. On 26 December early hours she moved again, but the continuous monitoring was called off. A month's data revealed that she moved about 30 km^2.

As pointed out by Barlow (2009), in the Bangladesh Sundarbans, a female tiger, occupying a home-range of 10-14 km^2, moves on average 1.65-1.72 km/day with a maximum of 10-11.3 km/day. Distances moved include traversing both terrestrial habitat and waterways; a SRF female tiger crosses a *khal* 10-15 times/month (equivalent to one crossing every two to three days. This frequent rate of waterway crossing suggests that a monitoring survey, based on track counts along creek banks, would have a high chance of detecting tiger presence.

Nylon-Net Fencing

Nylon net fences have been raised in all the tiger straying sensitive areas of Indian Sundarban. Tigers used to negotiate these fences primarily by three means viz. (i) by jumping over the fence, (ii) through creeks and (iii) through poorly maintained nylon net fences. Therefore, it has been decided to raise the height of the nylon net fence from 6 feet to 7 feet to prevent jumping over cases. Another fact which has come to light is that the fencing post of Baen (*Avicennia* sp.) has rotted at the base resulting in drooping of the fence, making it easier for the tigers to jump over. To overcome this problem, a decision has been taken to replace the Baen posts with bamboo or with Garan (*Ceriops* sp.). The advantage of using bamboo is it helps in habitat protection and mangrove conservation. A new design of the nylon net fence has been implemented to prevent tiger straying from creeks. This new design with "Floaters" has been successful at Jharkhali. The creek part of the fence has two strands- one fixed and another with floaters, which moves up and down with the tides, ensuring that the creeks are fully fenced even during the high tides.

A stretch of 96 km was covered in STR- 21 km in NP (West), 35 km in Sajnekhali WLS and 40 km in Basirhat ranges. Now this fencing has been extended by 10 km. The 24-Parganas (South) FD currently manages a 65 km area where a nylon net fence has been erected. It is found to play an important role in preventing the straying out of tigers into village's areas from forest. However due to tidal rhythm and creeks, the maintenance of nylon net fences is difficult. Tiger, an intelligent predator, is also adapting fast to overcome the barriers. Cases of chewing of nylon-net fences by tigers to create big holes and swimming over the fences during high tides in creeks have been recorded.

A protocol for maintenance of the nylon-net fencing has been designed with an aim of carrying out thorough checking and proper maintenance. The Protocol includes involvement of local Stakeholders in FPC/EDC members also along with forest staff. In addition, the Primary Response Team (PRT) members are also voluntarily involved with the local FD staff for fence-patrolling and maintenance on a regular basis.

Following success in the Indian counterpart, now Bangladesh plans to fence 60 km of Sundarbans as human-tiger conflicts surge during 2023-2024.

Provision for sweet water

Tigers need small amounts of fresh water to drink. In the absence of standing water, some tigers can meet their need from the meat of their prey; they must kill frequently, even if they consume only a small portion of meat from each of their victims. In other words, they may be killing frequently to satisfy their thirst as well as their hunger. If 'killing to drink' does contribute to man-eating in the Sundarbans, it would be reasonable to expect all its tigers to be eating people, but they obviously are not. If they were, human deaths would run into many thousands a year.

For the wild animals 43 sweet water ponds have been dug in STR near the camps and many other areas away from the camp. The size of the ponds was normally 20 - 30 m x 40 - 60 m in size and about 4 to 6 m in depth. The excavated soil from the pond was used to make a strong *bundh* around the pond. The bund was many times planted with mangroves plants for stabilisation. These *bundhs* did not usually allow the salt water entry into the pond during high tide. The pond construction was done mainly after the rains.

It is evident from the field that there is no specific trend for regular visits of a tiger to a particular pond. This is also clear from the pug-mark identification. It appears that tigers do not restrict themselves to a small territory and they also do not depend on the sweet water pond for their regular requirement of water. The tiger, which is visiting a pond for some days, may stop visiting the pond and some other tiger may visit the pond. At times it is noticed that there are long gaps between visits of tigers at the sweet water ponds. So, the water hole method for estimation of tiger population and study of tiger behaviour cannot be undertaken in Sundarbans. It is also found that the number of tigers visiting a pond varies from year to year. The sweet water ponds are frequently visited by deer but not by wild boars. Observations reveal that several tigers have visited the same area but on different days or different times depending on their needs.

The rain water which gets accumulated after the first rains is normally pumped out as this water tends to get saline. From the second year the collected rainwater can be used by the animals for drinking purposes. These ponds tend to get silted up after five to seven years of their construction, so they are to be again desilted/re-dug. The bunds also get damaged because of the holes created inside by crabs etc. and this in turn allows entry of tidal water into the ponds. During the study period it was found that the ponds are not frequented by animals including tigers during the rainy period. The tigers only visited the ponds during the months of December, January and February. It is found that the number of tigers visiting per year per pond varies from 0 to 6 in numbers. The tiger uses the sweet rain water which is collected in small puddles during the rainy season. As rains are distributed beyond the rainy season in the Sundarbans so the tigers fulfil their need of sweet water from the collected rain water in puddles etc.

However, it appeared that the poachers used these remote areas for hunting. So, to prevent this, a conscious decision was made by the management to close the water holes which were away from the camps. Fruit trees like Keora (*Sonneratia apetala*) and *Zizyphus* sp. which are preferred by the wild animals especially deer and macaques can be planted near the sweet water ponds.

The Chital deer survives due to excess salt excretion in the form of lachrymal secretion and the tiger has adapted to drinking waters of the saline creeks, yet the sweet water is an attraction and preferred to the saline water as evident from the pugmarks found near the water holes dug artificially all over the forests (Chaudhuri and Chakrabarti, 1980; Chaudhuri and Choudhury, 1994). When the area is swept over by severe storms along with high current, the wild animals have to cling hard to the forest floor for survival.

Monitoring of tigers

Monitoring of tigers, which is popularly known as the "Tiger Census" is in fact an exercise for the estimation of tiger population. The data is collected by moving in boats in creeks through islands as well as in water along the shore. Camera traps are also laid in 2 km^2 grids by moving in rivers and creeks. No camera is laid in grids in the interior of islands as it is unsafe for staff to move on foot. Based on monitoring results the necessary changes are made in the management practices to make them more effective. Of late, Bengal tigers have been seen more often in the Sundarbans, which means that the number of tigers in the forest has increased.

Community-based Voluntary Village Tiger Response Teams (VTRTs)

An initiative was taken to build a VTR team, who will instantly respond to any kind of human-tiger conflict incidents inside the village and keep the role as primary manager. The first team was established in 2007 in the Chandpai range of Bangladesh Sundarbans. At that time only two such teams were formed in that area because of the high frequency of tiger straying incidents. Those two teams immediately got accepted by the local people by showing their capability of handling conflict incidents in their area. The success story of the two groups was used to motivate people in other villages to establish other such groups. By 2010, 29 teams were established in the villages adjacent to Sundarbans area, and by the end of 2012 a total of 49 teams were established covering 80% of the border villages in four ranges of Sundarbans. Currently, VTRTs have 340 members including 20 women.

From 2007 to 2018, VTRTs successfully facilitated to rescue and set free three tigers, and more than 349 other wild animals (deer, wild boar, fishing cat, python, crocodiles, turtle, otter, bird, monkey, Bengal monitor, Water monitor, wild fox etc.). They effectively managed 30 tiger straying incidents and sent them back to the forest, conducted 140 patrolling inside villages, and recovered 27 dead bodies of tiger victims. They also provided emergency first-aid to seven people injured for tiger attacks, and conducted a total of 2,949 village meetings to raise awareness for conflict management. Apart from that, VTRTs helped the FD and firefighters to manage 12 fire incidents in the Bangladesh Sundarbans. India's Rapid Response Team (RRT) assisted the FD to rescue an adult male tiger strayed into Kultali Village near Garankanthi forests of western Sundarbans on 23rd December, 2021. The tiger was tranquilised and translocated to the Bonnie Forest Camp using the RRT's boat and after a day's monitoring, it was finally released in Dhulibhasani block of Ramganga Range.

Protection of aquatic corridors

The Sundarbans mangrove forest has been isolated for centuries from the nearest tiger occupied habitats due to absence of any wildlife corridor. With the Bay of Bengal on one side, the Sundarbans is criss-crossed by innumerable water channels of various widths and surrounded by human settlement and agriculture on its three sides. The mangrove tigers are often found swimming across 50-70 m wide water channels. In the west, the river Matla acts as an interdivisional corridor and on the eastern side, the river Raimangal acts as an international corridor as far as tiger movements are concerned. A male tiger was seen to cross the Gaji River at noon from Pirkhali 6 to 5 in STR on 3rd December, 2016. It transpires that the radio-collared tigers in general avoid crossing channels wider than 600 m. Tigers were found more active with a bimodal peak during morning (06:00 hrs-09:00 hrs) and late evening (18:00 hrs-20:00 hrs). Activity is less in midday hours (11:00 hrs- 16:00 hrs). The peak of tiger activity was observed at the same period as activity of spotted deer and wild pig peaks during the first morning hour. This is along expected lines as they are the primary prey of tigers in the Sundarbans.

The constant movement of boats acts as a potential barrier to the isolated tiger's dispersal between the islands within Sundarban. The cargo movement through the mangrove forests of Sundarbans has been stopped on the Indian side since 2011. Now, the ships and barges take alternate water routes, avoiding the Indian Sundarbans, to reach Bangladesh. But such a movement continues in the Bangladesh part.

Though rare, there are occasional reports of tigers swimming across the big rivers. Their movement corridors between islands have narrowed. Every day, hundreds of vessels ply across the multiple water channels of the mangrove and disturb their shifting. These minimised crossings of wide water channels need to be identified so that the various islands harbouring tigers remain genetically connected. Corridors are used by the prey species and other wildlife too. Without corridors the tiger habitat will remain fragmented and the isolated tiger populations become vulnerable to localised extinction.

A study report (2018) recommended that India and Bangladesh should carefully use the water-channels in the Sundarbans in a bid not to disturb the free movement of tigers between the two countries, which may ultimately affect their gene flow. Within the Sundarbans tiger population, low to medium level of genetic diversity has been reported. The primary concern regarding threats to genetic heterogeneity of this population is the probable isolation induced by wide rivers. There are reports of fine-scale genetic structure and clustering within the tiger population of Bangladesh Sundarbans that could largely be attributed to river systems. Increased continuous use of these water channels inside the forest as a conduit for commercial boat traffic has transformed the rivers to barriers to tiger movement. Despite efforts by FDs of both the countries, joint patrolling and joint management activities remain a non-starter. Boat traffic during the active phase of the tiger's aquatic movement needs to be controlled.

Conservation agencies

In Bangladesh only one agency, the FD, is solely responsible for the conservation of the Sundarban, while in India there are quite a few agencies at the state and national level involved in conservation activities. These include NTCA, SBR and Sundarban Development Board (SDB), notified on 7th March, 1973, under the department of Sundarban Affairs for promoting social, economic and cultural advancement of people residing in the Sundarban areas. SDB has jurisdiction over 19 community blocks. There are often overlapping roles and responsibilities creating confusion.

Forest Management in impact areas

Management boundaries differ between the two countries. In Bangladesh, while a 10-km wide band was declared an ecologically critical area (ECA), forest conservation essentially focuses on the RF and WLSs. Uses and pressures on the SRF originate with human populations residing in the so-called Sundarbans Impact Zone (SIZ), a band extending 20 km outside the SRF boundary comprising five districts and 17 upazilas thereof, which are as follows.

1. Bagerhat: Rampal, Mongla, Morrelganj, Sharankhola
2. Khulna: Daccop, Koyra, Paikgacha, Batiaghata
3. Satkhira: Shymnagar,Assasuni, Kaliganj
4. Pirojpur: Matbaria, Bhandaria
5. Barguna: Patharghata, Bamna, Barguna Sadar.

There are a few uncoordinated activities by government and non-government agencies, no real initiative has been taken to manage and develop the surrounding areas which house the population dependent on the Sundarban so far.

On the other hand, SBR consists of uninhabited PAs as well as inhabited transitional areas and there are quite a few programmes and initiatives undertaken by the government, non-government and research agencies. As a result, in Bangladesh planning and management has typically been focused on the physical resources. In SBR, the socio-economic issues affecting forest change, dependency and interlinkages on forest resources are better addressed.

Joint Forest Management (JFM)

Community based forest management has been introduced in the Indian Sundarbans since 1991 involving the fringe villagers in the protection works through JFM committees and has started to give dividends in the form of reduced human-wildlife conflict situations since last decade. Whereas up to 1998, 10 Forest Protection Committees (FPCs) and 12 Eco-development Committees (EDCs) consisting of 10,000 families have been formed in STR and 21 FPCs consisting of 8,300 families in undivided 24-Parganas FD. At present, 51 FPCs and 14 EDCs have been registered in SBR. It appears that in STR, 8,548 families of 14 EDCs and 11 FPCs protect about 252 km² of forests, falling under WLS and RFs (buffer area). In 24-Parganas (South) FD, 40 FPCs composed of 26,519 families protect about 390 km² of forests.

Prior to the 1990s and in the 1990s due to the frequent problem of tiger straying into the villages adjacent to TR, and the problem of cattle-lifting by the strayed tigers, many tigers were killed by the villagers. After implementation of JFM in Sundarban, human-wildlife (mainly tiger) conflicts have been reduced considerably in the villages. Formation of FPCs and EDCs as well as women self-help groups definitely helped curb the retaliatory killings of tigers by the fringe villagers and it also helped spread awareness among the people regarding the conservation of tigers. Rather than killing the straying tiger, villagers are driving it back into the forest by brandishing flaming torches and setting off firecrackers. If this fails, they call the nearest forest office to get a swat team on the ground with a tranquilliser to sedate the tiger so that it can be released back to the forest. FPCs and EDCs not only helped save the straying tigers, capture and release them in the wild, but also many other straying animals, even during the flash floods due to cyclones.

EDCs around TR are also eligible for 25% of the total revenues earned from tourism, which is then used for development works in the villages. The participants derive direct benefits from sustainable use of forest resources.

Numerous eco-development works like construction of jetties, brick paths, solar lighting, providing alternative livelihood support, skill development, plantation etc. have been benefitting the communities in great ways. But it has been alleged that manipulation, corruption, etc. by the village elites and politicisation of JFM benefits are depriving the real needy and deserving villagers from enjoying the usufructs.

In Bangladesh Sundarbans, the co-management activities like forest protection, monitoring, meeting role, training and organisational support are largely implemented through local stakeholders who are members of a number of institutions: the Village Conservation Forum (VCF), the People's Forum (PF) and the Community Patrol Group (CPG) at the local level, and the Co-management Committee (CMC) at the higher level, in which local, district and national stakeholders are involved. Stakeholders reported both positive and negative views on these co-management activities. While the positive views were mostly related to the benefits obtained from forest co-management practices (e.g., increased knowledge, awareness and ability to obtain entry permits), the negative views were related to the obstacles frequently faced by the local stakeholders, such as limited harvesting opportunities, poor meeting support and strict rules on eligibility for VCF/PF membership.

Border Security Force (BSF)

BSF guards the nearly 300 km riverine border with Bangladesh in Sundarbans. It patrols amid the dense mangrove forests or where the rivers meet the Bay of Bengal. Border guards on a fast patrol boat in the estuarine areas. A floating BOP (Border Outpost) stands guard past the last land BOP in Shamshernagar, North 24-Parganas. BSF has three floating BOPs deployed along the border with Bangladesh in the Sundarbans. 30-40 people are deployed on these floating BOPs that move at a speed of around 8 knots.

Chapter 13
Critical Conservation Issues

"As humans, we are fortunate to fight to secure so many rights: men's rights, women's rights, freedom of speech. But animals like tigers are losing their right to live. It is time we all support their most fundamental right- the right to exist!" – Gyalwa Dokhampa

Aziz *et al.* (2013) identified 23 threats; four were linked to tigers, two to prey and 17 to habitat. Of the identified threats, the highest ranked included poaching of tigers, poaching of prey, sea-level rise, upstream water extraction/divergence, wood collection, fishing, and harvesting of other aquatic resources.

Though the Sundarbans has been in the 'very good' category in all five assessment cycles so far, its overall ranking has been consistently declining over the last two decades, from second in 2006 to the current 31st, the Management Effectiveness Evaluation (MEE) of TRs 2022 reports showed. Incidentally, the tiger census report showed that the Sundarbans tiger number increased from 88 to 100 (including four cubs) between 2018 and 2022.

Lack of adequate manpower and the vulnerability of the location to climate change and submergence from sea level rise have been identified as the major challenges.

The evaluation describes key threats to the mangrove ecosystem ranging from agriculture and aquaculture, biological resource use (such as honey collection, hunting, and poaching of tigers), climate change (such as enhanced salinisation due to sea-level rise, the severity of cyclones and drought), natural system modifications (dams, barrages, embankments), pollution (including accidental spills during oil transportation), as well as residential and commercial development (continued mangrove clearing).

Here, landfalls of storms, cyclones and super-cyclones are an annual recurrence. Sidr, Aila, Fani, Bulbul, Amphan and now Yaas have proved yet again how vulnerable the Sundarbans have become to the vagaries of climate change. On 15 November 2007, cyclone Sidr struck Bangladesh with winds up to 240 km per hour. Flushed out of their natural habitat and deprived of food by cyclone and floods, the famed Royal Bengal tiger that roams these mangrove forests is desperately on the prowl. Villagers beat drums and light fires all night to keep it away. Some wildlife activists anticipated that some big cats, after being flushed out, might have crossed over to the Indian Sundarbans possibly through the Jhilla point from Bangladesh where cyclone Sidr destroyed much of their habitat, causing a rise in tiger attacks in SBR. A tiger had sneaked into a farmer's kitchen in the Sundarbans, triggering fear among the villagers.

Dozens of the tigers are feared to have drowned in Aila's storm surge. As of 27 May 2009, one tiger has been found alive in the Jamespur village; it was found crouching in the waterlogged cowshed of Pintu Mirdha following the cyclone's landfall. Mirdha managed to shut the cowshed door and informed the FD. But forest guards had to wait for the water to recede to get close to the animal. Neighbours were asked to evacuate as the animal paced up and down the locked cowshed. At around 1 pm, when the water level went down during low tide, the male tiger was tranquilised. Additionally the forest remains under an estimated 2.4 m of water. On 27 May, conservationists began a search for the tigers throughout the forest. The search teams were supplied with fresh drinking water for the tigers as their natural water source was inundated with salt water from Aila's storm surge. Besides the straying tiger, two spotted deer were rescued.

The eight post-Amphan tiger-attack cases- that also coincided with return of migrant labourers- were mostly reported from Pirkhali, Jhila and the forests of STR. Amphan damaged almost the entire nylon-fencing that had been erected to prevent tigers from entering human habitation in the Sundarbans, triggering recurrence of man-animal conflict in the fringe villages. Since the fencing was restored, the number of attacks by tigers on humans had almost come down to zero.

An analysis of cyclonic events in the Bay of Bengal over a period of 120 years indicates a 26% rise in the frequency of high to very high intensity cyclones over this time period. Salinity has gone up by 20% in the Sundarbans since 1990. Water level too has been rising by 3.14 mm on an average every year at Sagar Island while at Pakhiralay, the rate has been 5 mm/year. Sighting records confirm the scientists' claims. Recent evidence of accelerating global sea-level rise should also be taken into account. Also, with the sea swallowing up land, humans and tigers are being squeezed into an ever-shrinking space, leading to more human-animal conflicts. One of the major actions needed now is to invest in the planned retreat of millions of people and increase mangrove cover as a bio-shield against future cyclones and sea level rise.

The majority of the Indian part of Sundarbans is located in the South 24-Parganas district. In recent times, South 24-Parganas is devastated by more cyclones than any other district of India. Over the last year, 10 cyclonic disturbances have swept the country and two of them (Cyclone Yaas and Jawad) were in the Sundarbans. Sundarbans has now become the cyclone capital of India. According to the recently released Climate hazards and vulnerability Atlas of India by the India Meteorological Department, Pune, the return period of cyclonic storms in South 24-Parganas was 1.67 years on a scale of 1.5 to 60 years. The Shorter return periods indicate more frequent cyclones. To put it in context, India's return period for severe storms is 2.61 years. The researchers studied cyclonic storms of the past 60 years, especially those passing within 90 kilometres off the coastal districts. Around 70 per cent of the cyclones were of severe categories in the Sundarbans area. The median landfall location has also shifted eastward from northern Odisha toward the Sundarban area. These findings are in sync with the developments during the last three years when Sundarbans was impacted by one major cyclone after another- from Bulbul in 2019, Amphan in 2020 to Yaas and Jawad in 2021. As climate change is forcing more and more extreme weather events, life in the Sundarbans is going to get more challenging in the coming years.

The key to saving the Sundarbans is the preservation and plantation of mangroves. This will check salinity and prevent displacement of tigers, thereby bringing down the man-animal conflict.

The staff strength was about 50% when it was evaluated. The lack of enough staff is affecting many areas. For example, the performance of eco-development committees can be improved with a micro-plan, which will ensure better livelihood opportunities for people and discourage them from entering the forest area. Moreover, area development committees need to become functional and more management coordination is required between India and Bangladesh for the Sundarbans forest areas.

The deltaic region faces a substantial amount of accretion and erosion every year as was observed in STR.

Comparative area (in ha.) statistics of STR

Block name (compartments)	Old recorded area	Satellite imagery-based area
Arbesi (1-5)	15048.58	14037.8862
Bagmara (1-8)	29404.86	31089.0364
Chamta (1-8)	22077.33	23137.3178
Chandkhali (1-4)	15596.76	15477.774
Chhotohardi (1-3)	17573.68	16779.8086
Gona (1-3)	13908.91	15027.135
Gosaba (1-4)	17179.76	17485.2928
Harinbhanga (2-3)	11691.5	11805.1192
Jhila (1-6)	12313.62	13159.292
Khatuajhuri (1-3)	13246.56	11959.1852
Matla (1-4)	17636.84	19154.185
Mayadwip (1-5)	27346.96	30910.504
Netidhopani (1-3)	9303.643	8221.0344
Panchamukhani (1-5)	17672.87	17872.6304
Pirkhali (1-7)	19171.85	19973.2352
GRAND TOTAL	259173.709	265689.4362

Source: Tiger Conservation Plan for STR (2017-18 to 2023-27)

Area increased by 16515.7272 ha or say 165.16 km²; increased in ten blocks-Bagmara, Chamta, Gona, Gosaba, Harinbhanga, Jhila, Matla, Mayadwip, Panchamukhani and Pirkhali and decreased in five block- Arbesi, Chandkhali, Chhotohardi, Khatuajhuri and Netidhopani.

Recently, whereas the Kendo (Bhangaduni) Island located in the southernmost part of STR is reported to be reduced from 37.8 km² in 1991 to 22.7 km² in 2021, the Haliday Island WLS in 24-Parganas (South) FD also shrunk due to continuous erosion, landslides and severe cyclonic storms (Justin & Ghosh, 2022). A new *char* (accretion platform) land has developed on the river Haldi on the south and widening of the river Dutta on the north were also marked. Submergence of the Dobanki Island and associated *char* areas under Sajnekhali WLS is also observed. On the contrary, the eastern and north-eastern RF areas adjoining Bangladesh are being elevated in comparison to the NP West and Sajnekhali WLS.

Bera and Maiti (2019) observed that accretion increases landward from the coast in almost all channels. They observed a sharp difference of erosion/accretion between reclaimed moujas and islands covered by dense mangrove swamp. 54% moujas having embankments are retreating at an average rate of 6.14 metre/year; contrary to this only 35% mangrove islands are moving landward at a much slower rate (4.78 m/yr.) in comparison to habitat areas. Process of new land formation is also very steady in mangrove areas (4.78 ha./yr.) than habited areas (6.84 ha./yr.) which indicates channel stability and erosion resistance provided by mangroves.

Loss and degradation of the mangrove habitat has caused severe decline of the tiger population. Tiger conservation within and beyond the PAs is based on the ethical issues- "biodiversity, aesthetic values and integrity" as well as management of the mangrove ecosystem challenged by development works and anthropogenic activities.

The Sundarbans has been doing well in containing major problems like poaching and tiger-human conflict. Steps like putting nylon nets around the forest areas of the Sundarbans to stop straying of tigers in human habitats, has given a lot of dividends to the pristine forest area. There is scope for improvement and the state government took a policy decision to expand the TR area of the Sundarbans. To improve the overall management, we have decided in principle to include three ranges from the South 24-Parganas FD, namely Matla, Raidighi and Ramganga, into the STR, and expect NTCA to give clearance. These ranges are tiger habitats and hence can be better managed once they become part of the TR area. Once approved, the STR will be the biggest TR in the country. Presently, STR is around 2,585 km², which may become about 3,600 km² once the TR gets expanded; and will push back Nagarjunasagar TR of Andhra Pradesh into second place, which is currently the biggest TR in the country with around 3,296 km² area.

The Sundarbans tiger is an important component of global tiger recovery programme and its present stable population and, being a trans-boundary tiger habitat, its contiguity holds promising long term conservation. Shrinkage of the habitat in use by the tiger and decline in its population as well as increasing human-tiger conflict, particularly in the forest-village interface, due to extensive anthropogenic threats combined with rapid environmental degradation and climate change, are posing acute management challenges. Under the circumstances, the main management goals are to increase the current tiger population density, maintain sufficient prey and habitat, maintain connectivity between populations in the western and eastern Sundarbans to avoid gene pool stagnation, address challenges to reduce human-tiger conflict, relocate the tigers to balance the ratio of male and female tigers and monitor their movements, improve conservation capacity, improve law enforcement, build capacity and proper mechanism for awareness and education programmes as well as community involvement, build capacity to conduct tiger conservation research and monitoring, and encourage collaboration.

Mukul *et al.* (2019) recommended enhancing terrestrial protected area coverage, regular monitoring, law enforcement, awareness-building among local residents among the key strategies needed to ensure long-term survival and conservation of the Bengal tiger in the Sundarbans.

Chapter 14
Prospects

"We must protect tigers from extinction. Our planet's future depends on it." – Michelle Yeoh

To ensure the long-term survival of tigers in the Sundarbans, a multi-faceted approach is needed, including protecting and expanding tiger habitats, preserving population connectivity, minimising human-tiger conflicts, and combating threats like habitat loss, poaching, and illegal trade. It's important to restore habitats, increase ungulate populations, and plan reintroduction of tigers in low density areas to tackle conflict issues.

The involvement of various stakeholders, such as governments, NGOs, local communities, and businesses, is crucial. Strategies like increased patrolling, monitoring, and law enforcement, focus on "Other Effective Area-based Conservation Measures (OECM)" along with promoting eco-tourism and sustainable livelihoods for local communities, can help achieve this goal.

Though the mangrove forest of the Sundarbans is considered as vulnerable to endangered under Red List of Ecosystems driven by historical clearing and diminishing wildlife populations, there is cause for 'cautious future optimism' (Sievers *et al.*, 2020). The reasons stated are-

(i)the trend of historically high rates of mangrove clearing and degradation has since slowed down and

(ii)tiger population in the Sundarban mangroves has slightly increased and is stable. Of late, tigers have been spotted at certain locations where they were not seen before.

The increase in density and abundance of tigers can be attributed to some positive improvements of tiger habitat management. Specifically, SMART patrol has restrained some illegal resource collectors and poachers leading to the improvements of tiger's status.

Despite efforts to conserve tigers, there are still several challenges that need to be addressed. The ever-increasing biotic interference in the form of livelihood forest explorations, fishing, palm and timber, fuelwood and NTFP extractions, and growing national and international waterways make the Sundarbans landscape and the tiger population vulnerable. The ongoing threats such as reduced freshwater and sediment supply must be effectively monitored and managed. Temperature, salinity, freshwater flow, nutrients, and tidal amplitude are key drivers of mangrove ecosystem productivity and diversity.

There lies the necessity of increasing the notified core areas to prevent man-animal conflict. In order to intensify tiger management in the mangrove landscape, STR is likely to be expanded by about 1,000 km^2, as proposed by the FD, by merging three more forest ranges, namely Matla, Raidighi, and Ramganga under 24-Parganas (South) FD to STR, provided formal clearance from NTCA is received. According to All India Tiger Estimation 2022, this proposed extended area comprises roughly 30 tigers. Once STR gets expanded from 2,585 km^2 to about 3,600 km^2,the habitat would become the largest one in India. But the real problem is that most of the posts of field staff are lying vacant.

Further increase of tiger population in the greater Sundarbans may not be impossible in future if the habitats are adequately protected by filling up of the vacant posts because in Bangladesh Sundarbans protection and preservation was being hampered due to the FD's shortage of manpower and logistics with 344 out 1,172 posts lying vacant in 2018. As a result, there was only one guard for every nine km^2 of the forest. Thus, the FD's activities in the Sundarbans, including averting smuggling of forest resources and poaching, are being severely hindered. Even sometimes the boatmen have to play the role of forest guards. The vacancies are being increased with the retirement of the existing staff. In Indian Sundarbans, about 50% of posts are lying vacant.

It is important that the trans-boundary tiger population is managed as a single population through joint patrolling. To evaluate the situation in its entirety, extensive monitoring on spatial as well as temporal scale is needed. To preserve the ecological integrity of the area, cross-border collaboration and knowledge exchange between India and Bangladesh are imperative.

To UNESCO, the future of the Sundarbans lies in "biodiversity, aesthetic values and integrity" and management of ecological balances challenged by development works and anthropogenic activities.

It would be more useful if the entire SBR and SRF are sampled at one time. This should also be done simultaneously with the Bangladesh part of the Sundarbans to arrive at the best estimate for the Sundarbans as a whole. The two forest forces came together for tiger enumeration in 2018. The modalities of enumeration through camera traps have been standardised across the border. The Sundarbans delta is a single ecosystem, which is cleaved by the man-made national boundaries of India and Bangladesh. Tigers migrate between the two countries, as do other animals. Floral species on either side, too, are almost the same. Under the heightened impacts of climate change, trans-boundary cooperation and knowledge sharing between India and Bangladesh are important to maintain ecological integrity of the landscape. Declaration of a trans-boundary "peace park" is also being considered in order to jointly manage the unique ecosystem, which is vulnerable to accelerated climate change. But the decision is pending for long. Similar peace parks are also being considered for eco-sensitive zones along the border regions of Nepal, Bhutan and Sri Lanka.

We can ensure a future where these majestic tigers roam freely in their natural environment by working together on conservation, sustainable tourism, and community engagement.

Selected Bibliography

Ahmad YS. 1981. *With the Wild Animals of Bengal*. Dacca: Y.S. Ahmad.

Ahmed S. 2002. Losing battle for royal cats. *The Daily Star*, 6th May 2002, Internet edition, http://www.thedailystar.net.

Akash M, Chowdhury U, Khaleque FTZ, Reza RN, Howladar DC, Islam MR, Khan H. 2021. On the reappearance of the Indian grey wolf in Bangladesh after 70 years: what do we know? *Mammalian Biology* 101: 163-171. https://doi.org/10.1007/s42991-020-00064-4.

An L, Bohnett E, Battle C, Dai J, Lewison R, Jankowski P, Carter N, Armstrong EE, Khan A, Taylor RW, Gouy A, Greenbaum G, Thiéry A, Kang JT, Redondo SA, Prost S, Barsh G, Kaelin C, Phalke S, Chugani A, Gilbert M, Miquelle D, Zachariah A, Borthakur U, Reddy A, Louis E, Ryder OA, Jhala YV, Petrov D, Excoffier L, Hadly E, Ramakrishnan U. 2021. Recent Evolutionary History of Tigers Highlights Contrasting Roles of Genetic Drift and Selection. *Molecular Biology and Evolution* 38(6): 2366-2379. https://doi.org/10.1093/molbev/msab032.

Aziz MA. 2018. Notes on population status and feeding behaviour of Asian small-clawed otter (*Aonyx cinereus*) in the Sundarbans mangrove forest of Bangladesh. *IUCN Otter Spec. Group Bull.* 35(1): 3-9.

Aziz MA, Barlow ACD, Greenwood CG, Islam MA. 2013. Prioritizing threats to improve conservation strategy for the tiger *Panthera tigris* in the Sundarbans Reserve Forest of Bangladesh. *Oryx* 47(4): 510-518.

Aziz MA, Paul AR. 2015. Bangladesh Sundarbans: Present Status of the Environment and Biota. *Diversity* 7: 242-269.

Aziz MA, Tollington S, Barlow A, Greenwood C, Goodrich JM, Smith O. 2017a. Using non-invasively collected genetic data to estimate density and population size of tigers in the Bangladesh Sundarbans. *Global Ecology and Conservation* 12: 272-282

Aziz MA, Tollington S, Barlow A, Goodrich J, Shamsuddoha M, Islam MA, Groombridge JJ. 2017b. Investigating patterns of tiger and prey poaching in the Bangladesh Sundarbans: Implications for improved management. *Global Ecology and Conservation* 9: 70-81.

Aziz MA, Smith O, Barlow A, Tollington S, Islam MA, Groombridge JJ. 2018. Do rivers influence fine-scale population genetic structure of tigers in the Sundarbans? *Conservation genetics* 29(5): 1137-1151. https://link.springer.com/article/10.100.

Aziz MA, Islam MA, Groombridge J. 2020. Spatial differences in prey preference by tigers across the Bangladesh Sundarbans reveal a need for customised strategies to protect prey populations. *Endangered Species Research* 43: 65-74.

Aziz MA, Smith O, Jackson HA, Tollington S, Darlow S, Barlow A, Islam MA, Groombridge JJ. 2022. Phylogeography of *Panthera tigris* in the mangrove forest of the Sundarbans. *Endangered Species Research* 48: 87-97.

Badam GL, Sathe VG. 1991. Animal depictions in rock art and palaeoecology-A case study at Bhimbetka, Madhya Pradesh, India. In: Pager SA, Swartz BK, Willcox AR (Natal) (eds.), *Rock art-The way ahead: South African rock art research association first international conference proceedings* (pp. 196-208).

Bahuguna NC, Mallick JK. 2010. *Handbook of the mammals of South Asia with special emphasis on India, Bhutan and Bangladesh*. Natraj Publishers, Dehradun, India.

Baker EB. 1887. *Sport in Bengal and how, when, and where to seek it*. London: Ledger, smith & co.

Baker S. 1890. *Wild beasts and their ways, Reminiscences of Europe, Asia, Africa, and America. Vol. I*. London: Macmillan.

Bandyopadhyay S, Kar NS, Dasgupta S, Mukherjee D, Das A, 2023. Island area changes in the Sundarban region of the abandoned western Ganga– Brahmaputra–Meghna Delta, India and Bangladesh, *Geomorphology* 422: 108482. ISSN 0169-555X, https://doi.org/10.1016/j.geomorph.2022.108482.

Barlow ACD. 2009. The Sundarbans Tiger Adaptation, Population Status, And Conflict Management. *PhD Thesis*. The Faculty of the Graduate School of the University Of Minnesota.

Barlow ACD, Gani MO, Ahmed MIU, Rahman SKM, Hossain A, Hossain ANM, Islam T,, Saha UK, Smith JLD. 2006. *Sundarbans Tiger Project Activities and Results 2005-2006 Final report to the USFWS and the Save the Tiger Fund*.

Barlow ACD, Ahmed MIU, Rahman MM, Howlader A, Smith AC, Smith JLD. 2008. Linking monitoring and intervention for improved management of tigers in the Sundarbans of Bangladesh. *Biological Conservation* 141: 2031-2040.

Barlow ACD, Maza´k J, Ahmad IU, Smith JLD. 2010. A preliminary investigation of Sundarbans tiger morphology. *Mammalia* 74(3): 329-331.

Barlow ACD, Ahmad I, Smith JLD. 2013. Profiling Tigers (*Panthera tigris*) to Formulate Management Responses to Human-Killing in the Bangladesh Sundarbans. *Wildl. Biol. Pract.* 9(2): 30-39.

Barlow ACD, Smith JLD, Ahmad IU, Hossain ANM, Rahman M, Howlader A. 2011. Female tiger (*Panthera tigris*) home range size in the Bangladesh Sundarbans: the value of this mangrove ecosystem for the species' conservation. *Oryx* 45(1): 125-128.

Bera R, Maiti R. 2019. Quantitative Analysis of Erosion and Accretion (1975-2017) Using DSAS – A Study on Indian Sundarbans. *Regional Studies in Marine Science* 28: 100583.

Bernier F. 1916. *Travels in the Mogul Empire A.D. 1656-1668*. New York: Oxford University Press.

Bhattacharya S, Dutta B, Mondal U, Mukherjee J, Mitra M. Helminthiasis in a Bengal tiger (*Panthera tigris tigris*)- A case report. *Explor. Anim. Med. Res.* 2(2): 184-188.

Biswas SR, Choudhury JK, Nishat A, Rahman MM. 2007. Do invasive plants threaten the Sundarbans mangrove forest of Bangladesh? *Forest Ecology and Management* 245(1-3): 1-9.

Brahmachary RL, Dutta J. 1981. On the pheromones of tigers: Experiments and theory. *Am. Nat.* 118: 561-567.

Brahmachary RL, Dutta J. 1987. Chemical communication in the tiger and leopard. In: Eds. Tilson RL, Seal US (eds.) *Tigers of the World*. Noyes Publication, USA, 296-302.

Burger B, Bekker M, Bekker J, Le Roux M, Fish N, Fourie W, Weibchen G. 2008. Chemical Characterization of Territorial Marking Fluid of Male Bengal Tiger, *Panthera tigris*. *Journal of chemical ecology* 34: 659-671. 10.1007/s10886-008-9462-y.

Carter N, Levin, S, Barlow, A, Grimm, V. 2015. Modeling tiger population and territory dynamics using an agent-based approach. *Ecological Modelling* 312: 347-362. 10.1016/j.ecolmodel.2015.06.008.

Carter N, Wilson E, Gurung B. 2023. Social networks of solitary carnivores: The case of endangered tigers and insights on their conservation. *Conservation Science and Practice* 10.1111/csp2.12976: 1-12.

Chakrabarti K. 1987. Sundarban mangrove-biomass productivity and resource utilization: an in-depth study. *Indian Forester* 113(9): 622-628.

Chakrabarti K. 1992. *Man eating tigers*. New Delhi, India: Darbari Publication. 142 p.

Chakrabarti R. 2009. Local People and the Global Tiger: An Environmental History of the Sundarbans. *Global Environment* 3: 72-95.

Chakrabarti R. 2010. Prioritizing the tiger: A History of human–tiger conflict in theSundarbans.

Current Conservation 4(4): 44-47.

Chatterjee A. 2019. Hunting as a Leisurely Sporting Activity in Colonial Bengal: A Historical Discourse. *Vidyasagar University Journal of History* 7(2018-2019): 36-43.

Chatterjee S. 2023. Rising trend of man-tiger conflict at man-nature interface of Indian Sundarbans: study towards traditional understanding and challenging livelihood of Sundarbans people. *Safety in Extreme Environments* 5(3): 35-46.

Chaudhuri AB. 2007. *Biodiversity of mangroves*. New Delhi: Daya Publishing House, 332 pp.

Chaudhuri AB, Chakrabarti K. 1972. Wildlife biology of Sundarban forests-observations on tigers. *Cheetal* 15(1): 65-68.

Chaudhuri AB and Chakrabarti K. 1980. Wildlife biology of Sundarban forests:further study on the habitat and behaviour pattern of Sundarban tiger. In: Director, Zoological Survey of India (Ed.), *Proceedings of the workshop on wildlife ecology, Dehra Dun, January 1978*, p. 27-33.

Chaudhuri AB and Choudhury A. 1994. *Mangroves of the Sundarbans. Volume 1: India*. Bangkok, Thailand: IUCN, 247 pp.

Chauhan NPS. 2011. Man–eating and cattle–lifting by tigers and conservation implications in India. 8 th European Vertebrate Pest Management Conference. *Julius–Kühn–Archiv* 432:178-179.

Choudhury SR. 1979, Olfaction ecology of peak cubbing in Simlipal's tigers. *Indian Forester* 103: 577-588.

Chowdhury A, Naz A, Sharma SB, Dasgupta R. 2023. Changes in Salinity, Mangrove Community Ecology, and Organic Blue Carbon Stock in Response to Cyclones at Indian Sundarbans. *Life (Basel)* 13(7): 1539. doi: 10.3390/life13071539.

Cooper DM, Dugmore, AJ, Gittings, BM, Scharf AK, Wilting A, Kitchener AC. 2016. Predicted pleistocene-holocene range shifts of the tiger (*Panthera tigris*). *Diversity and Distributions* 22(11): 1199-1211.

Cott HB. 1940. *Adaptive colouration in mammals*. London: Methuen.

Cotton E. 1907. *Calcutta, Old and New: A Historical and Descriptive Handbook of the City*. Calcutta: W. Newman.

Curtis SJ. 1933. *Working Plan for the Sundarbans Division for the period from April 1st 1931-March 31st 1951*. Calcutta: Bengal Government Press.

Das AK. 1980. Observations on the reproductive behaviour of the tiger *Panthera tigris tigris* Linn. in captivity. *J. Bombay nat. Hist. Soc.* 77(2): 46-53.

Das CS. 2011. Tiger straying incidents in Indian Sundarban: statistical analysis of case studies as well as depredation caused by conflict. *European Journal of Wildlife Research* 58(1): 205-214.

Das NC. 1896. *Ancient Geography of Asia compiled from Valmiki-Rāmāyana*. Bharat-Bharati.

Das SK, Sarkar PK, Saha R, Vyas P, Danda AA, Vattakavan J. 2012. *Status of Tigers in 24-Parganas (South) Forest Division, West Bengal, India*. World Wide Fund for Nature-India, New Delhi.

De Rathindranath. 1991. *The Sundarbans*. Kolkata: Oxford University Press.

Dendup P, Lham C, Wangchuk W, Jamtsho Y. 2023. Tiger abundance and ecology in Jigme Dorji National Park, Bhutan. *Global Ecology and Conservation* 42: e02378.

Deuti K, Roy Choudhury B. 1999. Possible causes for the straying of tigers in the Sundarbans and some suggested remedies. In: Guha Bakshi DN, Sanyal P, Naskar KR (Eds.). *Sundarbans Mangal* Calcutta: Naya Prakash, p. 460-465.

Dey TK. 2007. *Deer population in the Bangladesh Sundarbans*. Chittagong: The Ad Communication. 112 pp.

Dey TK, Kabir MJ, Ahsan MM, Islam MM, Chowdhury MMR, Hassan S, Roy M, Qureshi Q, Naha D, Kumar U, Jhala YV. 2015. *First phase Tiger status report of Bangladesh Sundarbans*. Wildlife Institute of India and Bangladesh Forest Department, Ministry of Environment and Forests. Government of the People's Republic of Bangladesh, Dhaka.

Dhar SB, Mondal S. 2023. Nature of human-tiger conflict in Indian Sundarban. *Trees, Forests and People* 12: 100401. https://doi.org/10.1016/j.tfp.2023.100401.

Dinerstein E, Loucks C, Wikramanayake E, Ginsberg J, Sanderson E, Seidensticker J, Forrest J, Bryja M, Heydlauff A, Klenzendorf S, Leimgruber P, Mills J. 2007. The Fate of Wild Tigers. *Bioscience* 57(6): 508-514. doi: 10.1641/B570608.

Dunbar Brander AA. 1923. *Wild Animals in Central India*. London: E. Arnold.

Ferdiousi J, Khan MMH. 2021. Human-wildlife conflict along the edge of the Sundarbans mangrove forest in Satkhira, Bangladesh. *Jahangirnagar University J. Biol. Sci.* 10(1 & 2): 59-70.

Gani MO. 2002. A study on the loss of Bengal Tiger (*Panthera tigris*) in five years (1996-2000) from Bangladesh Sundarbans. *Tigerpaper* 29(2): 6-11.

Garga DP. 1948. How far can a tiger swim? *Journal of the Bombay Natural History Society* 47(3): 545-546.

Ghosh A. 1999. Pride of Bengal. In: WWF Tiger Conservation Programme: *Three Years and Beyond*. New Delhi: WWF TCP.

Ghosh J. 2020. The Bengal Tigers of India. *JOJ Wildlife and Biodiversity* 2(5): 55-58.

Ghosh M, Saha U, Roy S, Talukdar B. 1992. Subrecent remains of great one-horned rhinoceros from southern West Bengal, India. *Current Science* 62(8): 577-580.

Goodrich J, Lynam A, Miquelle D, Wibisono H, Kawanishi K, Pattanavibool A, Htun S, Tempa T, Karki J, Jhala Y, Karanth U. 2015. *Panthera tigris*. The IUCN Red List of Threatened Species 2015:e.T15955A50659951.http://dx.doi.org/10.2305/IUCN.UK.2015-2.RLTS.T15955A50659951.en.

Gour DS, Bhagavatula J, Bhavanishankar M, Reddy PA, Gupta JA, Sarkar MS, Hussain SM, Harika S, Gulia R, Shivaji S. 2013. Philopatry and Dispersal Patterns in Tiger (*Panthera tigris*). *PLoS ONE* 8(07): e66956. doi: 10.1371/journal.pone.0066956.

Gupta AC. 1966. Wildlife of lower Bengal with particular reference to the Sundarbans. *West Bengal Forests, Centenary Commemoration Volume*. Calcutta: Forest Directorate. Government of West Bengal. pp. 233-237.

Hait AK, Behling H. 2019. Responses of the mangrove ecosystem to Holocene environmental change in the Sundarban Biosphere Reserve, India. *Acta Palaeobotanica* 59(2): 391-409.

Halder NK. 2011. Scientific approach for tiger conservation in the Sundarbans. *Tigerpaper* 38(4): 5-9.

Haque MZ, Reza MIH, Rahim SA, Abdullah MP, Elfithri R, Mokhtar MB. 2015. Behavioral change due to climate change effects accelerate tiger human conflicts: A study on sundarbans mangrove forests, Bangladesh. *International Journal of Conservation Science* 6(4): 669-684.

Hasan MK, Aziz MA, Alam SMR, Kawamoto Y, Jones-Engel L, Kyes RC. 2013. Distribution of Rhesus Macaques (*Macaca mulatta*) in Bangladesh: inter-population variation in group size and composition. *Primate Conservation* 26(1): 125-132.

Hazra S, Ghosh T, Das Gupta R and Sen G. 2002. Sea level and associated changes in the Sundarbans. *Science and Culture* 68(9–12):309-321.

Hazra S. 2010. *Temporal change detection (2001-2008) of the Sundarban*. Unpublished Report. WWF-India.

Hendrichs H. 1975. The status of the tiger *Panthera tigris* (L.) 1758 in the Sundarbans mangrove forest (Bay of Bengal). *Säugetierkundliches Mitt*. 23(3): 161-199.

Hossain ANM, Lynam AJ, Ngoprasert D, Barlow A, Barlow CG, Savini T. 2018. Identifying landscape factors affecting tiger decline in the Bangladesh Sundarbans. *Global Ecology and Conservation* 13: e00382.

Hu J, Westbury MV, Yuan J, Wang C, Xiao B, Chen S, Song S, Wang L, Lin H, Lai X, Sheng G. 2022. An extinct and deeply divergent tiger lineage from northeastern China recognized through palaeogenomics. *Proceedings of the Royal Society* B.2892022061720220617. http://doi.org/10.1098/rspb.2022.0617.

Hunter WW. 1858. *A History of British India Vol.1*. London: James Madden.

Hunter WW. 1876a. *A statistical account of Bengal: Volume 7 Districts of Maldah, Rangpur and Dinajpur*. London: Trübner & Co.

Hunter WW. 1876b. *A statistical account of Bengal: Volume 9 Districts of Murshidabad and Pabna*. London: Trübner & Co.

Inskip C, Fahad Z, Tully R, Roberts T, MacMillan D. 2014. Understanding carnivore killing behaviour: Exploring the motivations for tiger killing in the Sundarbans, Bangladesh, *Biological Conservation* 180: 42-50. https://doi.org/10.1016/j.biocon.2014.09.028.

Inskip C, Carter N, Riley S, Roberts T, MacMillan D. 2016. Toward Human-Carnivore Coexistence: Understanding Tolerance for Tigers in Bangladesh. *PLoS ONE* 11(1): e0145913. doi:10.1371/journal.pone.0145913.

Islam SN, Gnauck A, Voigt Hans-Jürgen. 2011. Fourier polynomial approximation of estuaries water salinity in the Sundarbans region of Bangladesh. *Int. J. Hydrology Science and Technology* 1(3/4): 207-223.

Islam SN, Reinstädtler S, Gnauck A. 2019. Invasive Species in the Sundarbans Coastal Zone (Bangladesh) in Times of Climate Change: Chances and Threats. In: *Impacts of Invasive Species on Coastal Environments*. 0.1007/978-3-319-91382-7_2.

Jack JC. 1918. *Bengal District Gazetteers: Bakarganj*. Calcutta.

Jackson P. 1990. *Endangered species: Tigers*. London: Apple Press.

Jaroš MF. 2012. The ecological and ethological significance of felid coat patterns (Felidae). *PhD Thesis*. Prague: Department of Philosophy and History of Science, Charles University.

Jalais A. 2010. *Forest of tigers: People, politics and environment in the Sundarbans*. New Delhi: Routledge.

Jayanthi M, Duraisamy M, Kabiraj S, Thirumurthy S, Samynathan M, Panigrahi A, Muralidhar M. 2023. Are the Sundarbans, the World's largest mangroves region under threat?-An ecosystem-based geospatial approach to assess changes past, present, and future in relation to natural and human-induced factors. *Land Degradation & Development* 34(1): 125-141. https://doi.org/10.1002/ldr.4448.

Jhala YV, Qureshi Q, Gopal R, and Sinha PR (eds.). 2011. *Status of the Tigers, Co-predators, and Prey in India*. New Delhi: National Tiger Conservation Authority, Govt. of India; Dehradun: Wildlife Institute of India.

Jhala YV, Qureshi Q, Gopal R. 2015. *The status of tigers in India 2014*. New Delhi: National Tiger Conservation Authority/Dehradun: Wildlife Institute of India. 28 p.

Jhala YV, Dey TK, Qureshi Q, Kabir J, Md, Bora J, Roy M. 2016. *Status of tigers in the Sundarban landscape Bangladesh and India*. Bangladesh Forest Department; National Tiger Conservation Authority, New Delhi, & Wildlife Institute of India, Dehradun. TRNO -2016/002.

Jhala YV, Sadhu A. 2017. Additional file 1: Field Guide for Aging Tigers. In: Sadhu *et al.* Demography of a small, isolated tiger (*Panthera tigris tigris*) population in a semiarid region of western India. *BMC Zoology* 2(16): 1-13.

Jhala YV, Qureshi Q, Nayak AK. (eds.) 2020. *Status of tigers, copredators and prey in India, 2018*. National Tiger Conservation Authority, Government of India, New Delhi, and Wildlife Institute of India, Dehradun.

Jhala YV, Gopal R, Mathur V, Ghosh P, Negi HS, Narain S, Yadav SP, Malik A, Garawad R, Qureshi Q. 2021. Recovery of tigers in India: Critical introspection and potential lessons. *People and Nature* 3(2): 281-293.

Ji A, Johnson MT, Walsh EJ, McGee J, Armstrong DL. 2013. Discrimination of Individual Tigers (Panthera tigris) from Long Distance Roars. *Journal of the Acoustical Society of America* 133(3): 1762-1769.

Justin SJ, Ghosh D. 2022. A Contemporary Study of Environmental Challenges in Sundarban Tiger Reserve: A Unesco World Heritage Site. *EPRA International Journal of Multidisciplinary Research* (IJMR) 8(12): 253-260.

Karanth KU, Sunquist ME. 1995. Prey Selection by Tiger, Leopard and Dhole in Tropical Forests. *Journal of Animal Ecology* 64(4): 439-450.

Karanth KU, Nichols JD. 1998. Estimation of tiger densities in India using photographic capture and recaptures. *Ecology* 79(8): 2852-2862.

Karanth KU, Nichols JD. 2000. Ecological Status and Conservation of Tigers in India. *Final technical report to the Division of International Conservation, US Fish and Wildlife Service, Washington, DC, and Wildlife Conservation Society, New York, USA; Centre for Wildlife Studies, Bangalore, India.*

Karanth K, Nichols J, Seidensticker J, Dinerstein E, Smith J, Mcdougal C, Johnsingh A, Chundawat R, Thapar V. 2003. Science deficiency in conservation practice: The monitoring of tiger populations in India. *Animal Conservation* 6:141-146. 10.1017/S1367943003003184.

Karanth KU, Nichols JD, Kumar NS, *et al.* 2004. Tigers and their prey: predicting carnivore densities from prey abundance. *Proc Natl Acad Sci.* 101(14): 4854-4858.

Karanth KU, Goodrich JM, Vaidyanathan S, Reddy GV. 2009. *Landscape scale, ecology-based management of wild tiger populations.* Washington, D.C.: Global Tiger Initiative, World Bank, and Wildlife Conservation Society.

Karim A. 1994. Physical environment and vegetation. In: Hussain Z and Acharya G. (eds.). *Mangroves of the Sundarbans.* Vol.2: Bangladesh. Bangkok, Thailand: IUCN.

Kellner CJ, Brawn JD, Karr JR. 1992. What is habitat suitability and how should it be measured? Pp.476-488. In: McCullough DR, Barrett RH (eds.) *Wildlife 2001: Populations.* New York: Elsevier Applied Science.

Kelly A, Goosen J, Venter M, Younus A. 2020. Management of Bengal tiger attacks- A case report and literature review. *Interdisciplinary Neurosurgery* 22: 100824, ISSN 2214-7519, https://doi.org/10.1016/j.inat.2020.100824.

Khan MMH. 2004. Ecology and conservation of the Bengal tiger in the Sundarbans mangrove forest of Bangladesh. *PhD Thesis.* Cambridge: University of Cambridge.

Khan MMH. 2007. *Project Sundarbans Tiger: Tiger density and Tiger-Human Conflict Final Technical Report.* USA: Save the Tiger Fund, National Fish and Wildlife Foundation.

Khan MMH. 2008. Prey selection by tigers (*Panthera tigris*) in the Sundarbans West Wildlife Sanctuary of Bangladesh. *J Bombay Nat Hist Soc* 105(3): 255-263.

Khan MMH. 2009. Can domestic dogs save humans from tigers *Panthera tigris*? *Oryx* 43(1): 44-47.

Khan MMH. 2012. Population and prey of the Bengal Tiger *Panthera tigris tigris* (Linnaeus, 1758) (Carnivora: Felidae) and their prey in the Sundarbans, Bangladesh. *Journal of Threatened Taxa* 4(2): 2370-2380.

Khan MMH, Chivers D. 2007. Habitat preferences of tigers *Panthera tigris* in the Sundarbans East Wildlife Sanctuary, Bangladesh, and management recommendations. *Oryx* 41(4): 463-468. doi:10.1017/S0030605307012094.

Kitchener AC, Breitenmoser-Würsten C, Eizirik E, Gentry A, Werdelin L, Wilting A, Yamaguchi N, Abramov AV, Christiansen P, Driscoll C, Duckworth JW, Johnson W, Luo S-J, Meijaard E, O'Donoghue P, Sanderson J, Seymour K, Bruford M, Groves C, Hoffmann M, Nowell K, Timmons Z, Tobe S. 2017. A revised taxonomy of the Felidae: The final report of the Cat Classification Task Force of the IUCN Cat Specialist Group *Cat News* Special Issue 11: 66-68.

Kolipakam B, Singh S, Pant B, Qureshi Q, Jhala YV. 2019. Genetic structure of tigers (*Panthera tigris tigris*) in India and its implications for conservation. *Global Ecology and Conservation* 20: e00710.

Kurup GU. 1980. Some parameters of tiger survival in India. In: Director, Zoological Survey of India (Ed.), *Proceedings of the workshop on wildlife ecology, Dehra Dun, January 1978*, p. 87-98.

Lahiri RK. 1974. *Report on the Investigation of Depredation Caused by Tiger in Parbatirpur Mouza, under Basanti Police Station*. Forest Directorate, West Bengal. 5 pp. (mimeo).

Lewis J, Alam MM, Dey TK, Barlow ACD. 2012. *Bangladesh Tiger Action Plan Manual: Wild tiger capture and immobilisation*. Wildlife Trust of Bangladesh.

Liu, Y-C, Sun X, Driscoll C, Miquelle DG, Xu X, Martelli P, Uphyrkina O, Smith JLD, O'Brien SJ, Luo S-J. 2018. Genome-wide evolutionary analysis of natural history and adaptation in the world's tigers. *Current Biology* 28(23): 3840–3849. doi:10.1016/j.cub.2018.09.019.

Luo S-J, Johnson, W, Smith J, Brien S. 2010. What Is a Tiger? Genetics and Phylogeography. *Tigers of the World*. 10.1016/B978-0-8155-1570-8.00003-7.

Mahmood H, Ahmed M, Islam T, Zashim Uddin M, Ahmed ZU, Saha C. 2021. Paradigm shift in the management of the Sundarbans mangrove forest of Bangladesh: Issues and challenges, *Trees, Forests and People* 5: 100094. https://doi.org/10.1016/j.tfp.2021.100094.

Mallick JK. 2002. *Biodiversity Resource Assessment and Management of Buxa Tiger Reserve through GIS*. Wildlife Wing, Directorate of Forest. Govt. of West Bengal. Kolkata. Mimeo. 60 pp+maps.

Mallick JK. 2010a. GPS tracking of the tigers in Sundarban Tiger Reserve. *Annual Report 09-'10* (WWF, West Bengal State Office, Kolkata). 1: 21-22.

Mallick JK. 2010b. Past and present status of the Indian Tiger in northern West Bengal, India: an overview. *Journal of Threatened Taxa* 2(3): 739-952.

Mallick, J. K., 2010c. Endangered fishing cats in human-dominated landscape. *Environ* 10(3): 40-43.

Mallick JK. 2011. Status of the mammal fauna in Sundarban Tiger Reserve, West Bengal- India. *Taprobanica* 3(2): 52-68.

Mallick JK. 2012. Mammals of Kalimpong Hills, Darjeeling District, West Bengal, India. *Journal of Threatened Taxa* 4(12): 3103-3136.

Mallick JK. 2013. Ecology, status and aberrant behaviour of Bengal Tiger in the Indian Sundarbans. In: Gupta VK, Verma AK (eds). *Animal Diversity Natural History Conservation* Vol.2. pp. 381-454. New Delhi, India: Daya Publishing House.

Mallick JK. 2015. Transformation of Sundarban tiger to opportunistic omnivore. *Tigerpaper* 42(2): 17-21

Mallick JK. 2019a. *Panthera tigris*: range and population collapse in Northern West Bengal, India. *Biodiversity Int J.* 3(3): 110-119.

Mallick JK. 2019b. Primates in the Sundarbans of India and Bangladesh, 110-123. In: Nowak K, Barnett AA, Matsuda I (Eds.) Primates in Flooded Habitats: Ecology and Conservation. Cambridge University Press, UK.

Mallick JK. 2021. *Shadow in the Swamp: The Aberrant Tiger of Sundarbans.* https://roundglasssustain.com/.

Mallick JK. 2023a. *Biodiversity of the Sundarbans: Ethics of Conservation Ecology (1770- 2022).* Ethics International Press Ltd, UK. Pp. 356.

Mallick JK. 2023b. Changing mangrove forests in the Indian Sundarbans during the 21st century. In: Misra TK (ed.) *Research in emerging fields of biological science (Multidisciplinary aspect).* Chhattisgarh: Astitva Prakashan; 85-93.

Mallick N. 2007. Control of man–tiger conflict in Sundarban Tiger Reserve. *Banabithi* October: 41-45.

Majie AK, Mondal P, Ghosh SK, Banerjee DN, Roy Burman J. 2013. Management and Rehabilitation of a Sundarban Tiger with a Chronic Wound and Infective Arthritis. *Indian Forester* 139(10): 879-882.

Mandal RN, Saenger P, Das CS, Aziz A. 2019. Current Status of Mangrove Forests in the Trans-boundary Sundarbans. In: *The Sundarbans: A Disaster-Prone Eco-Region* (pp.93-131).

Mani F. 2012. Guns and shikaris: The rise of the sahib's hunting ethos and the fall of the subaltern poacher in British India, 1750-1947. *Graduate Theses, Dissertations, and Problem Reports* 594. https://researchrepository.wvu.edu/etd/594.

Manjrekar MP, Prabu CL. 2014. Status of otters in the Sundarbans Tiger Reserve, West Bengal, India. *IUCN Otter Spec. Group Bull.* 31(2): 61-63.

Mannan IA, Sujauddin M, Sohel MSI. 2023. Evaluating the Financial Effectiveness of Funded Projects on Tiger Conservation in Bangladesh. *Tropical Conservation Science* 16: 1-14.

Mathai MV. 1999. Habitat Occupancy by Tiger Prey Species Across Anthropogenic Disturbance Regimes in Panna National Park, Madhya Pradesh, India. *Ph.D Dissertation.* Dehradun: Wildlife Institute of India.

Mazak V. 1981. *Panthera tigris. Mammalian Species.* 152, 8 May 1981, Pages 1-8, https://doi.org/10.2307/3504004.

McDougal C. 1977. *The face of the tiger.* London: Rivington Books.

McDougal C. 1999. You can tell some tigers by their tracks with confidence. p.383. In: Seidensticker J, Christie S, Jackson P (eds). *Riding the Tiger.* Cambridg: Cambridge University Press.

Miller CS, Hebblewhite M, Petrunenko YK, Seryodkin IV, Goodrich JM, Miquelle DG. 2014. Amur tiger (*Panthera tigris altaica*) energetic requirements: implications for conserving wild tigers. *Biol. Conserv.* 170: 120-129.

Miquelle DG, Goodrich JM, Kerley LL *et al* . 2010. Science-based conservation of Amur tigers in the RussianFar East and Northeast China. Pages 403-423 In Tilson R, Nyhus P. (eds.) *Tigers of the world: the science, politics, and conservation of Panthera tigris.* 2nd ed. Elsevier/AcademicPress, Oxford, U.K.

Mishra A. 2022. Geo Spatial Approach for Tiger Habitat Suitability Modeling A Case Study of Bandhavgarh National Park Madhya Pradesh. *PhD Thesis.* Bhopal: Department of Applied Geology, Barkatullah University.

Mitra S, Chanda A, Das S, Ghosh T, Hazra S. 2022. Salinity Dynamics in the Hooghly-Matla Estuarine System and Its Impact on the Mangrove Plants of Indian Sundarbans. In: Mukhopadhyay A, Mitra D, Hazra S. *Sundarbans Mangrove Systems: A Geo-Informatics Approach,* pp. 305-328. UK: CRC Press.

Mondol S, Karanth KU, Ramakrishnan U. 2009. Why the Indian subcontinent holds the key to global tiger recovery. *PLOS Genet* 5: e1000585.

Montgomery S. 2008. *Spell of the tiger: The man–eaters of Sundarbans*. Chelsea Green Publishing Company, 240 pp.

Montgomery S. 1995. *Spell of the Tiger: The Man-Eaters of Sundarbans*. Houghton Mifflin, Boston.

Mottram JC. 1915. Some observations on pattern-blending with reference to obliterative shading and concealment of outline. *Proceedings of the Zoological Society of London* 1915: 679-692.

Mukherjee S. 2004. Ecological investigations on mangroves of the Sundarban Tiger Reserve in West Bengal (India) with special reference to effective conservation through management practice. *Ph.D. Thesis*. University of Kalyani.

Mukherjee S. 2006. Rhesus monkey and chital in the Sundarbans. *Banabithi, Wildlife Issue,* October 2006: 33-35.

Mukherjee S and Sen Sarkar N. 2013. The Range of Prey Size of the Royal Bengal Tiger of Sundarbans. *Journal of Ecosystems* 2013(1): 1-7.

Mukul SA, Alamgir M, Sohel MSI, Pert PL, Herbohn J, Turton SM, Khan MSI, Munim SA, Ali Reza AHM, Laurance WF. 2019. Combined effects of climate change and sea-level rise project dramatic habitat loss of the globally endangered Bengal tiger in the Bangladesh Sundarbans. *Science of The Total Environment* 663: 830-840.

Naha D. 2015. Human tiger conflict ranging pattern and habitat use by tiger in Sundarban tiger reserve India. *PhD. Thesis.* Department of Wildlife Science, Saurashtra University.

Naha D, Jhala YV, Qureshi Q, Roy M, Sankar K, Gopal R. 2016. Ranging, Activity and Habitat Use by Tigers in the Mangrove Forests of the Sundarban. *PLoS One*. 11(4):e0152119. doi: 10.1371/journal.pone.0152119. PMID: 27049644; PMCID: PMC4822765.

Neumann-Denzau G. 2006. The tiger as scavenger: Case histories and deduced recommendations. *Tigerpaper* 33(1): 1-9.

Neumann-Denzau G, Denzau H. 2010. Examining the extent of human-tiger conflict in the Sundarbans forest, Bangladesh. *Tigerpaper* 37(2): 4-7.

NEWS. 1996. *Collaborative study on problems of tiger straying in the villages of the Sundarbans.* Kolkata: Nature, Environment and Wildlife Society.

O'malley, LSS. 1911. *Bengal District Gazetteers: Midnapore*. Calcutta, Bengal Secretariat Book Depot.

O'malley, LSS. 1912. *Bengal District Gazetteers: Jessore*. Calcutta, Bengal Secretariat Book Depot.

O'Malley LSS. 1914. *Bengal District Gazetteers: 24-Parganas*. Calcutta, Bengal Secretariat Book Depot.

Nigam P, Muliya SK, Srivastav A, Shrivastava AB, Mathur VC, Malik PK. 2016. Patterns of Mortality in Free Ranging Tigers. *Technical Report*. Wildlife Institute of India – National Tiger Conservation Authority. TR No. 2016/012, pp. 93.

Pandit PK. 2012. Sundarban Tiger Reserve (*Panthera tigris tigri*s)- A new prey species of estuarine Crocodile (*Crocodylus porosu*s) at Sundarban Tiger Reserve, India. *Tiger Paper* 39(1): 1-5.

Pandit PK and Guha A. 2015. Peculiarities in food chain of Sundarban Tiger Reserve: Recent case studies. *Indian Journal of Biological Sciences* 21: 17-22.

Pandit RV. 1994. Osteology of Indian Tiger. *Technical Bulletin No. VI*. Directorate of Project Tiger, Melghat, Amravati 444 602.

Panwar HS. 1979. A note on tiger census technique based on pugmark tracings. *Tigerpaper* 6(2-3): 16-18.

Pariwakam M, Joshi A, Navgire S, Vaidyanathan, S. 2018. *A Policy Framework for Connectivity Conservation and Smart Green Linear Infrastructure Development in the Central Indian and Eastern Ghats Tiger Landscape. Volume 1*. Wildlife Conservation Trust (WCT), India.

Pennant T. 1798. *The view of Hindoostan, vol. 2: Eastern Hindoostan*. London, Henry Hughs.

Perry R. 1965. *The world of the tiger*. New York: Atheneum.

Pirazzoli PA. 1996. *Sea level changes: the last 20000 years*. New York: Willey, p 211.

Poddar-Sarkar M, Ray S, Chowdhury SP, Samanta G, Raha P, Brahmachary RL. 2013. On the Body Odour of Wild-Caught Mangrove-Marsh Bengal Tiger of Sundarban. In: East, M., Dehnhard, M. (eds) *Chemical Signals in Vertebrates12*. Springer, New York, NY. https://doi.org/10.1007/978-1-4614-5927-9_17

de Poncins E. 1935. A hunting trip in the Sunderbunds in 1892. *Journal of the Bombay Natural History Society* 37: 844-858.

Powell ANW. 1957. Call of the tiger. *Technical Report*, Robert Hale Ltd., London, pp. 1-237.

Rahman M, Islam M, Ahmed, J. 2009. Tiger bite: An unapprehended injury. *Journal of the College of Physicians and Surgeons-Pakistan* 19: 595-597.

Rahaman SMB, Sarder L, Rahaman MS *et al.* 2013. Nutrient dynamics in the Sundarbans mangrove estuarine system of Bangladesh under different weather and tidal cycles. *Ecol Process* 2:29 https://doi.org/10.1186/2192-1709-2-29.

Ray R, Majumder N, Das S, Chowdhury C, Jana TK. 2014. Biogeochemical cycle of nitrogen in a tropical mangrove ecosystem, east coast of India. *Marine Chemistry* 167: 33-43.

Reza AHMA. 2000. Ecology of the Bengal tiger, *Panthera tigris tigris,* in the Sundarbans. *M.Sc. Thesis.* Dhaka: Jahangirnagar University.

Reza AHMA, Feeroj MM, Islam MA. 2002. Man-tiger Interaction in Bangladesh Sundarban. *Bangladesh J. Life Sci.* 14(1&2): 75-82.

Reza AHMA, Islam MA, Feeroz MM, Nishat A. 2004. *Bengal tiger in the Bangladesh Sundarbans.* Dhaka: IUCN Bangladesh.

Rishi V. 1988. Man, Mask, and Maneater. *Tigerpaper* 15(3): 9-14.

Rishi, V. 1992. Making of a Cattle Lifter - A Sundarban Case Study. *Tigerpaper* 19(2):14-17.

Roy Chowdhury D, Das SK, Saha R. 2018. *Status of Tigers in the Sundarban Biosphere Reserve: Session 2015-2016.* World Wide Fund for Nature-India, New Delhi.

Roy M. 2019. Evaluating methods to monitor tiger abundance and its prey in Indian Sundarbans. *PhD Thesis.* Department of Wildlife Science, Saurashtra University.

Rudra K, Lahiri JN. 2021. *Bharatiya Sundarban.* Kolkata: Ananda Publishers Private Limited.

Saif S, Rahman HT, Macmillan D. 2018. Who is killing the tiger *Panthera tigris* and why? *Oryx*, 52(1): 46-54. doi:10.1017/S0030605316000491.

Salter RE. 1984. *Status and utilization of wildlife.* FAO, Rome, Italy, p. 56

Samanta S, Hazra S, Mondal PP, Chanda A, Giri S, French JR, Nicholls RJ. 2021. Assessment and Attribution of Mangrove Forest Changes in the Indian Sundarbans from 2000 to 2020. *Remote Sens.* 13: 4957. https://doi.org/10.3390/rs13244957.

Sanderson EW, Moy J, Rose C, Fisher K, Jones B, Balk D, Clyne P, Miquelle D, Walston J. 2019. Implications of the shared socioeconomic pathways for tiger (*Panthera tigris*) conservation. *Biological Conservation* 231: 13-23.

Sankhala K. 1978. *Tiger! The story of the Indian Tiger*. Calcutta: Rupa.

Sanyal P. 1999. Man-eating tigers of Sundarbans. In: Guha Bakshi DN, Sanyal P, Naskar KR (Eds.). *Sundarbans Mangal*. Calcutta: Naya Prakash, pp. 449-454.

Schaller GB. 1967. Indian Wildlife: The Deer and the Tiger. A Study of Wildlife in India. Chicago: University of Chicago Press. 376 pp.

Seidensticker J, Lahiri RK, Das KC, Wright A. 1976. Problem Tiger in the Sundarbans. *Oryx* 8(3): 267-273.

Seidensticker J, Hai MA. 1983. *The Sundarbans Wildlife Management Plan: conservation in the Bangladesh coastal zone: a World Wildlife Fund Report*. Gland, Switzerland: International Union for the Conservation of Nature and Natural Resources.

Seidensticker J and McDougal C. 1993. Tiger predatory behaviour, ecology and conservation. *Symposium of the Zoological Society of London* 65: 105-125.

Seidensticker J, Christie S, Jackson P. 1999. Preface. In: J. Seidensticker, S. Christie and P. Jackson (eds), *Riding the tiger: tiger conservation in human-dominated landscapes*. Cambridge University Press, Cambridge, UK.

Sharma RK, Qureshi, Q, Jhala YV. 2011. Home range size of a tigress in Sundarbans, India: preliminary results. *CATnews* 54: 13-16.

Sharma S, Jhala YV, Sawarkar V. 2003. Gender discrimination of tigers by using their pugmarks. *Wildlife Society Bulletin* 258-264. 10.2307/3784382.

Sharma S, Dutta T, Maldonado JE, Wood TC, Panwar HS, Seidensticker J. 2013. Forest corridors maintain historical gene flow in a tiger metapopulation in the highlands of central India. *Proc Biol Sci.* 280(1767):20131506.

Sharma S, Jhala YV, Sawarkar VB. 2005. Identification of individual tigers (*Panthera tigris*) from their pugmarks. J. Zool., Lond. 267: 9-18.

Siddiqi NA, Choudhury JH. 1987. Man-eating behaviour of tigers (*Panthera tigris* Linn) of the Sundarbans - twenty-eight years' record analysis. *Tigerpaper* 14(3): 26-32.

Sievers M, Roy Chowdhury M, Adame MF, Bhadury P, Bhargava R, Buelow C, Friess DA, Ghosh A, Hayes MA, McClure EC, Pearson RM, Turschwell MP, Worthington TA, Connolly RM. 2020. Indian Sundarbans mangrove forest considered endangered under Red List of Ecosystems, but there is cause for optimism. *Biological Conservation* 251:108751, ISSN 0006-3207, https://doi.org/10.1016/j.biocon.2020.108751.

Simson FB. 1886. *Letters (No.47) on sport in Eastern Bengal.* London: RH Porter.

Singh R, Pandey P, Qureshi Q, Sankar K, Krausman PR, Goyal SP. 2020. Acquisition of vacated home ranges by tigers. *Current Science* 119(9): 1549-1554.

Singh S, Mishra S, Aspi J, Kvist L, Nigam P, Pandey P, Sharma R, Goyal S. 2014. Tigers of Sundarbans in India: Is the Population a Separate Conservation Unit?. *PLoS ONE* 10: 1-24. 10.1371/journal.pone.0118846.

Singh SK. 2017. Conservation Genetics of the Bengal tiger (*Panthera tigris tigris*) in India. *Acta Univ. Oul. A* 689, 2017.

Smith JLD, McDougal C, Miquelle D. 1989. Scent marking in free-ranging tigers, *Panthera tigris. Animal Behaviour* 37(1): 1-10.

Somerville, A. 1924. *Shikar near Calcutta with a trip to the Sunderbunds.* W. Newman & Co., Limited, Calcutta: 87-88.

Somerville, A. 1933. *At midnight comes the killer.* Calcutta: Thacker Spink.

Sramek J. 2006. Face Him like a Briton: Tiger Hunting, Imperialism, and British Masculinity in Colonial India, 1800-1875. *Victorian Studies* 48(4): 659-680.

Soutyrina SV, Riley MD, Goodrich JM, Seryodkin IV, Miquelle DG. 2013. *A population estimate of Amur tigers using camera traps.* Sikhote-Alin State Nature Biosphere Reserve, Wildlife Conservation Society, Pacific Geographic Institute, Russian Academy of Sciences Far Eastern Branch, Dalnauka, Vladivostok.

Sunquist ME. 1981. The social organization of tigers (*Panthera tigris*) in Royal Chitawan National Park, Nepal. *Smithson. Contrib. Zool.* 336: 1-98.

Sunquist M, Karanth KU, Sunquist F. 1999. Ecology, Behaviour and Resilience of the Tiger and its Conservation Needs, pp.5-18, In: Seidensticker J, Christie S, Jackson P. (eds.) *Riding the Tiger: Tiger conservation in human-dominated landscapes.* London: Cambridge University press.

Sunquist M, Sunquist F. 2002. *Wild cats of the world.* University of Chicago press, London. Pp 1 -443.

Thakur S, Maity D, Mondal I, Basumatary G, Ghosh P, Das P, De T. 2021. Assessment of changes in land use, land cover, and land surface temperature in the mangrove forest of Sundarbans, northeast coast of India. *Environment, Development and Sustainability* 168(2): 1917-1943. 10.1007/s10668-020-00656-7.

Tiwari A, Saran S, Avishek K. 2021. *Habitat suitability analysis of tigers using decision support system of Indian bio-resource information network (IBIN) portal.* The 42nd Asian Conference on Remote Sensing (ACRS2021) 22-24th November, 2021 in Can Tho University, Can Tho city, Vietnam.

Trivedi S, Zaman S, Ray Chaudhuri T, Pramanick P, Fazli P, Amin G, Mitra A. 2016. Inter-annual variation of salinity in Indian Sundarbans. *Indian Journal of Geo-Marine Science* 45(3): 410-415.

Uddin N, Enoch S, Harihar A, Pickles RS, Ara T, Hughes AC. 2022. Learning from perpetrator replacement to remove crime opportunities and prevent poaching of the Sundarbans tiger *Conserv. Biol.* 13997.

Uddin N, Enoch S, Harihar A, Pickles RSA, Hughes AC. 2023. Tigers at a crossroads: Shedding light on the role of Bangladesh in the illegal trade of this iconic big cat. *Conservation Science and Practice* 5(7): e12952. https://doi.org/10.1111/csp2.12952.

Variar AS, Anoop NR, Komire S, Vinayan PA, Sujin NS, Raj A, Prasadan PK. 2023. Prey selection by the Indian tiger (*Panthera tigris tigris*) outside protected areas in India's Western Ghats: implications for conservation. *Food Webs* 34: e00268, ISSN 2352-2496, https://doi.org/10.1016/j.fooweb.2022.e00268.

Verma M, Negandhi D, Khanna C, Edgaonkar A, David A, Kadekodi G, Costanza R, Singh R. 2015. *Economic Valuation of Tiger Reserves in India: A Value+ Approach.* Indian Institute of Forest Management. Bhopal, India. Wallace AR. 1870. Mimicry, and other protective resemblances among animals. In: *Contributions to the theory of natural selection. A series of essays,* 45-129. London: Macmilian.

Vidyanadhan R, Ghosh RN. 1993. Quaternary of the East Coast of India. *Current Science* 64(11&12): 804-816.

Vivekanandan J. 2021. Scratches on our sovereignty? Analyzing conservation politics in the Sundarbans. *Regions & Cohesion* 11(1):1-20.

Wikramanayake E, Dinerstein E, Seidensticker J, Lumpkin S, Pandav B, Shrestha M, Mishra H, Ballou, JD, Johnsingh AJT, Chestin I, Sunarto S, Thinley P, Thapa K, Jiang G, Elagupillay S, Kafley H, Pradhan NMB, Teak S, Jigme K, Cutter P, Aziz MA, Than U. 2011. A landscape-based conservation strategy to double the wild tiger population. *Conservation Letters* 4(3): 219-227.

Worldwide Fund for Nature–India (WWF-India). 2011. *Proceedings of the International Conference on Tiger Conservation and Global workshop on implementation of the Global Tiger Recovery Program (GTRP) March 28- 29, 2011.* New Delhi: Vigyan Bhavan,
79 pp.

Xiao W, Feng L, Mou P, Miquelle DG, Hebblewhite M, Goldberg JF, Robinson HS, Zhao X, Yoshimura H, Hirata S, Kinoshita K. 2021. Plant-eating carnivores: Multispecies analysis on factors influencing the frequency of plant occurrence in obligate carnivores. *Ecology and Evolution* 11(16):10968-10983. doi: 10.1002/ece3.7885. Erratum in: *Ecol Evol*. 2021 11(19):13618-13619. PMID: 34429895; PMCID: PMC8366844.

Zhou B, Wang T, Ge J. 2016. Estimating abundance and density of Amur tigers along the Sino–Russian border. *Integrative Zoology* 11(4): 322-332.

www.ingramcontent.com/pod-product-compliance
Lightning Source LLC
Chambersburg PA
CBHW030237230326
41458CB00092B/320